Second Edition

California Cultural LANDSCAPES

An Exploration of Spatial Patterns Over Time

Dean H.K. Fairbanks
California State University, Chico

Cover images © Shutterstock, Inc.

Kendall Hunt
publishing company

www.kendallhunt.com
Send all inquiries to:
4050 Westmark Drive
Dubuque, IA 52004-1840

Copyright © 2009, 2013 by Kendall Hunt Publishing Company

ISBN 978-0-7575-9354-3

All rights reserved. No part of this publication may be reproduced,
stored in a retrieval system, or transmitted, in any form or by any means,
electronic, mechanical, photocopying, recording, or otherwise,
without the prior written permission of the copyright owner.

Printed in the United States of America
10 9 8 7 6 5 4 3 2 1

CONTENTS

Acknowledgements ix
Preface xi

chapter 1
UNDERSTANDING THE CALIFORNIA LANDSCAPE: ELEMENTS OF GEOGRAPHY 1
Key Terms 1
Introduction 1
1.1 **Five Themes of Geographical Science** 2
    1.1.1 Location 3
    1.1.2 Place 5
    1.1.3 Movement 7
    1.1.4 Human-Environmental Interaction 8
    1.1.5 Region 8
1.2 **Culture and Cultural Landscapes: Basic Themes** 10
1.3 **California Geography: Basic Themes to Explore** 11
    1.3.1 An Approach to Understanding California 11
    1.3.2 Diverging Images of California 11

chapter 2
GEOLOGICAL HISTORY AND LANDFORMS 13
Key Terms 13
Introduction 13
2.1 **Basic Geology: The Building Blocks** 14
    2.1.1 Constructive Interaction 14
    2.1.2 Destructive Interaction 14
    2.1.3 Conservative Interaction 17
2.2 **The Rock Cycle** 18
2.3 **California's Brief Geologic History** 19
2.4 **Geologic Natural Hazards: Living on Shaky Ground** 22
2.5 **Natural Landform Regions** 25
    Geomorphologically-based Landscapes 25
    2.5.1 Sierra Nevada 25
    2.5.2 Klamath Mountains 27
    2.5.3 Peninsular Ranges 28
    2.5.4 Southern Cascades and Modoc Plateau 29
    2.5.5 Great Central Valley 29
    2.5.6 Coast Ranges 29
    2.5.7 Transverse Ranges 31
    2.5.8 Basin and Range 31
    2.5.9 Mojave Desert 33
    2.5.10 Colorado Desert/Salton Trough 33

chapter 3
CALIFORNIA CLIMATE: SCALES AND GENERAL CONTROLS 39
Key Terms 39
Introduction 39
3.1 **Global-Scale Controls** 40
    3.1.1 Latitude 40
    3.1.2 West Coast Position and Air Masses 40
    3.1.3 El Niño Events 42

3.2 **Regional Scale Controls—Influence of Topography** 44
3.3 **Small Scale Issues** 45
3.4 **Precipitation and Temperature: Spatial and Seasonal Patterns** 47
   3.4.1 Precipitation Controls 47
      3.4.1.1 Seasonal Precipitation Cycles 47
      3.4.1.2 Controls on Precipitation Spatial Patterns 47
   3.4.2 Temperature Controls 49
      3.4.2.1 Seasonal Temperature Cycles 49
      3.4.2.2 Controls on Temperature Spatial Patterns 49
3.5 **Climatic Variation by Landform Region** 54
   3.5.1 Sierra Nevada 54
   3.5.2 Klamath Ranges 54
   3.5.3 Peninsular Ranges 54
   3.5.4 Southern Cascades and Modoc Plateau 54
   3.5.5 Great Central Valley 54
   3.5.6 North and Central Coast Ranges 55
   3.5.7 Transverse Ranges 55
   3.5.8 Basin and Range 55
   3.5.9 Mojave 55
   3.5.10 Salton Trough/Colorado Desert 56

chapter 4

CALIFORNIA FLORA AND FAUNA: DIVERSITY, HAZARDS AND CONSERVATION 75
   Key Terms 57
   Introduction 57
4.1 **Bioregions of California** 58
4.2 **Determinates of Vegetation Patterns** 59
   4.2.1 Climate (Past and Present) 59
   4.2.2 Topography 60
   4.2.3 Geology and Soils 60
   4.2.4 Natural Disturbances—Fire 60
   4.2.5 Human Activity 61
4.3 **Plant Communities—General Summary** 61
4.4 **Plant Diversity** 63
4.5 **Floral History** 65
4.6 **California Animal Diversity** 68
4.7 **Endangered Species and Conservation** 72

chapter 5

CALIFORNIA NATIVE AMERICAN LANDSCAPES 75
   Key Terms 75
   Introduction 75
5.1 **Pre-contact California: California's First Peoples** 76
   5.1.1 Native Languages: A Tower of Babel Allegory 76
   5.1.2 Living with a Highly Variable Climate and without Agriculture 78
5.2 **California Native American "Tribes," Tribelets and Cultural Areas** 83
   5.2.1 Northwestern Cultural Area 84
   5.2.2 Central Cultural Area 86
   5.2.3 Southern Cultural Area 87
   5.2.4 Colorado River Cultural Area 88

5.2.5 Great Basin Eastern California Cultural Area 88
5.2.6 Northeastern Cultural Area 89
5.3 **Ecological Impacts of Spanish Contact** 89
5.3.1 Impacts of Direct Spanish Settlement 90
5.4 **Summary** 91

chapter 6

SPANISH EXPLORATION AND SETTLEMENT PATTERN IN ALTA CALIFORNIA 95

Key Terms 95
Introduction 95
6.1 **Indiana Jones and the Kingdom of Queen Califia! . . . Or Maybe Not . . .** 96
6.1.1 Northward Exploration and Settlement 96
6.2 **Spanish Settlement Pattern** 99
6.2.1 Missions 100
6.2.2 Presidios 106
6.2.3 Pueblos 107
6.3 **Postcards from a Decaying Empire** 109

chapter 7

MEXICAN RANCHO ERA: SETTLEMENT AND DEVELOPMENT BY "LOS CALIFORNIOS" 113

Key Terms 113
Introduction 113
7.1 **Trade and Economic Growth** 114
7.2 **Mexican Ranchos and Mission Secularization** 115
7.2.1 Rancho Diseños 118
7.3 **A Rigid Cultural and Economic Landscape in a Mixed Blood Land** 120
7.4 **Pre-Gold Rush Scenario** 122

chapter 8

GOLD RUSH: THE U.S. TAKES OVER—THE WORLD'S CULTURES MOVE INTO NORTHERN CALIFORNIA 125

Key Terms 125
Introduction 126
8.1 **Bear Flaggers, the Texas Game and a Game Changing Character—John Frémont** 126
8.1.1 Texas Game, a Precursor of Things to Come 126
8.1.2 John Charles Frémont, Manifest Destiny, and the Bear Flag Revolt 126
8.1.3 The Mexican-American War Reaches California 128
8.2 **Before the Gold Rush—California Before the Dawn** 130
8.3 **The Rush Is On: California the Distant Outpost** 132
8.4 **1849ers—The Argonauts and Rapid Growth in Northern California** 134
8.4.1 Early Mining Patterns 136
8.5 **California Life in the 1850s** 137
8.5.1 Cultural Diversity and Ethnic Conflict 138
8.5.2 Urban Growth vs. Gold Rush Communities 144
8.6 **U.S. Western Settlement Procedures—Homesteadin'** 144
8.6.1 Homestead Act of 1862—The Great Grid of the West 144
8.7 **Industrialization of Mining: Evolution of California Mining** 146
8.7.1 The Gold Rush Mining Law 146

8.7.2 Rise of Industrial Hydraulic Mining 147
8.7.3 Rise of Hard Rock Lode Mining 148
8.8 Gold vs. Golden Grain: The Downfall of California Gold Mining 149

chapter 9   THE BEGINNINGS OF MODERN CALIFORNIA AND THE RISE OF SOUTHERN CALIFORNIA 155

Key Terms 155
Introduction 155
9.1 Early Transportation and Communication 157
9.1.1 The Coming of the Central (Southern) Pacific Railroad 158
9.1.1.1 Chinese as Scapegoats: Racist Exclusion under the Yellow Peril 160
9.2 Search for the Ideal Landscape 164
9.2.1 Final Destruction of the California Native American Cultural Landscape 166
9.3 Southern California Development: 1880–1900—Development of the Myth 167
9.3.1 The Mission Myth: Romance Tourism 169
9.3.2 Southern California Vitality: Climate for Health, Freedom of Expression, and Real Estate Bubbles 170
9.3.3 "Oranges for Health, California for Wealth:" Expansion of the Citrus Industry 172
9.3.4 The Oil Boom: Southern California's "Black Gold Rush" 174
9.3.5 The Cultural Character of the 1880s Boom 175
9.4 The Progressive Era: Arts and Crafts Movement in Southern California 175
9.4.1 Craftsman Bungalow Architecture in Southern California 176
9.4.2 Mission Revival Design in Southern California 177
9.5 A Development Pattern Snapshot of California: 1880–1930 179

chapter 10   CALIFORNIA AGRICULTURE AND IRRIGATION AGRI-"CULTURES" 183

Key Terms 183
Introduction 183
10.1 California Agriculture and Reclamation 185
10.1.1 Water Doctrine in California: Prior Appropriation vs. Riparian Rights 186
10.1.1.1 Miller-Lux vs. Haggin: A Battle between Two Land Monopolists 187
10.1.2 Wright Irrigation Act 1887 187
10.2 Do You Know What We Grow? California Agriculture and Cuisine 189
10.2.1 Fruits, Tree Nuts, and Berries 190
10.2.2 Vegetables—Truck Crops 193
10.2.3 Field Pasture, Seed Crops, and Nursery/Floriculture 193
10.2.4 Dairy, Cattle, and Poultry 196
10.3 Irrigation Agri-"Cultures:" Chinese, Japanese, Filipinos, Indian Punjabis, and Mexicans 198
10.3.1 Early Chinese Labor and Agriculture: A Glimpse of the Future 198
10.3.2 Japanese Farm Labor 199

Contents vii

      10.3.3 Filipino Farm Labor 201
      10.3.4 Indian Punjabis in the Sacramento Valley: A Uniquely Concentrated Labor Group 201
      10.3.5 Mexican Farm Labor 201
  10.4 Factories in the Fields: Great Depression—Dust Bowl—Okies 204
      10.4.1 The Dust Bowl: Disaster from a Major Drought and Poor Land Management 204
      10.4.2 The "Okies" in Agriculture, an Agri-"Subculture" 206

chapter 11

URBAN-RURAL CALIFORNIA: WATER, LAND AND DESIGN 211

Key Terms 211
Introduction 212
  11.1 **Urban Water Imperialism: Stories from the South and North** 213
      11.1.1 Los Angeles: The Owens River Valley Saga 213
      11.1.2 San Francisco: Hetch Hetchy for a City with No Water Rights 215
  11.2 **Water Projects: Colorado River, the Central Valley Project and the State Water Project** 217
      11.2.1 Controlling the Colorado River and the "Accidental" Creation of the Salton Sea 217
      11.2.2 Colorado River Compact and Boulder Dam 218
      11.2.3 Central Valley Water Project: A New Deal in Water Delivery 220
          11.2.3.1 Irrigated Acreage Limitations: A Big Problem in California 224
      11.2.4 State Water Project: World's Largest Water Works 224
  11.3 **California's Perennial Water Woes: What's Next for the Dry State?** 226

chapter 12

THE RISE OF THE MODERN CALIFORNIA LANDSCAPE 229

Key Terms 229
Introduction 229
  12.1 **World War II: Industry, Infrastructure, and Immigration** 233
      12.1.1 Non-Military Industry and Military Lands 234
  12.2 **Modern Cultural Conflict and Migration** 235
      12.2.1 Japanese Internment 235
      12.2.2 Mexican-Americans and the Zoot-Suit Riots 236
      12.2.3 African-Americans: From Rural to Uniquely Urban 237
  12.3 **Housing: Suburbanization—The Rise of the Tract Home** 237
      12.3.1 Transportation and Economy 240
  12.4 **California's Era of Expanding the Mind and Restless Youth Energy** 243

chapter 13

CALIFORNIA: PARADISE OR GRITTY REALITY? 249

Key Terms 249
Introduction 249
  13.1 **Modern Geographical Cultural Patterns** 252
      13.1.1 Native Americans 252
      13.1.2 African-Americans 253
      13.1.3 Asians 253
      13.1.4 Hispanics 256

13.2 California's Modern Cultural Regions Revealed 259
    13.2.1 Urban Empires on the Pacific
        13.2.1.1 Southern California 259
        13.2.1.2 The San Francisco Bay Area 261
    13.2.2 The Great Central Valley 264
        13.2.2.1 Sacramento Metro-Delta Corridor 264
        13.2.2.2 San Joaquin Valley 265
        13.2.2.3 Sacramento Valley 265
    13.2.3 Mountainous, Arid-scapes, and Between the Urban Empires 266
        13.2.3.1 Central Coast 266
        13.2.3.2 The Wet Northwest 267
        13.2.3.3 The Lonely Southern Cascade and Northeast 267
        13.2.3.4 The Sierra Nevada 267
        13.2.3.5 The Mojave Desert and Trans Sierra 268
        13.2.3.6 The Colorado Desert and Salton Sea 269
13.3 California: Eden or Wasteland? 271
    13.3.1 NorCal vs. SoCal or Coastal vs. Interior? 272
13.4 Living the "La Vida Loca" in Mediterranean California 274

ACKNOWLEDGEMENTS

Student textbooks are comprehensive summaries and integrations of a vast amount of primary and secondary source material. I must strongly acknowledge that this textbook could not have been written without the availability of all the wonderful and insightful materials based in books, journal articles, online and online lecture presentations by lecturers and professors from other institutions around the state for teaching California geography. Besides how we teach the course for undergraduates needing a General Education (GE) requirement at California State University Chico (CSUC), it was insightful to visit various colleagues' websites at other universities and community colleges around the state to see how they approach teaching California geography. This textbook is written based on a combined point of view to help support their teaching about the geographical nature of California. It is a combination of their material and our lecturing group's approach at CSUC that I have tried to truthfully capture. In a final analysis, however, it is the access to so much wonderful material that makes this textbook work. Without their sources, insights and interpretations, this textbook would not have been completed. I deeply acknowledge and respect their fine efforts. I am merely a vehicle to repackage and present it to an undergraduate student audience to help support the various incarnations of this course around the state. All the references for each chapter are the backbone to this textbook, and I strongly encourage interested students to look at the referenced material for a deeper understanding. I hope I have translated each author's thoughts and interpretations in my own writing style, to you the student, in a meaningful way.

The spark that led to this textbook began with three people. First, a big thanks to Jim Monaco, who has been lecturing the California geography course at CSUC since the early eighties. Upon my arrival at CSUC, Jim mentored me in this course as he had been mentored by the late Professor David Lantis. Jim often lamented to me how there was no appropriate textbook for the course. In his words, the one or two textbooks that were available are inappropriate for GE (especially with an ethnic diversity designation) and use an older style for presenting regional geography. Regional geographic studies are not just about descriptions of regions or focuses on broad themes, but must aim at unpacking the geographic pattern and process dynamics that human-environment dominated regions have developed over time. Second, I would like to thank my wife Portia Ceruti, who one day in early 2004 came home from the used bookstore with an old (1920s) textbook on California geography by Harold Fairbanks, who upon investigation turned out to be related to my family line. I was surprised to find that his textbook was the standard geography textbook on California adopted statewide for all public K–12 schools in the state during the 1920s. This find provided me the next nudge down the road to writing this textbook now in its second edition. Thirdly, I would like to thank my grandfather Charles Hurlbut (1920–2012), who also constantly asked when I was going to write a textbook and why not do it on California geography. He was originally a teacher of geography and California history in middle schools and high schools from 1948–1985. Even well into his retirement years he still had a fond love of California geography and history. His generation greatly benefited from the post- World War II GI education bill that made the California State University system one of the largest and finest university systems in the country. In addition, a thank you to colleagues, students, friends, and family members who provided valuable feedback in the first edition and especially the second edition, including Frank Bayham, Guy King, Nori Sato, Steve Stewart, Robert Pierce, Jon Libby, Adrienne Harrold, Jake Palazzo, and the 600 students from GEOG 105 that provided valuable feedback for the second edition.

Student textbooks are not just about the author, but they are created by the author and a talented team of people. The folks at Kendall Hunt are a hard-working, stellar bunch, who were a pleasure to work with. A big thank you to acquisition editor, Dr. Frank Forcier for coming to my office and convincing me to write this textbook. My production editors, Angela Puls (first edition) and Linda Chapman (second edition), are frankly the best editors I have ever worked with on a writing project. They both kept me on task and relatively on time and were so helpful with all my questions. Permissions editor Caroline Kieler was diligent with locating my image material. Cover designer, Jeni Chapman, connected with my design ideas and came up with an awesome jacket layout, and finally the copyediting and interior designer group in Execustaff Composition provided me with excellent chapter edits to ponder and a great chapter layout design for students to follow. New with this second edition is a companion website, which are increasingly appreciated by students in this digital era. I would like to thank the folks at Great River Technologies and especially Samantha Bartholomew for developing the book's valuable website. Thanks to everyone! I would also like to thank emeritus Professor Jerry Williams from my department for allowing me to fully integrate and package his student *California Atlas* with this textbook. The combination makes this a valuable reference for faculty and students.

Finally, any errors in omission or commission of information, mis-interpretations, typographical errors, and failure to paraphrase information properly are mine.

Cheers,

Dean Fairbanks

PREFACE

This textbook was written for the community college or university student who might be a bit curious about, California, the state they are studying in or might have been raised. Its aim is to support a course in California geography by introducing students to the physical setting and historical cultural geography of California's ever changing cultural landscapes.

In writing *California Cultural Landscapes: An Exploration of Spatial and Temporal Patterns*, my goal was to create a textbook for the student taking California geography as a General Education requirement. This type of course is taught in almost all of our community colleges across the state as well as in many of our California State University (CSU) campuses, including mine at CSU Chico. The study of California, its diversity over time and space is tied to aspects of economics, politics, transportation, human behavior, cultural identity, natural resources, and the physical setting of the state. Using geographical methods to provide students the opportunity to develop an understanding of cultural and social institutions is a powerful approach. In the many years I have taught California geography at California State University, Chico students have constantly commented on how they can "see" the differences between places and want to know why one place can be so different from another in the state. Why did it become that way? Is there really a difference between *SoCal* and *NorCal*? What about coastal versus interior, while physically different, is their a cultural difference? If so, why?

While the physical geographic setting is the first layer of landscape view humans come in contact with, focus quickly shifts to the cultural elements of the human built landscape followed by an assessment of economic class, ethnic and gender issues in a place. This textbook was not only written to provide a guide to understanding why California has evolved into such a vastly different region than other parts of the U.S., but also why inside its political boundaries physical spaces and cultural places can be strikingly different or similar but separated by significant distance. As such, this textbook uses social science methods and perspectives and historic as well as contemporary perspectives and influences to explain and explore California. It is different than other California geography textbooks as rather than taking a theme or a regionalization based approach it uses an understanding of the physical geographic setting to approach historical geography with the traditional regional geography themes of population/migration, cultures, economics, and natural resources integrated into the spatial historical chronology. California's landscapes hold memory and a "sequence of occupation" approach to understanding their contemporary pattern has proven to be a more fulfilling approach to regional geographical studies. This approach to studying California is one based on a co-evolutionary understanding: environment, cultures, and human institutions interacting over space to develop new pathways to emergent patterns that we can see geographically and experience. Therefore *California Cultural Landscapes* allows students to demonstrate learning in the development and variation of California cultural and social institutions; and how California's cultural and social development and variations have affected ethnic groups, institutions and human behavior. With this in mind *California Cultural Landscapes* is designed to help students with the following student learning objectives:

1. Develop a geographic understanding of California's changing cultural landscapes, including:
   a. the physical processes that shape the patterns on California's surface,
   b. how these physical processes affect human cultures,
   c. how various cultures have modified the physical environment,
   d. the patterns of human migration and settlement throughout the region, and
   e. the role of interethnic relations in shaping these migrations and settlement patterns.

2. This textbook comes with a full color student *California Atlas* and a companion website that is integrated within the text for increased comprehension and exploration.

*To my wife, Portia,
and our children Sibaya and Sabin.
Thank you for your love and support.*

CHAPTER 1

Understanding the California Landscape
Elements of Geography

Key terms

Absolute location	Geography	Movement
Cultural characteristics	Hearth	Parallel
Cultural elements	Help-hinder	Physical characteristics
Cultural geography	Human-environment	Place
Cultural landscape	Immigration/emigration	Possibilism
Cultural sequence	Integration	Push-pull
Culture	Isolated	Regions
Determinism	Latitude	Relative location
Diaspora	Location	Scale
Diffusion	Longitude	Synthesis
Formal region	Meridian	Toponym
Functional region	Migration	Vernacular region

Introduction

The discipline of geography is both a fascinating yet an often overwhelming approach to understanding our world. The numerous physical and environmental processes and interconnected cultural landscapes often come together to create complex spatial patterns that can change over time. A student can find the myriad concepts from many supporting and complimentary fields of study required to understand the patterns based on the many processes to be daunting.

California to even the casual observer, represents one of the many diverse regions of the world, physically and culturally. California's natural dynamic processes and landscapes and its human cultural history, its people and their landscapes, are well connected in interesting ways and apparent to anyone willing to explore them. Therefore, understanding both the physical and cultural evolution of landscapes in California (See pages 3, 5, and 27 in *California Atlas*) requires comprehension of some basic concepts grounded in the fields of economics, anthropology, sociology, political science and the natural sciences. In this chapter, we will briefly cover these investigative tools to help explore California over space and time.

A geographical understanding of California's past, present and future will come from understanding the underlying processes that allowed for the dramatic physical environments and the diversity of cultural patterns to develop where they did across the state. The cultural landscape features always evolving in California are the roadways that were once Native American trails, the suburban planned community subdivisions spreading across rural California farmland, the landscapes of water poverty in Southern California that became ones of abundance through large scale water transfers from distant places, and the way 'minority' Hispanic, Asian and mixed population groups are becoming the 'majority' in the state's demographic profile. In addition, what allows California to have the eighth largest economy in the world (2009 ranking), which is almost double the economic size of the next state, Texas, in the internal U.S. ranking? These are the geographical patterns, trends and questions this textbook will attempt to answer.

The above introductory discussion really begs the question: What is a student supposed to learn geographically about this state? Does it really just come down to Nor Cal/So Cal, Coastal Cal/Valley Cal/Mountain Cal? Are there truly hard distinctions? What is all the fuss with the bumper stickers one sees declaring Nor Cal or So Cal as rightful entities on the landscape? What do these statements say to you about the regionalization of California? Are these distinct regions that warrant a discussion? In this textbook you will come away with a better understanding of how this region/state functioned in the past and how it functions in the present to produce the geographical diversity it displays to those who live here or who visit. While a popular tourism commercial says "What happens in Las Vegas, stays in Vegas," this is not the case for California. For what happens in California always tends to leave California and have a profound effect on the rest of the U.S., and sometimes even on the world. There is the American dream, but there is also the California dream. Out of all the many regions the U.S. contains, California holds something special and profound in the minds of most citizens and foreigners. This cannot honestly be said about, say, studying the geography of Kansas! As the online catalog description of the California geography course at the University of California at Berkeley says: "California may be a 'state of mind'—as bumper stickers say—but it is also the most powerful place in the most powerful country in the world. Its wealth and diversity in both human and natural resources has contributed to its extraordinary resilience, making it a center of technological and cultural innovation."

1.1 FIVE THEMES OF GEOGRAPHICAL SCIENCE

We can categorize most fields of knowledge as part of the humanities or as part of science (social and physical), two avenues of understanding the world that differ only in their medium. Those trained in the humanities view the world by expressing their deepest feelings and beliefs through the medium of canvas, clay, dance, theater, music or language. Those trained in the sciences view the world by expressing their deepest feelings and beliefs by solving problems, exploring patterns and relationships, and testing ideas through objective observation and then building a model of how the world works.

The word **geography** literally means 'description of the earth.' The goal of the geographical sciences is to explain how and why things, both physical and cultural, spatially and temporally differ over the surface of the earth. In short, geographers study "what is where, why, and so what." Physical geography differs from human or cultural geography only in the relative emphasis given to natural phenomena (features, processes and materials of the atmosphere, hydrosphere, lithosphere and biosphere) compared to the phenomena of human society (features, processes and materials created by humans—cultural elements). Geography is further distinguished from the other natural and social sciences by virtue of the attention geographers give to three major themes:

1. **Scale** refers to the recognition that how we view and interpret phenomena depends on the spatial extent and the spatial resolution of the measurements we collect.

2. **Integration** refers to the recognition that the landscape functions differently as a whole than one would predict by adding up the individual effects of climate, water, landforms, plants and animals and human cultural elements.

3. **Synthesis** refers to the attention geographers give to understanding the environmental opportunities (resources) and environmental constraints (hazards)

upon human activities and upon human impacts on the environment.

The Five Themes of geography are one recognized way to organize our knowledge of the earth and our relationship to it. This textbook will use these tools of understanding to explore the range of issues California faced in the past, is facing in the present and will face in its future.

1.1.1 LOCATION

Absolute location defines a point on the earth's surface by specifying coordinates on a Cartesian mathematical grid. The latitude and longitude of a point define its location. These X, Y pairs represent our globally recognized address system for identifying the location of places (Figure 1.1).

- **Latitude** (north-south direction, or the Y-axis on a Cartesian grid) is measured along lines that run east and west **parallel** to the equator (0° latitude) called parallels. The image that this should conjure in your mind is that of slicing an onion, where the slices become smaller in diameter as one slices from the middle (equator) to the ends (poles). The north and south poles respectively represent 90° and –90°.

- **Longitude** (east-west direction, or the X-axis on a Cartesian grid) is measured along lines that run north and south through the poles called **meridians**. The image that this should conjure is that of slicing orange wedges, i.e., slicing down from pole to pole. The prime meridian (0° longitude) is internationally recognized as running through Greenwich, England, with the opposite side of the circle representing the International Date Line (180° longitude).

For example, note the latitudes and longitudes of San Francisco and Los Angeles, California in Figure 1.1.

To understand their world geographic 'addresses' we can use what was stated earlier about latitude and longitude to see that the earth, being spheroid, is divided into 360°. Therefore, using the analogy of a clock: 1° = 60′ (minutes) and 1′ = 60″ (seconds). The first 'address' above states that San Francisco is 37° (degrees) and 45′ (minutes) north of the equator, and 122° (degrees) and 26′ (minutes) west of the prime meridian (0° longitude line). Los Angeles, however, is a little closer to the equator at 34° and 03′ north of the equator, and a little closer to the prime meridian at 118° and 14′ west of longitude 0°. If you were given a location for somewhere in Chile the latitude would have a negative in front of the number or an 'S' for south of the equator after the complete latitude depiction of degrees, minutes and sometimes seconds. Likewise, if you were given a location for somewhere in Iraq, while it would be north of the equator, its longitudinal position would be east of the prime meridian and therefore be a positive number or have an 'E' after the complete longitude depiction.

Examining the state (see page 4 in *California Atlas*) we note that California extends from the subtropics (32° 30′ N) at the southern border with Mexico into mid-latitudes (42° N) at the northern border with the state of Oregon. In all, California has been legally defined with a northwest-southeast elongated trending shape that covers about 9° 30′ of latitude and a little more than 10° of longitude on earth.

Lastly, with the advent of road maps and now Global Positioning Systems (GPS) in our cars and cell phones, most of us are use to the concept of **relative locations**. These are described by landmarks, time, distance or direction from one place to another and may associate one place with another. For example, the distance from the point on California's northern coastal border with Oregon to the point on the southern coastal international border with Mexico encompasses ~2,027 km (1,260 miles) of jagged coastline. Or the flight distance from San Francisco flying east to South Lake Tahoe for a snowboard holiday is ~240 km (150 miles).

Knowing a few basic rules about working with latitude and longitude values can allow you to calculate the distance from one location to another. Looking at the longitude values again for San Francisco and Los Angeles we can calculate how far apart they are in a west-east direction. To do this, we subtract 122° 26′ from 118° 14′ and get 4° 12′.

- **Rule 1:** Since 1° = 85 km (53 miles); we can therefore calculate the distance by multiplying and thus get 357 km (223 miles). Remember, however, that while you should get 340 km (212 miles) for the 4°, there are 12′ to work with. Therefore, since 1° = 60′, then 12′/60′ = 0.2; or in other words 12′ is 20% of the full 60′ and you should therefore multiply 0.2 with 85 km (53 miles) to derive the fractional distance (17 km or 11 miles) and then add it to the result from the whole degrees multiplication (340 + 17 = 357 km).

4 CHAPTER 1 Understanding the California Landscape

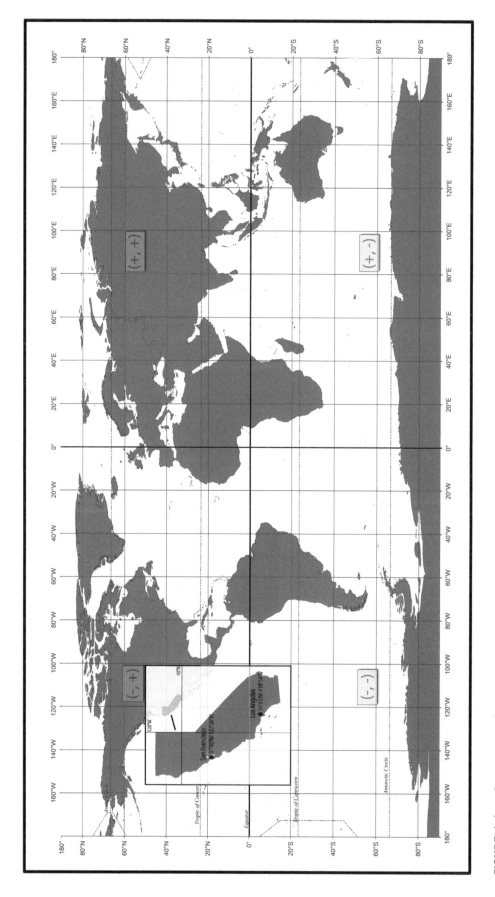

FIGURE 1.1 *Cartesian coordinate system placed on the planet with California denoted and the locations and latitude/longitude coordinates illustrated for San Francisco and Los Angeles. (Courtesy Curtis Page)*

■ **Rule 2:** This rule involves carrying over values, recalling that 1° = 60'. If the values had been slightly different, such as 122° 14' from 118° 26', then you would not bring over 19 to subtract in the minutes column but rather 60', thereby reducing 122° to 121° and adding 60' to 14' to get 74', which then is subtracted from 26'. The answer is now 3° 48'.

How is any of this applied in everyday life? Following are some examples. We can use geographic coordinates to search for places in Google Earth. U.S. Federal Law (the Patriot Act) requires your cell phone to have a GPS device to locate your position in case you dial a 911 call. Next time you travel on an airplane look in at the cockpit. There you will see a large digital display of the aircraft's current latitude-longitude location with a corresponding latitude-longitude location for its final destination or for landmarks it passes over along its flight path. The aircraft's computers are calculating the air miles using latitude-longitude as you fly to your holiday destination.

One last element of a location is its situation, which provides the context for a location's regional position in relation to other locations. Many things can be said of California's situation. It is situated on a tectonic plate boundary (the Pacific rim of fire), which leads to earthquakes and volcanoes. Its West Coast continental position between latitudes 32° 30'–42° N provides a Mediterranean climate. For another example, revisit Figure 1.1 and follow straight lines of latitude from California's northern and southern borders east to Europe and you will see that the states latitudinal range is the same as Southern Europe, the Mediterranean Sea and North Africa. It was also historically **isolated** geographically. Native American populations in California were separated from other populations via deserts and mountain ranges. These same physical barriers, as well as, long distance ocean voyages from historical sources of settlements made for hindered European exploration and settlement, and later American migration movements. Finally, it has a strong historical and modern Latin American connection, and a modern connection with the Asian Pacific Rim economies. Later chapters will explore all of these topics to provide the context for the current situation California finds itself in and as a means of providing ways to explore this region's future.

1.1.2 PLACE

A place is any location and its immediate surroundings deemed unique enough to name. This includes inhabited and many uninhabited places. A **toponym** is the name given to a place on Earth. A place may be named for a person, be related to religion or the origin of its settlers or derive from features of the physical environment, to name just a few situations. The tremendous ethnic variety of the pioneers lured to California by the prospects of finding gold in the 1850s left many identifying and picturesque names of cultural diversity on the landscape. These include French Gulch, Dutch Flat, Kanaka Bar, Mormon Bar, Goldpan Canyon, Rough and Ready, and Beggars Bar to name a few. Or consider the number of coastal counties with Spanish Catholic saints' names, for example, San Diego, Santa Barbara, Santa Cruz and so forth. Each is connected with the extent of settlement of the Spanish mission era.

Generally, a place is a geographical area that we can define by its geographical characteristics and location, its historic events and by physical or cultural change over time. Places are described by their:

■ **Physical characteristics**—such as topography, landforms, climate, geology, relation to water and native plants and animals, natural hazards, etc.

■ **Cultural characteristics**—such as the ethnic character of the population, the history of settlement and human artifacts (building types, settlement patterns, roads) and the socio-economic/political character of the population (demographic age breakdown, Democrat/Republican/Libertarian, clothing, food habits, etc.). These usually result in changes to the natural environment through cultural tradition or development of agreed upon policy.

The image people have of a place is based on their experiences, both intellectual and emotional. People's descriptions of a place reveal their values, attitudes and perceptions. For example, John Muir in 1901 described the Sierra Nevada as such "All the world lies warm in one heart, yet the Sierra seems to get more light than other mountains. The weather is mostly sunshine embellished with magnificent storms, and nearly everything shines from base to summit. . . . Well may the Sierra be called the Range of Light, not the Snowy Range; for only in winter is it white, while all the year it is bright."

Physical Characteristics

California's physical characteristics as a whole represent a place of exceptional natural landscapes with high diversity. For example:

- Highest point (Mt. Whitney (36° 34′ 43″N, 118° 17′ 36″W), 4421 m/14,505 feet and lowest point (Death Valley (36° 27′ 20″N, 116° 52′W), −85.5 m/ −282 feet in the continental U.S.

- Deserts (Mojave and Colorado) and temperate rainforests (coastal redwood forests)

- Large topographic variations (i.e., Klamath, Cascades, Sierra Nevada mountains), broad valleys (i.e., Great Central Valley, Salton Trough), and long narrow coastal valleys (i.e., Santa Clara, Salinas and Lompoc valleys).

Overall, California possess impressive and enchanting scenic beauty.

California also represents a region of unique environments driven by the strong oceanic influences of the cold southward flowing California current. This allows for the development of the Mediterranean climate California enjoys throughout the year, defined by cool winter rains and warm summers of drought. This climate only occurs in five areas of the world: California, central Chile, southwest South Africa, southwest Australia, and the Mediterranean Sea region. This uniqueness leads to a productive and species-rich marine and terrestrial environment, as well as to excellent agricultural land (if supplemental water can be found to apply to crops). In addition, the unique historical geological environment that is part of being on the Pacific and North American plate tectonic boundary has left California with abundant mineral wealth, a key feature to its geographical and historical development.

Another way of viewing California's physical characteristics is by observing its severe constraints and natural hazards. So, while having a Mediterranean climate type is unique, it also is prone to a highly variable rainfall regime and a very long summer drought, which leads to a hazardous fire regime. In addition, the topographic barriers caused by the numerous mountain ranges in the state lead to large arid and semi-arid regions to the east of their ridge lines (see pages 6 and 7 in the *California Atlas*). All of these elements can lead to increased concerns over water supply, both for the natural environment and for human needs.

California is a land of dangerous, unstable landscapes. Earthquakes, volcanoes, floods and steep erodible slopes lead to mudslides; and active coastlines and seasonal wildfires contribute to a sense of uneasiness and chaos. The unpredictable nature of many of these natural hazards and the seasonal foreboding created by the others are all facts of life for those who choose to live on the West coast of North America.

Cultural Characteristics

Along with the state's physical traits, the people of California have always been represented by diversity and dynamic changes, and they are connected to the environment in profound ways. To provide some clear examples, however, we need to see the strong impact of cultural groups on histories and on the built environment. In fact, since the arrival of the first Native Americans in the region, California has always had a highly diverse human population. We will see this illustrated later on in Chapter 5 by the large number of Native American language groups, through the arrival of the Spanish and Russian explorers, the people of exceptional character and diverse ethnicities who arrived during the Gold Rush, the melting pot dynamics found in the region after WWII and finally in the strong focus upon modern immigration to keep the state supplied with various levels of skilled and unskilled labor.

This textbook will examine relationships between humans in general and upon specific cultural groups with nature in particular. For example, human development in California is strongly tied to living in areas where land is flat, water is abundant (or if not, it can be transferred from another source) and the climate is mild. The majority of people have avoided living in the mountains, deserts and areas with harsh climates. When looking at cultural groups specifically, we can see that their spatial living choices are linked more directly to economy: agriculture, manufacturing, services and past political policies. For example, why do African-American communities congregate in inner city areas within the two main urban centers, San Francisco Bay area and Los Angeles County? Why were the Okie-Arkie Dust Bowl migrants of the 1930s particularly associated with Bakersfield, also known as Nashville West?

1.1.3 MOVEMENT

Humans move themselves, their possessions and their ideas/cultural heritage from place to place. The movement of people, the import and export of goods and the advent of mass communication have all played major roles in shaping our world. People everywhere interact in some way. They travel from place to place and they communicate. People interact with each other through movement. Humans occupy places unevenly on Earth because of the environment but also because we are social beings (i.e., ~80 percent of the U.S. population is located within fifty miles of the coast). We interact with each other through travel, trade, information flows (e-mail, cell phone text messaging, webcams) and political events.

Geographical science studies the **migration/immigration/emigration** of people (as well as plants and animals), the *transportation* of goods, the *access* to services and the *diffusion* of ideas and technologies. Migration (Figure 1.2) is linked to the situational terms of:

- **Push-Pull**—factors or situations that (a) force migrants to leave their place of origin and (b) attract them to a new destination. Push factors represent the negative attributes of a migrant's place of origin (i.e., unemployment, wars and environmental disasters). Pull factors are positive attributes of a place of destination. Many pull factors are simply the reverse of push factors, such as employment opportunities, good quality services, and safety. In any case, the most powerful push and pull factors are economic ones.

- **Help-Hinder**—factors or situations (obstacles) that might help or hinder one from reaching a new place (i.e., international immigration policy, specialized employment, poor communications, family divorce, death of a family member, etc.)

Diffusion is the process by which a characteristic spreads across space from one place to another over time. The interaction of sophisticated communications and transportation networks in the world is complex. The ability to upload and share on the World Wide Web has encouraged rapid diffusion, because Web surfers throughout the world have access to the same material simultaneously and quickly.

The place from which an innovation originates is called a **hearth**, and the innovation diffuses from there to other places. A hearth can emerge via a cultural group willing to try something new by being able to allocate resources to nurture the innovation. At the same time the group must also have the technical ability to achieve the desired idea and the economic structures to facilitate its implementation. The technical culture of the Santa Clara valley in the '60s and '70s is a case of a hearth from which the personal computer emerged. In 1973, the Intel 8080 microprocessor was developed in the valley, which eventually led to the development of the first Apple II computers and later to the IBM personal computers and their compatibles. Today the Silicon Valley is synomous with the personal computer and the software industry—and now the global World Wide Web via famous companies such as Google and Facebook.

Last, we must confront the term **diaspora**, which refers to any population that shares a common cultural or ethnic identity and that was either forced to leave or voluntarily left its place of origin and became residents in often completely different areas. While we could connect this term with the discussion on migration/immigration above, and it has its associated push/pull and help/hinder elements, the context is different. It represents a mass movement of people from one area to another typically because of the following conditions: strife in the homeland (i.e., the Chinese male diaspora in the nineteenth century to the California Gold Rush); economic pressure (i.e., the African-American diaspora in the 1940s leaving the U.S. South for large California cities; and the Mexican-Hispanic diaspora since the 1970s to California); environmental disaster (i.e., Dust Bowl events with the migration of Okies-Arkies to California in the 1930s); and finally

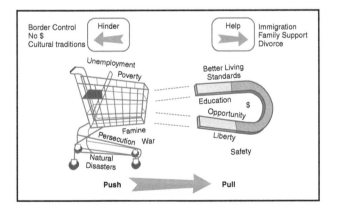

FIGURE 1.2 *Conceptual illustration of the migration issues faced by families and the cultural elements that get carried or left behind during geographical transitions.*

ethnic cleansing (i.e., Native Americans in California escaping from the coast and Sierra Nevada foothills and eventually receiving placement within the Rancheria reservation system). Each of these groups has displayed a different course from its original place of settlement in California's cultural development.

1.1.4 HUMAN-ENVIRONMENTAL INTERACTION

As people move about and settle in different places on the Earth they interact with the physical environment. Humans and the environments they encounter affect each other. Humans change the environment to suit their needs, and then sometimes a location's natural processes change it back—for example, in California there are floods, tornadoes, hurricanes, and especially earthquakes, mudslides and fires.

The concepts involved in human-environment interaction cover adaptation, modification and dependence with two larger general themes usually dominating our understanding:

- **Geographical determinism**—is a measure of the degree to which the physical environment influences human cultural development (i.e., crop-climate constraints). Humans tend to live in flat areas, with abundant water, mild climates, and so forth. Native Americans once established their densest settlements where there were abundant water and food resources. Modern California's population density clearly shows this as the majority of people reside in coastal and inland coastal valleys and plains as opposed to the mountainous regions of the state. Of course this has changed over time. For instance, during the Gold Rush, which lasted from 1849 into the 1880s, the miners had to live near their gold claims, so the Sierra Nevada foothills and mountains became well established population centers. Then, with the demise of gold mining, mass abandonments of these settlements took place in favor of the Central Valley towns and coastal cities. Determinism is becoming an issue again in the era of climate change as agricultural crops will be forced to move to new, more favorable areas for optimal production.

- **Possibilism**—describes the degree to which a particular culture is able to modify its physical environment. We heat and cool our buildings for comfort. We choose suitable clothing for the diversity of climate regions and summer or winter conditions. We use large scale engineering to transfer massive amounts of water from regions of abundance to regions of poverty. Natural factors, as a whole, no longer play major roles in determining the concentrations of California's population.

Since the arrival of the Spanish in California there has been a continual need to modify and control nature. We can see this largely in the development of large water projects to transfer water from wetter parts of the state to drier parts. For example, large water transfer schemes are used to 'green' inland valleys for large-scale agricultural production. The southern San Joaquin valley is considered desert-like in its climatic condition, but irrigated agriculture has severely modified this, as large canals bring water from well watered places like the Klamath Mountains and the Cascade Range in the northern region of the state.

In addition, the development of a large-scale energy grid and highway transportation network allows for the large urban areas to quickly spread east away from the coastal conveniences and into the inland valleys. In the case of Southern California, this penetration is now taking place in the harsher climates of the Mojave (Antelope valley) and Colorado (Coachella and Imperial valleys) deserts. Large scale land developments continue to encroach into the Great Central Valley, some even moving back into the Sierra Nevada Foothills in areas that had depopulated directly after the Gold Rush era ended (i.e., Grass Valley). But it is in these urban fringe areas, like the chaparral covered mountains of Southern California or the pine forests of the Sierra Nevada foothills, that society is facing natural processes that may keep expansion in check, i.e., wildfires, mudslides, flooding and encounters with mountain lions or bears. Nature can still control people at some stage within this realm.

1.1.5 REGION

A region is the basic traditional unit of study in geography. A region is any contiguous area of the earth's surface where all places within it share a particular set of physical and/or cultural characteristics. Regions are human constructs that we can map and analyze.

The Sierra Nevada mountain range is a physical region defined by geologic history and rock type. Cultural regions in California include such examples as the vineyard growing and wine making areas within the coast range valleys or the Okie-Arkie settlement region in the San Joaquin valley.

There are three basic types of regions (Figure 1.3):

■ **Formal regions** are those defined by governmental or administrative boundaries (i.e., the United States, California or Los Angeles County). These regional boundaries are not open to dispute; therefore physical regions fall under this category (i.e., the Sierra Nevada, the Great Central Valley and the Mojave Desert).

■ **Functional regions** are those defined by a function (i.e., San Jose City limits, the State Water Project, air quality monitoring basins, the Bay Area Rapid Transit [BART] or a newspaper service area—San

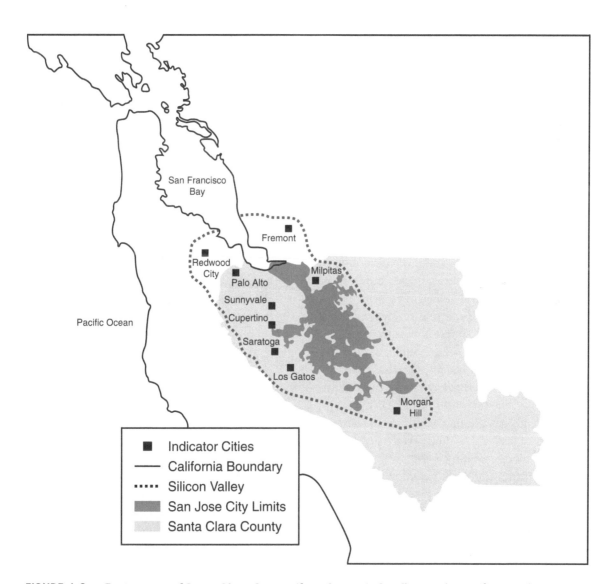

FIGURE 1.3 *Regions map of Santa Clara County (formal region), San Jose city limits (functional region) and Silicon Valley (vernacular region). (Courtesy Curtis Page)*

Diego Tribune). If the function ceases to exist, the region no longer exists.

- **Vernacular regions** are those loosely defined by people's perceptions (i.e., East LA, The Hood (South Central LA), The Silicon Valley, The Valley (San Fernando), the Tenderloin District (San Francisco), the North Beach district (San Francisco), the Inland Empire (Riverside), Emerald triangle (Marijuana growing region of Mendocino, Humboldt, and Trinity Counties), etc.

1.2 Culture and Cultural Landscapes: Basic Themes

Culture refers to patterns of human activity and the symbolic structures that give such activities significance and importance. We can understand cultures "as systems of symbols and meanings that even their creators contest, that lack fixed boundaries, that are constantly in flux, and that interact and compete with one another." Sociologists often define these as "learned behavior" that consists of several critical elements (cultural elements):

language	clothing
religion	housing styles/buildings
race and ethnicity	settlement pattern
gender	roads
diet	politics

We can say several things about the **elements of culture**, mainly that they are specific, located in space (i.e., they originate from or are found in a specific place), purposeful, rule-following and rule-making, and they rely on communicating and interacting with people.

By further narrowing, we can address the term **cultural geography**. This consists of specific cultural traits contained in an area that has its own cultural history and ecology. The focus can be on urban or rural environments and also on popular culture and vernacular architecture (as long as they pertain to an area). Therefore, cultural geography is the study of cultural products and norms and their variations across and relations to spaces and places and time. It focuses upon describing and analyzing the ways language, religion, economy, government, built environment and other cultural phenomena vary or remain constant from one place to another and upon explaining how humans function spatially. The study of a cultural geographic area is usually done over time to understand the **cultural sequence** of who had been in an area and what remains, and how the cultural elements retain or lose their representation among successional cultural groups or internal cultural group evolution.

In an analytical sense, cultural geography can be the study of cultural areal differentiation. This represents, for example, a study of the differences in ways of life and encompasses ideas, attitudes, languages, practices, institutions, and structures of power, which include the whole range of cultural practices in geographical areas.

Last, we come to what Carl Sauer (late geographer from U.C. Berkeley) developed as the ultimate aim of understanding cultures and societies, which not only develop out of their environmental landscape (determinism), but also shape them too (possibilism). This interaction between the 'natural' landscape (topography, vegetation, climate, etc.) and humans (structures, settlement patterns, etc.) creates the cultural landscape. His classic definition of a **cultural landscape** reads as follows:

"The cultural landscape is fashioned from a natural landscape by a cultural group. Culture is the agent, the natural are the medium, the cultural landscape is the result."

The United Nations World Heritage Committee has defined cultural landscapes as distinct geographical areas or properties uniquely "represent[ing] the combined work of nature and of humans . . ." This concept has been adapted and developed within international heritage arenas (UNESCO) as part of an international effort to reconcile "one of the most pervasive dualisms in Western thought—that of nature and culture."

The World Heritage Committee has identified and adopted three categories of cultural landscape. These range from (i) those landscapes most deliberately 'shaped' by people, through (ii) an "organically evolved landscape" that may be a "relict (or fossil) landscape" or a "continuing landscape," to (iii) those least evidently 'shaped' by people (yet highly valued because of the "religious, artistic or cultural associations of the natural element"). In the case of California, the region has two natural landscapes (category iii) as part of the World Heritage List: Redwoods National and state parks and Yosemite National Park.

1.3 California Geography: Basic Themes to Explore

1.3.1 An Approach to Understanding California

This textbook was not designed to take the reader on a traditional geographer's systematic topical overview of the state, such as covering individual chapters on physical geography topics, population/migration, ethnic groups, industry, economy, energy, urbanization, etc., or conducting a region to region survey that simply describes each landform section of California (i.e. its physical and cultural makeup). While some of the systematic topics are important and we can consider California's twelve physiographic regions a structured approach to studying environmental and cultural differences around the state, the geographical dynamics brought out over a geospatial cultural-historical approach best informs us where California started and where it has the potential to go. We can explain so much of the geographical diversity found in the state through related and connected forces and landscapes approached within a space and time framework. In this respect, it is best to first describe and explain the physical environment that is California and then to 'roll the tape forward' by populating the environment with the diversity of cultural groups that have been part of the state, from 15,000 b.p. (i.e., Native Americans) to Spanish 'discovery', the Mexican period's Californios, the U.S. takeover/Gold Rush, the resulting ethnic conflict, the rise of Southern California, irrigation agri-cultures, the Depression era, the WWII era and beyond to the present 21st century. To understand where this state currently 'is' and where it might be headed it is important to know that landscapes hold physical and cultural memory and thus California is an integration of its past and future dreams, and nightmares. Using the approach of historical geography interspersed with a systematic topical approach to examining important geographical themes and processes is a more fulfilling method to understand the multi-faceted complexities of California's cultural landscapes.

1.3.2 Diverging Images of California

This textbook examines the many myths of California—the dreams and the nightmares. The state has had many boosters; even before Spanish 'discovery' it was woven into a mythical land and written about as a romantic fantasy while sought as a real landscape by its early European explorers—an earthly paradise loaded with gold and ruled over by Queen Califia—the powerful Amazonian. The 1849 Gold Rush was a powerful push factor motivating young men from far flung places to 'strike it rich' in what was perceived as an easy manner, and the West was never the same again. These people of exceptional character and diverse ethnic background believed in the power of the golden state to make their dreams of an easier life come true. For others, however, California represented the final frontier, the manifest destiny of a young nation striving for representation on both continental coasts. The westward expansion of farming families and business entrepreneurs into a state brimming with substantial resource wealth allowed for a public personality type to develop in the state, one of people open to experience. In many respects, since the Gold Rush California has been a state known as the land of unlimited possibilities. This has all led to tremendous economic productivity and national specialization in such varied fields as winemaking, computer/software engineering, biotechnology and the entertainment industry. Finally, the varied geographical history, the stunning physical features and the mild climate have produced a whole public relations industry geared towards presenting California as the land of gracious and affluent living. From California cuisine, patio/BBQ life, hot tub culture to beach and surfing culture, to hip hop dance culture ("krumping") in the inner cities, and Western rodeos in the rural regions, the state's image of itself reflects an idealized past and an idealized present.

Unfortunately for many, the promised land has its dark side; there is a California of nightmares. Since Spanish 'discovery' an unending pressure of environmental and cultural damage has occurred. Native American groups have suffered under introduced diseases and the supposedly helping 'reforming' hands of the mission period, as well as further loss under racist brutality and the holocaust-like conditions brought on by the anarchy of the Gold Rush. By the early 1900s California's once unique and diverse Native American groups had been nearly exterminated, leaving less than 10,000 individuals out of an original estimated 350,000. The reality is that California fostered the worst slaughter of Native American peoples in U.S. history. In many ways the madness of the Gold Rush itself brought on numerous problems by rapidly bringing together more diverse immigrants than occurred in any other part of the U.S. in a short time period and concentrated to a defined area. While many of these young men never meant to stay, just to get their share of the gold and then go home, mining failure was a common fact. Frustrations ran high among the miners stuck in impossible conditions, which led to swindling, corruption and exploitation among the many conflicting ethnic groups. Non-Caucasians and non-Americans alike lost out to racist attitudes and brutality through scapegoating, and many of these attitudes have continued into present day California. Swindling and exploitation have played a role up to the present day's bust of the Dot-com economy (1995–2001) and the unsupportable home and commercial real estate market (2001–2007), brought on by boosters, developers, and realtors. This last event, with respect to the housing boom and eventual bust, had a similar scene witnessed in Southern California during the 1880's for the same reasons.

In addition, the Gold Rush was the start of many severe pollution problems inflicted on the state's waterways (i.e., mercury contamination and increased sedimentation) and eventually on air quality through the ripple effect of rising urbanization in valleys atmospherically controlled by thermal inversions. The final culmination of a turbo-charged economy with many players wanting to make it big in the California dream is the crass materialistic culture found largely in the state's urban centers. Many outside the state see California's urban and planned community suburbia landscapes as shallow, cultureless wastelands. These are ones that Hollywood film and TV studios constantly exploit (i.e., *Desperate Housewives*, *The OC*, and *Beverly Hills 90210*).

All the above elements play into five major issues in California's recent past and soon to be future nightmares: political myopia through the liberal/conservative divide, immigration friction, coastal versus inland values and economic livelihoods, urban/suburban sprawl, and finally the ongoing problems with water resources. Towards the end of this textbook we will address these issues after we understand that California's future has a unique geographic past.

Bibliography

Atkinson, D., Sibley, D., Jackson, P., and N. Washbourne (2005). *Cultural geography: a critical dictionary of key concepts*. I.B.Tauris.

Crang, M. (1998). *Cultural Geography*. Routledge, New York, NY.

Durrenberger, R.W. and R.B. Johnson (1976). *California Patterns on the Land*. 5th edition, Mayfield Publishing Company, Mountain View, CA.

Gudde, E.G. (1969). *California Place Names: The Origin and Etymology of Current Geographical Names*. University of California Press, Berkeley, CA.

Jordan, T., Domosh, M. and L. Rowntree (1994). *The Human Mosaic: A thematic introduction to cultural geography*. HarperCollins, New York, NY.

Lantis, D.W., Steiner, R., and A.E. Karinen (1989). *California: The Pacific Connection*. Creekside Press, Chico, CA.

Michaelson, J. (2008). *Geography of California*. Course at UC Santa Barbara, Dept. of Geography. [online] http://www.geog.ucsb.edu/~joel/g148_f08/.

Miller, C.S. and Hyslop, R.S. (1983). *California: The Geography of Diversity*. Mayfield Publishing Company, Mountain View, CA.

National Geography Standards, Geography Education Standards Project (1994). *Geography for Life: The National Geography Standards*. National Geographic Society Committee on Research and Exploration, Washington D.C.

National Geographic Society (2009). *Geography Standards*. [online] http://www.nationalgeographic.com/xpeditions/standards/

Pannell, S. (2006) *Reconciling Nature and Culture in a Global Context: Lessons from the World Heritage List*. James Cook University, Cairns, Australia.

Selby, W. (2006). *Rediscovering the Golden State: California Geography*. 2nd edition, John Wiley Press, New York, NY.

Starr, K. (2005). *California: A History*. The Modern Library, New York, NY.

UNESCO (2005). *Operational Guidelines for the Implementation of the World Heritage Convention*. UNESCO World Heritage Centre. Paris, France.

CHAPTER 2

Geological History and Landforms

Key terms

Antler orogeny	Intensity (Mercalli) scale	Sacramento Valley
Basalt	Juan de Fuca Plate	San Andreas fault
Basin and Range	Kettlemen Hills	San Joaquin Valley
Batholith	Klamath Mountains	Sedimentary
Cistmontane	Lake Tulare	Serpentine
Coast Ranges	Landform regions	Sierra Nevada
Colorado Desert/Salton Trough	Magnitude scale	Sonoma orogeny
Conservative	Mendocino Fracture Zone	Southern Cascade
Constructive	Metamorphic	Strik-slip
Destructive	Modoc Plateau	Subduction
Earthquakes	Mojave Desert	Sutter Buttes
East Pacific Rise	Mt. Lassen	Topography
Farallon Plate	Mt. Shasta	Transmontane
Franciscan mélange	Nevadan orogeny	Transverse motion
Geomorphology	Orogeny	Transverse Ranges
Glaciation	Peninsular Ranges	Tsunami
Granitic	Plate tectonics	Volcanoes
Great Central Valley	Rock cycle	
Igneous	Sacramento-San Joaquin Delta	

Introduction

The diversity of California's natural landscapes is truly extraordinary. In no other state is the variety of physical environments so wide, ranging from warm deserts to temperate rain forests, from below sea level basins to glaciated peaks, and from coastal salt marshes to alpine meadows. This **topographic** variation, which is one of the key factors responsible for landscape diversity, as well as California's continental West coast position and its large latitudinal range, are unmatched elsewhere in the U.S. The diversity and contrasts begin along the coastline then continue eastward over a series of mountain ranges that front the coast. From there they cross the broad flat valleys, rise over the glaciated crests of the Sierra Nevada Range and the volcanic landscapes of the Cascades and then move into the diverse desert areas of eastern California. These desert areas include the Colorado and Mojave Deserts in the south and the Basin and Range in the central and north (see page 5 in *California Atlas*). The cumulative effects of plate tectonic driven crustal plate deformations wrenching mountains higher, and occasionally allowing volcanic activity to come to the surface under great pressures, should remind us that sometimes violent geologic processes are shaping California. Millions of years of these events coupled with continued weathering-erosion from cycles of rain, snow, and heat including occasional episodes of glaciation have combined to create the varied landscape regions we have come to recognize and appreciate in this state.

2.1 Basic Geology: The Building Blocks

Understanding the **constructive** and **destructive** processes responsible for the physical landscapes that are emerging and changing over time in California will help us place current landscapes in proper perspective. This section provides a brief introduction to plate tectonics and the rock cycle to understand the evolution of landscapes, and the geological development and distribution of mineral resources that were historically significant to the economic development of the state.

To understand California's physical landscape is to understand **plate tectonics**. The Earth's upper crust consists of discrete plates (Figure 2.1) riding on top of the plastic-like asthenosphere of the Earth's upper mantle. Radioactive decay at the Earth's core creates convective movement of heat that rises through the lower and upper mantles, allowing the asthenosphere to flow gradually. The rising heat drags the brittle plates along, causing their surfaces to break and bend. There are two types of plates: light granitic-based continental plates representing older crustal formations and denser basalt-based oceanic plates representing younger crustal material. Over hundreds of millions of years plates have drifted around the planet. Most geologic or tectonic activity occurs along plate boundaries. There are three major types of active plate boundaries occurring in California (Figure 2.2): **constructive** interaction-divergent spreading zones, **destructive** interaction-converging and subducting plates, and **conservative** interaction-transform sliding plate boundaries.

2.1.1 Constructive Interaction

This motion occurs when two plates are spreading away from each other, thinning the crust between them and causing a gap where new material can emerge in the form of magma to fill the gap. In the oceanic basins this situation creates mid-ocean ridges, like the most famous one that snakes down through the center of the Atlantic Ocean (Figure 2.1). In California, this type of seafloor spreading occurs north of Cape Mendocino, where the Gorda plate is moving to the east away from the Pacific plate. Similar spreading is occurring in the Gulf of California with the Baja peninsula moving west away from the Mexican mainland and causing a widening of the gulf that continues on up into California's Salton Trough (see page 5 in *California Atlas*). In and around the southern Salton Trough the crust is thinning, which is called a continental rift zone. This is similar to what is happening in the Great Rift Valley of East Africa, although on a smaller scale. Evidence for thinning in the Salton trough comes from a series of small active mud volcanoes in the region, which are formed by a shallow magma chamber superheating the groundwater that rises to the surface with clays. This region is noted for its geothermal power generation because of this spreading phenomenon.

2.1.2 Destructive Interaction

This type of plate activity consists of two types of converging zones: one that subducts and one that compresses. In the case of **subduction**, one plate is usually heavier, allowing the other lighter plate to ride up over the heavier plate (Figure 2.3). This forces the heaviest plate to plunge back into the asthenosphere causing the destruction of the plate as it melts back into magma; it is one way of viewing our planet's recycling program.

There are two different ways subduction can occur, and each creates a different landscape outcome. One occurs when the subduction happens between two

CHAPTER 2 Geological History and Landforms 15

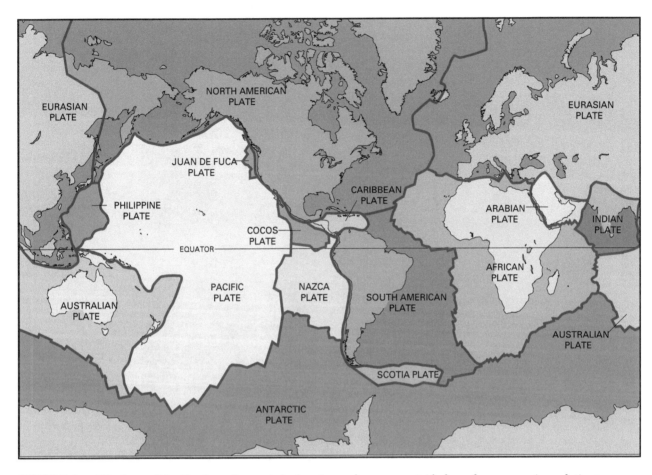

FIGURE 2.1 *The layer of the Earth we live on is broken into a dozen or so rigid plates that are moving relative to one another. (Courtesy U.S. Geological Survey, http://pubs.usgs.gov/gip/dynamic/slabs.html)*

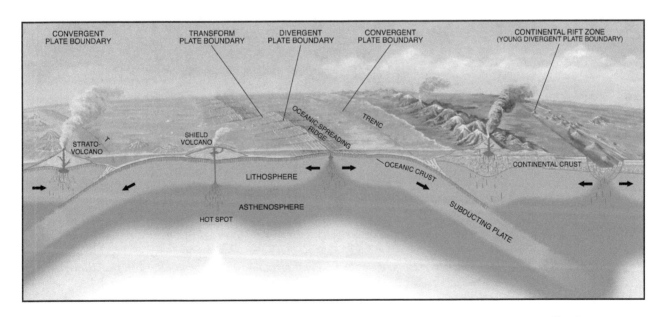

FIGURE 2.2 *Conceptual view of the main types of plate tectonic boundary processes occurring within the Earths crust. (Courtesy U.S. Geological Survey, http://pubs.usgs.gov/gip/dynamic/Vigil.html html)*

16 **CHAPTER 2** *Geological History and Landforms*

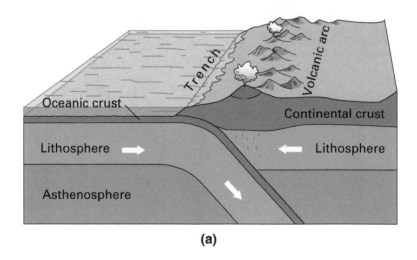

(a)

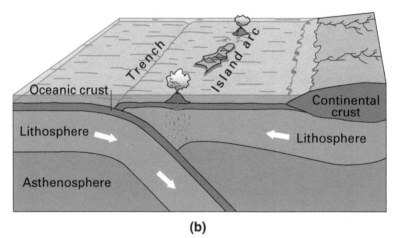

(b)

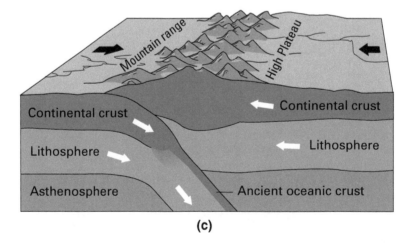

(c)

FIGURE 2.3 *Illustration of: (a) oceanic-continental convergence with subduction and the formation of continental volcanoes; (b) oceanic-oceanic convergence with subduction and the formation of volcanic island arcs; and (c) continental-continental convergence and the formation of compression-based folded mountains. (Courtesy U.S. Geological Survey, http://pubs.usgs.gov/gip/dynamic/understanding.html#anchor6715825)*

oceanic plates. The extra magma melt from the one plate erupts under the other oceanic plate to form volcanic island arcs. Japan, New Zealand and Alaska's Aleutian islands are the most famous forms of this interaction (Figure 2.3a). In the other case, the denser oceanic crust subducts under the lighter continental crust, thus melting and causing the extra magma to erupt on the surface of the continental plate creating a volcanoe or chain of volcanoes (Figure 2.3b). The Cascade Range of **volcanoes** represents this type of interaction. The southern end of the Cascade Range is in California, and its two most prominent volcanoes are **Mt. Lassen** and **Mt. Shasta**.

Finally, there is the case where compression exists between two plates. This is not unlike two cars having a frontend collision with each other. In the case of plate activity, we usually see it happen when two continental plates smash into each other, and instead of subduction the crustal material becomes compressed upward, creating folded/faulted mountains (Figure 2.3c). In California, the best examples of this type of interaction include the coastal ranges, the Transverse ranges and the Peninsular ranges. An example of the mountain-building power of this compression occurred during the Paso Robles earthquake of 2003, which caused some portions of the southern end of the Santa Lucia Mountains of Big Sur to rise almost one foot in elevation.

2.1.3 CONSERVATIVE INTERACTION

The infamous **San Andreas Fault** system represents the interaction between two plates that slide past one another (Figure 2.4). The **transverse motion** that this type of plate boundary represents on continents is called **strike-slip** faulting. In California, the Pacific plate is sliding northwest past the much slower North American plate at the rate of two inches per year (Figure 2.5). While the North American plate is also moving northwest, it is moving slower, so it appears as if the plates are going in opposite directions along their slip boundary. This is not unlike the relative motion one witnesses when passing another car on the freeway, even though both cars are going in the same direction. An example of how much these two plates have moved has

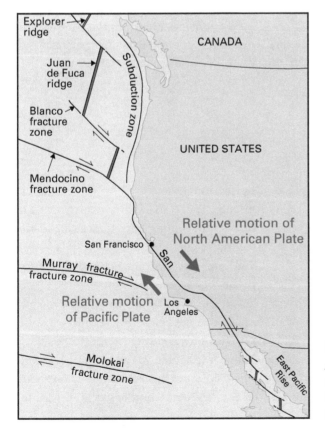

FIGURE 2.4 *The Mendocino and Murray fracture zones are some of the ocean floor transform faults. The San Andreas is one of the few transform faults exposed on land. (Courtesy U.S. Geological Survey, http://pubs.usgs.gov/gip/dynamic/understanding.html#anchor6715825)*

18 **CHAPTER 2** *Geological History and Landforms*

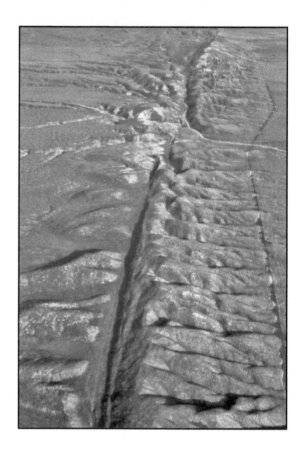

FIGURE 2.5 *Aerial view of the San Andreas fault slicing through the Carrizo Plain east of the city of San Luis Obispo. (Courtesy U.S. Geological Survey, http://pubs.usgs.gov/gip/ dynamic/San_Andreas.html)*

been documented using the distance between the San Gabriel Mountains in the Transverse Ranges (east of Los Angeles) and the Orocopia Mountains in the Colorado Desert (east of northern end of the Salton Sea). They once were together, but now they are separated by 130 miles. One other infamous transverse motion is occurring off Cape Mendocino, known as the **Mendocino Fracture Zone**. Geologists call the slipping along plates, either oceanic or continental, a transform fault.

2.2 THE ROCK CYCLE

In order to trace California's geological history, it requires a brief introduction to the **rock cycle**, which informs us of the formation, classification and names of rock types. Following Figure 2.6, there are three major rock formation categories: **igneous**, **sedimentary** and **metamorphic**.

Igneous rocks are derived directly from magma sources in two different ways—by extrusive or intrusive display. Geologists consider extrusive igneous landforms and the rocks that come from this activity the more exciting of the two. In this case, magma pushes up to the surface of the crust and extrudes onto the surface either by developing a volcano or by flooding the surface of a landscape. Extruded igneous rocks are developed from rapid cooling once they erupt on the surface, resulting in smaller mineral crystals. Black **basalt** and andesites represent the two most common igneous extrusive rock types.

Intrusive igneous rocks also derive from a magma chamber, but one that does not break the surface crust. Instead, the magma moves slowly towards the surface, which then allows it to cool slowly and eventually solidify into rocks with larger mineral crystals. **Granitic** rocks are the best representation of intrusive igneous

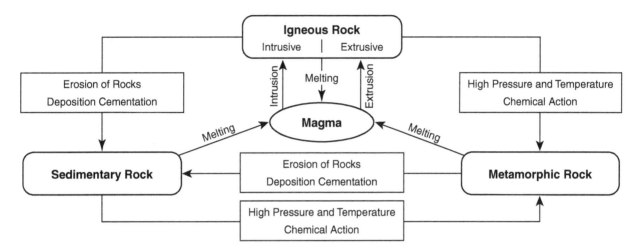

FIGURE 2.6 *The rock cycle.*

rock types. They usually form a **batholith**—a large emplacement of igneous intrusive rock that forms from cooled magma under the crustal surface.

Once new material has made it to or near the surface via either of the igneous processes, the cycle of weathering, erosion and transport of broken-down rock material can occur. The transport of rock material from one area to eventual deposition in another, its compaction over time through the accumulation of overlying material and eventually the cementation via chemical reactions and great pressure from the weight leads to the formation of sedimentary rock types. Sandstone, mudstone, shale and limestone represent examples of sedimentary rocks.

Metamorphic rocks represent the last rock category. In special situations, further application of heat and pressure that does not totally melt the original rock can change igneous or sedimentary-derived rocks. The heat and pressure allow further chemical reactions to occur that changes the original rock chemistry into new rock types. For example, limestone becomes marble or sandstone becomes quartzite. Plate tectonic activity in California has created large pockets of metamorphic rocks, such as schists and gneisses, which tend to be problem areas for road building or general construction. In fact, our state rock, **serpentine**, is a type of schist that swells with water and is very slippery during the winter rainfall season, causing numerous landslides around the state. Next time you are driving Highway 101 through the central coast (especially around the San Luis Obispo area) or you are on Highway 1 through Big Sur look for the green rock at the road cuts. Then look for the Caltrans road repair crews.

2.3 A Brief Geologic History of California

While scientists estimate the Earth is 4.6 billion years old, they estimate that the oldest rocks still visible in California are around 1.8 billion years old. These rocks are largely metamorphically derived and located in the Basin and Range province, in the Mojave Desert and in the San Gabriel Mountains, which are part of the Transverse Ranges. Geologists consider the time between 1.8 billion and 400 million years b.p. a quiet period in the region where California eventually became located. This means that no major mountain building processes were at work. For the most part, the western edge of North America consisted of shoreline running from what is now southern Idaho through central Nevada to southeastern California.

Between 700 and 300 million years b.p. California lay largely underwater as a broad marine shelf, deeper in the north and shallower in the south, as marine sedimentation studies have observed. Therefore, the then western coastal edge of North America was eroding onto the marine continental shelf and creating large sedimentary deposits. These deposits represent largely limestone or carbonate rocks still found in the Klamath Mountains, parts of the higher southern Sierra Nevada and in thicker layers in mountains east of the Sierra Nevada, i.e., Inyo and White Mountains.

The first mountain building (**orogeny**) era began sometime between 400 and 65 million years b.p. The first major disruption is called the **Antler orogeny**, which lasted from 400 to 250 million years b.p. During this period there was complex tectonic activity from roughly the latitude of Monterey northwards (but more prominently off the northern Sierra Nevada and Klamath regions) in the form of a set of volcanic island arcs, typical of an oceanic-oceanic subduction zone (Figure 2.3b). As the North American continent rode over these subducted rocks, material became welded onto its western edge, and the volcanic island arc joined the region of what is now the Sierra Nevada foothills and the central Klamath Mountains during what is called the **Sonoma orogeny** (250 to 210 million years b.p.). These are the volcanoes that would later contribute to the wealth of gold found in the Sierra Nevada and Klamath Mountains.

During the 210–100 million years b.p. timeframe the collisions along the last of the subducting **Farallon plate** resulted in the **Nevadan orogeny**, one of the greatest mountain-building episodes in California's geologic history. The subducted material melted and rose in huge magma chambers that never made it to the surface. Instead they slowly cooled into batholiths of granite that would form the future Sierra Nevada and Peninsular Ranges and the Salinian granites of Monterey/Carmel and northern Big Sur.

The uplift of the ancestral Nevadan mountains from the collision pushed the sea farther to the west and created a barrier along what is now the eastern Great Central Valley. This represented the western boundary of the North American continent. As things rise so must they erode, and during the 100–65 million years b.p. period the Nevadan mountains weathered, eroded and deposited their sediments on the continental shelf, slope and trench. The sedimentary rocks that these ancient mountains created in their erosion phase occur scattered throughout California west of today's Sierra Nevada range. Most importantly, this material is the modern basis for the Great Central Valley sediments, which in some areas are up to six miles thick.

Rise of modern landforms—The last 65 million years b.p. represent relatively current processes occurring in California, which formed the landforms upon which Californians find themselves living. This period illustrates that most of California consists of a series of pieces of the Earth's oceanic crust that have been thrust up and pasted onto the western edge of the North America continent. Focusing on the period of 65–25 million years b.p. reveals many processes that continue today. The ancient Nevadan mountains were gone, and further sedimentation was creating the Franciscan mélange of oceanic sediments and mantle material in the remaining trench created by the subduction of the Farallon plate. The **Franciscan mélange** represents the sedimentary deposition of material for the future Coast Range rocks (an important wine grape growing region). At the end of this period the Pacific plate arrived circa 25 million years b.p., causing a conservative interaction to replace the previous several hundred million years of destructive interaction. This was the time of the formation of the **San Andreas Fault** system (Figure 2.7).

During the formation of the San Andreas Fault system the complexity of the processes heightened. While the arrival of the Pacific plate caused the **East Pacific Rise** to meld with the western edge of North America and create a largely conservative strike-slip system, there were also constructive processes produced in the south that gave rise to the Salton Trough and the Gulf of California through spreading and rifting. In fact, the East Pacific Rise, which represents a mid-oceanic ridge, terminates at the Salton Trough from its start down near Antarctica. The evidence for the spreading and rifting having occurred is with us today, including deep basins in the coastal regions, the rise of the modern Sierra Nevada (5–7 million years b.p.) and the Basin and Range mountains, as well as the last remnants of volcanism in the trans-Sierra and Mojave Desert.

The collision of the Pacific plate with the North American plate creates much folding/faulting within the San Andreas system and the coastal regions. Everything from San Diego, including the Peninsular Ranges, the Transverse Ranges, and the Coast Ranges to Cape Mendocino, owes its formation to this period. At Cape Mendocino the Mendocino Transform Fault continues westward until it meets up with the East Pacific Rise, which continues north towards Alaska, with the Juan de Fuca plate as its eastern neighbor. The **Juan de Fuca plate** is currently subducting under the North American plate from Cape Mendocino to just north of

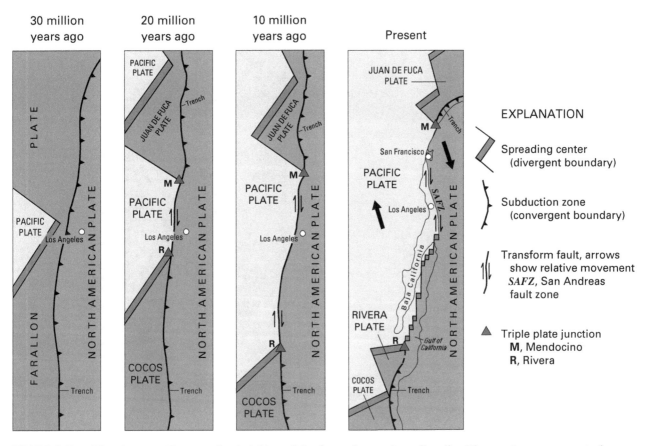

FIGURE 2.7 *This diagram illustrate the shrinking of the formerly very large Farallon Plate, as it was progressively consumed beneath the North American Plate, leaving only the present-day Juan de Fuca and Cocos Plates as remnants. Large solid arrows show the present-day sense of relative movement between the Pacific and North American Plates. (Courtesy U.S. Geological Survey, http://pubs.usgs.gov/gip/dynamic/Farallon.html)*

Vancouver Island, Canada (Figure 2.4). The subduction of this plate is responsible for the volcanic activity that has created the Cascade Ranges and the Modoc Plateau. While scientists are still evaluating the evidence, they consider the Sutter Buttes in the Sacramento Valley as the southern terminus of the Cascade activity and Mt. Lassen and Mt. Shasta as confirmed products of the subduction process (Figure 2.3a).

Glaciations (Figure 2.8) represent the last major geologic process to provide large scale landscape features in California. The Klamath Mountains and Cascade-Sierra Nevada Range, have undergone repeated periods of glaciation in the past 1.5 million years b.p. While today we are in an interglacial period marked by a much warmer climate, we can see evidence for glaciation in the spectacular canyons in the Sierra Nevada. This includes Yosemite Valley as well as the jagged features of the Trinity Alps of the Klamath. The most extensive glaciation, however, occurred in the Sierra Nevada during what is called the Sherwin glaciation about one million years ago, creating the modern Sierran landscape. Currently there are an estimated one hundred small glaciers remaining in the Sierra Nevada and on Mt. Shasta. In this era of enhanced global warming, however, the glaciers in the Sierra Nevada are rapidly melting due to unnaturally elevated greenhouse gases.

Another aspect of past glaciation is the melting period that created abundant Ice Age waters that covered much of the Basin and Range and the Mojave Desert provinces. In these areas large pluvial lakes covered the basins, including Death Valley (formerly Lake Manly), Lake Edwards (now Edwards Air Force Base), Lake Russell (now reduced in size to Mono Lake) and Lake Mojave, to name a few. In addition, 600,000 years b.p. the Great Central Valley was a large lake, known as Lake Clyde, which during a deglaciation period finally over topped the Coast Ranges by San Francisco Bay and drained through to the Pacific to form the Sacramento-San Joaquin Delta region.

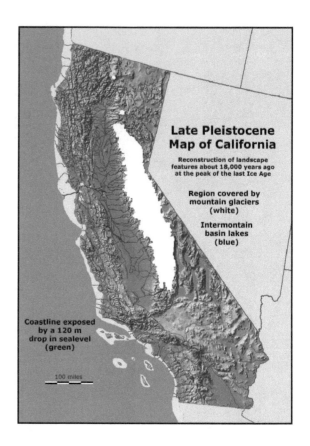

FIGURE 2.8 *Late Pleistocene map of California showing glaciation and large pluvial lakes. (Courtesy U.S. Geological Survey, http://education.usgs.gov/california/maps/glaciers2.htm)*

2.4 GEOLOGIC NATURAL HAZARDS: LIVING ON SHAKY GROUND

Humans notice tectonic activity in the form of geologically-based natural hazards. Tectonic natural hazards include **earthquakes**, **tsunamis** and **volcanic** activity. These natural hazards provide constraints on human activities by their processes and create features that affect human landscape development. Natural hazards have distinct spatial patterns, so it follows that geographers would want to describe, explain and predict the spatial distributions of natural hazards.

Earthquakes occur when rocks suddenly slide past one another along faults (Figure 2.9). Most earthquakes occur along faults near plate boundaries, releasing the energy built up over tens, hundreds or thousands of years, during which the plates tried to move but remained stuck. Geologists are still unable to predict *when* an earthquake is likely to occur, but they are very good at predicting *where* earthquakes are likely to occur (see page 8 in *California Atlas*).

Scientists can measure the size of an earthquake in two ways: by **intensity** (**Mercalli**) and by **magnitude scales**. The intensity of an earthquake measures its effect on people, objects and structures. Reports by people who experience an earthquake determine the intensity. It is based on a scale from I – X+ (nothing felt to extreme shaking/heavy damage). The measurement of magnitude (referred to by major media as the "Richter scale") is based on a seismograph's record of the amount of shaking (energy released) during an earthquake. Using a logarithmic scale, an earthquake registering a magnitude of 6 is 10 times stronger than a magnitude of 5, 100 times stronger than a magnitude of 4, 1,000 times stronger than a magnitude of 3, and 10,000 times stronger than a magnitude of 2. Scientists classify magnitudes greater than 7 as major earthquakes, capable of causing mass destruction and death (e.g., probably a >IX on the intensity scale).

In all earthquakes, energy is released as the two sides of the fault slide past one another. Seismic waves carry this energy through rock, generating the ground shaking that causes much of the damage during earthquakes. Seismic waves come in two forms: body waves and surface waves. Body waves move through the Earth's

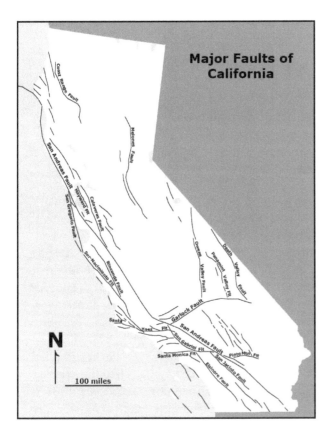

FIGURE 2.9 *Generalized map of major earthquake faults in California. (Courtesy U.S. Geological Survey, http://education.usgs.gov/california/maps/faults_names2.htm)*

interior and travel much more quickly than surface waves. Surface waves move over the surface of the Earth and cause much of the destruction during earthquakes.

The size of an earthquake depends mostly on the size of the fault that slipped. California's infamous San Andreas Fault system has created at least two great historical earthquakes: the 1857 Fort Tejon earthquake (between Paso Robles and Kettleman city on Highway 41; magnitude 7.9, intensity IX) slipped along a segment more than 300 km (200 miles) long; and the infamous 1906 San Francisco earthquake (magnitude 7.8, intensity XI) was caused by a slip along 430 kilometers (267 miles). Most recently, California has experienced the 1989 Loma Prieta earthquake (magnitude 6.9), the 1994 Northridge earthquake (magnitude 6.7) and the 2003 Paso Robles earthquake (magnitude 6.6). All were in or near populated areas.

Finally, there are the rare but potentially devastating **tsunamis**—large ocean waves that are an indirect effect of offshore slippage or underwater landslides caused by earthquakes. The most infamous one to affect California resulted from the 1964 Alaska Aleutian Islands earthquake (magnitude 9.2), which created a tsunami that crossed the Pacific but especially affected Crescent City in Del Norte County. A twenty foot wave destroyed much of the harbor, waterfront and downtown.

Also included as rare, but devastating, is the volcanic activity in the state (Figure 2.10). The last serious eruption by one of the state's major volcanoes came from Mt. Lassen in 1915, but much of the damage was contained to the eastern slope of the mountain and in the surrounding forested area. Otherwise, scientists consider both Mt. Lassen and Mt. Shasta currently dormant but not dead. Another major volcanically active area is the Long Valley Caldera and the associated Mono-Inyo Craters volcanic chain to the east of Mammoth Mountain between Mono Lake and Bishop (running parallel with Highway 395). A caldera is a cauldron-like volcanic feature formed by the collapse of land following a volcanic eruption. This area is still active with numerous hot springs and a geothermal power plant, and geologists continuously monitor it for shifts in its magma chamber.

California In Focus

SAN FRANCISCO EARTHQUAKE 1906

The first historical recording of earthquakes in California started in 1769, when Gaspar de Portola's expedition was jolted by a large earthquake while they camped on the Santa Ana River near the present city of Orange in Orange County. While earthquakes were later noted by several of the Spanish mission priests, and several missions were destroyed by them, earthquake reports were practically non-existent until the 1850s, when the Gold Rush population surge brought increased interest in the local geology.

It was the April 18, 1906 San Francisco earthquake that brought full attention to this type of hazard in California and with it the birth of modern seismology. The 1906 earthquake, and the fires that followed, resulted in the greatest seismic disaster in California history. Every city along the San Andreas Fault, from Hollister in San Benito County to Fort Bragg in Mendocino County, was violently shaken (representing a 300 kilometer fault displacement). The magnitude was recorded at 7.8, which resulted in 3,000 deaths and $524 million in property loss. All from an earthquake that was estimated to last one minute. The key to understanding the damage that San Francisco received relies on the city's geography and bedrock type. Buildings on solid rock foundations had minimal damage (the Presidio, Telegraph Hill, Bernal Heights), but buildings on land that had been recovered from the bay or filled in over old swamps and riverbeds had the greatest damage. These areas included the Marina district, the Embarcadero, China basin, south of Market, city hall plaza, Union Square and the Mission district. Uncompacted fill in these areas jolted and settled unevenly. It broke foundations, swayed buildings and broke natural gas lines, which led to great fires that spread across the city.

Aftermath of the San Francisco earthquake and fire of 1906. (Courtesy Library of Congress)

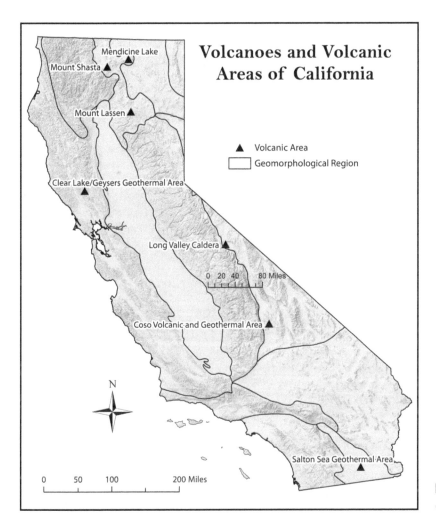

FIGURE 2.10 *Volcanic hazard areas of California. (Courtesy Kirstyn Pittman)*

2.5 Natural Landform Regions

GEOMORPHOLOGICALLY-BASED LANDSCAPES

The geological complexity of California's varied landscapes can be addressed by examining the major topographic features. The use of natural **landform regions** (Figure 2.11), recognized by state authorities, provides an organized means of explaining the state's geomorphologic diversity (see page 5 in *California Atlas*). **Geomorphology** is the geographical science that deals with the history of the Earth's landscapes as created by plate tectonics, the rock cycle, and informed by weathering-erosion and glaciation.

One of California's most prominent major topographic features is the continuous high mountain axis consisting of several mountain systems that divide the state into a moister western slope and a very arid eastern slope. The term **cismontane** describes the portion to the west of the mountain axis, which includes the Cascade Range, the Sierra Nevada Range, the Transverse Range and the Peninsular Ranges. The western area represents approximately 70 percent of the state. The remainder of the state that is east of the mountain crests is **transmontane** California. Within each of these areas are additional physiographic regions and locally important sub-regions.

2.5.1 SIERRA NEVADA

The **Sierra Nevada** represents an elongated northwest trending large-scale fault block with a gently sloped western side and a steep sloped eastern side (i.e., looks like a door stop from the side). The range represents

26 **CHAPTER 2** *Geological History and Landforms*

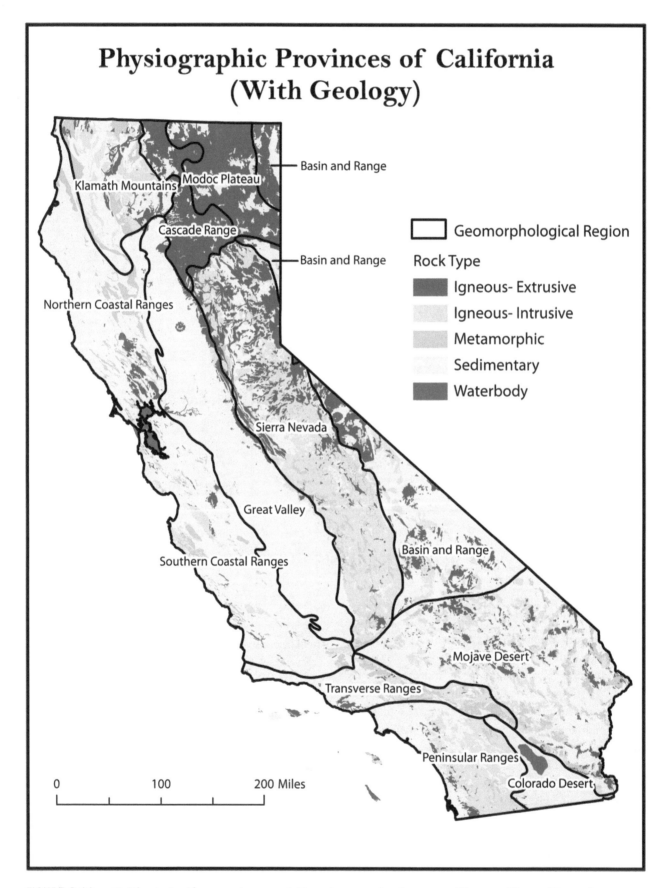

FIGURE 2.11 *California landform provinces overlaid on the general geology map. (Courtesy Kirstyn Pittman)*

largely granite rock (a large batholith) with volcanic deposits found in its northern half. Its length is 660 km long (400 miles) and extends from the Feather River canyon (Highway 70) to Tejon Pass where the Tehachapi Mountains meet the Transverse Ranges. The Sierra Nevada displays a relatively uniform skyline (Figure 2.12) and alpine upland with lower elevations in the north and its highest elevations (including Mt. Whitney) in the south. Large rivers with deep-cut canyons characterize the northern Sierra, which is generally extensively dissected, while the southern Sierra represents an unbroken massif not yet cut by erosion. From the central Sierra to its northern end, the rivers have cut through the volcanic deposits (originally a volcanic arc island) and have exposed gold-bearing gravels. These gold gravels have been redeposited along the river banks in modern placer (alluvial sand, gravel and valuable minerals) deposits, which miners discovered in 1848.

2.5.2 KLAMATH MOUNTAINS

The **Klamath Mountains** are located in the northwestern corner of California adjacent to Oregon. They represent a complex geologic history somewhat comparable to the Sierra Nevada. They are among the oldest geological features in the state, composed largely of several distinct plates representing ancient metamorphic and granitic rock material. They are an uplifted

FIGURE 2.12 *(a) Portion of the Sierra Nevada range showing Mt. Whitney as viewed from the eastern side of the range, along Highway 395. In the foreground are the Alabama Hills. (Courtesy Dean Fairbanks) (b) Yosemite Valley in Yosemite National Park.*

and folded mountain group like the Coast Ranges, but they have a more rugged and difficult terrain than the latter (Figure 2.13). The region has high mountain elevations (greater than 2,500 m; 8,000 ft) dissected by canyons and valleys from some of the state's largest rivers, which also harbor the state's largest trees—the coastal redwoods.

2.5.3 PENINSULAR RANGES

The **Peninsular Ranges** are a series of northwest trending mountain ridges aligned parallel with the San Andreas Fault. Only a small portion of the northern section exists in California; the rest of the 900 mile long unit stretches to the tip of Baja California. They are

FIGURE 2.13 *A view of the Trinity Alps near Granite Lake in the Klamath Mountains. (Shutterstock Paula Cobleigh, 2009.)*

FIGURE 2.14 *Aerial view of the Cuyamaca mountains in the Peninsular ranges of San Diego County, note the exposed granites that the entire range is composed. (© Shutterstock Skworking, 2012)*

made of primarily granitic rock (a large batholith) similar to the Sierra Nevada that was uplifted through the activity of a series of parallel fault zones associated with the San Andreas Fault (Figure 2.14). They occur largely in San Diego County as westward-tilting fault blocks and occasional basins (i.e., San Jacinto) and include elevated coastal plains cut by estuaries.

2.5.4 SOUTHERN CASCADES AND MODOC PLATEAU

One can view the **Southern Cascades** and the **Modoc Plateau** as part of the same system, but they have vastly different landscape attributes. The Cascade Range is a chain of volcanoes that extends from Canada to northern California (Figure 2.15). The largest volcanoes of the southern Cascades in California are **Mt. Shasta** (4,317 m; 14,154 ft) and **Mt. Lassen** (3,187 m; 10,449 ft). These volcanic features dominating this region are associated with a shift in plate movement from transverse to subduction of the **Juan de Fuca** plate and therefore creating igneous extrusive processes of largely basalt rock flows. In addition, the Cascade Range does not present a continuous high elevation ridge system like the Sierra Nevada. Instead, from the Feather River Canyon north on the east side of the Great Central Valley there are volcanic deposits creating layered plateaus (Figure 2.15a) above which the volcanic peaks arise as isolated cones.

In comparison, the Modoc Plateau is a flat area covered by extensive basaltic lava flows (Figure 2.16). Geologists consider it a southern extension of the vast Columbia River lava floods that spread across the Pacific Northwest ~3 million years b.p. While the occasional small volcanic cone dots the plateau, it lies mostly between 1,000 and 2,000 m elevation (3,300–6,600 ft).

2.5.5 GREAT CENTRAL VALLEY

The **Great Central Valley** is a large-scale, flat valley filled with up to six miles of sediments in some places. Geologists divide the valley into two sub-valleys: In the north above the Sacramento and San Joaquin River Delta is the **Sacramento Valley**, and to the south is the **San Joaquin Valley**. The Central Valley is essentially an unbroken lowland plain that lies between the Coast Ranges and the Sierra Nevada and Cascade Ranges. Many major rivers dominate the area, and the southern portion has rivers with internal drainage that at one time created large shallow lakes, i.e., **Lake Tulare** (Figure 2.17), which farmers have drained for large-scale agriculture. Two topographic features occur in the valley landscape. In the Sacramento Valley portion lies the **Sutter Buttes** (Figure 2.18), which breaks the plain, while in the southern San Joaquin Valley the **Kettleman Hills** (Figure 2.17) represent a folded dome outlier of the Coast Ranges. The **Sacramento-San Joaquin Delta** region forms the boundary between the two valleys, essentially representing the drainage area for the entire Central Valley. The rich organic, peat-dominated land here is at or below sea level, with many islands.

2.5.6 COAST RANGES

The **Coast Ranges** (Figure 2.19) are a series of northwest-southeast trending parallel ridges extending from Humboldt County to Santa Barbara County. They are broken into the following sections: a northern (Marin County north) and southern or central coast ranges (San Francisco Bay south). The San Andreas Fault to the west bounds the region until it disappears at Tomales Bay in Marin County. From the coast, the Coast Ranges consist of a series of coastal and interior ranges separated by north-south trending valleys, including Sonoma, Napa, Santa Clara, Salinas, etc. The Coast Ranges represent an area of generally low elevations but complex topographic relief that is prone to landslides, i.e., the Big Sur region.

The San Andreas Fault system has contributed to this regions geological complexity. The parent materials include the Franciscan sedimentary mélange, metamorphics, and both intrusive and extrusive igneous rock. In the north around Lake, Sonoma and Napa Counties ancient remnants of volcanic activity intersperse with the Franciscan sediments. In some areas the underground heat from past events is still visible in the Calistoga and Geyserville geothermal works used to generate electricity (see Figure 2.10). In the south the intrusive Salinian granites separate western and eastern Franciscan regions in Monterey County and are especially evident as beautiful beach rock outcrops at Pacific Grove. Products of the folding and faulting in the landscape have led to large pockets of metamorphic rocks, including **serpentine** (California's state mineral and responsible for landslides), cinnabar (a rock containing mercury that was mined—quicksilver mining—from this area to support the Gold Rush era) and asbestos.

30 **CHAPTER 2** *Geological History and Landforms*

FIGURE 2.15 *(a) Basalt lava layer formations of the Southern Cascades as represented by the Mt. Lassen foothills (Courtesy Portia Ceruti); and (b) View of Mt. Shasta stratovolcano in the southern Cascades (© Jim Feliciano, 2012. Used under license from Shutterstock, Inc.).*

FIGURE 2.16 *Gently rolling basalt flow landscape of the Modoc Plateau near Tule Lake. (Courtesy Portia Ceruti)*

2.5.7 TRANSVERSE RANGES

The **Transverse Ranges** are a series of mountain ridges that are unique in that they run east-west. The Transverse Ranges extend from the Santa Ynez Mountains in the west to the San Gabriels and the San Bernardino Mountains in the east, and they include the northern Channel Islands. The San Andreas Fault controls the orientation, and numerous other faults contribute to the rapid upliftment of this set of ranges. The San Andreas Fault also separates the San Bernardino and San Gabriel Mountains (Figure 2.20).

The western end of the Transverse Ranges has a lower elevation, is largely composed of strongly folded sedimentary parent material (i.e., sandstones) and lies the closest to the coastal plain ending at Pt. Conception (in Santa Barbara County). The higher elevations of the ranges lie in the east, dominated by granitic and meatamorphic formations. Across the entire range, however, the mountains are very steep and rugged.

2.5.8 BASIN AND RANGE

The **Basin and Range** province is a region that largely lies outside the state boundary in Nevada, Utah and Idaho. It is a landscape dominated by evidence of crustal extension. Large scale block faulting has created a land of high mountains and deep valleys all parallel with each other and trending in a north-south orientation (Figure 2.21). The region lies to the east of the Sierra Nevada-Cascade Ranges and Modoc plateau. The basins vary

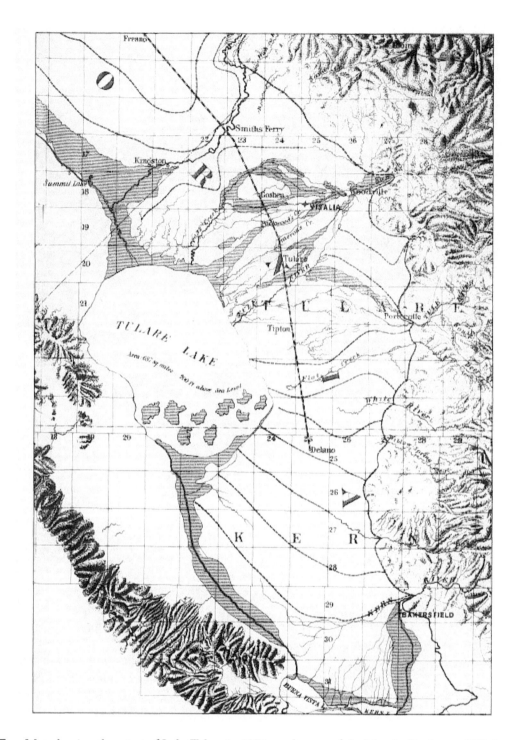

FIGURE 2.17 *Map showing the extent of Lake Tulare in 1874, to the west of the lake the Kettleman Hills have been drawn. (From Report of the Board of Commissioners on the Irrigation of the San Joaquin, Tulare, and Sacramento Valleys of the State of California; Washington: Government Printing Office, 1874)*

FIGURE 2.18 *Sutter Buttes situated in the Sacramento River Valley. In the foreground are the rice paddies and wetland wildlife reserves of Colusa County. (Courtesy Bruce King)*

in elevation from below sea level at Death Valley to an average of 2,500 m (8,000 ft).

2.5.9 MOJAVE DESERT

The **Mojave Desert** was created by the rain shadow that developed after the upliftment of the southern Sierra Nevada (the Tehachapi Mountains extension) and the Transverse Ranges. The western boundary is the San Andreas Fault, and the northern boundary is the Garlock Fault, creating a sharply bounded region (see Figure 2.9). The region is rather flat with broad valleys (dry lake beds exist in the centers of most valleys), but broken in various areas by low mountains (Figure 2.22) and past volcanic activity in the central Mojave. In the eastern Mojave higher fault block ranges exist with steep relief, also with some past volcanic activity.

2.5.10 COLORADO DESERT/SALTON TROUGH

The **Colorado Desert/Salton Trough** is bounded by major fault systems and represents a deep, down-faulted basin that was the northern extension of the Gulf of California before the Colorado River delta filled it in with sediments. The flat, lowland nature of the region is below sea level in some places and is the location of the Salton Sea (Figure 2.23). The region also represents the southern terminus of the slipping portion of the San Andreas Fault. There are rugged low lying mountains (i.e., the Turtle and Chocolate Mountains) spotting the area, but a key feature of this landscape is the large scale sand dune fields (i.e., Algodones dunes) that occupy areas to the east of the Salton Sea, as well as the large washes that drain into the Colorado River. The Colorado Desert extends into southern Arizona and northern Mexico.

FIGURE 2.19 *(a) Interior of the North Coast Range in the Mendocino National Forest; and (b) View of the Central Coast Range (Santa Lucia Mountains) along the Big Sur coastline.*

(c) Interior of the Central Coast Range in the Big Sur (Santa Lucia Mountains); and (d) Abandoned New Idria mercury (quicksilver) mine in the Diablo Mountains of the interior Central Coast Range. (All courtesy Portia Ceruti)

FIGURE 2.20 *View of the San Gabriel Mountains. as seen from Bel Air, in Southern California. (© Thomas Barrat, 2012. Used under license from Shutterstock, Inc.)*

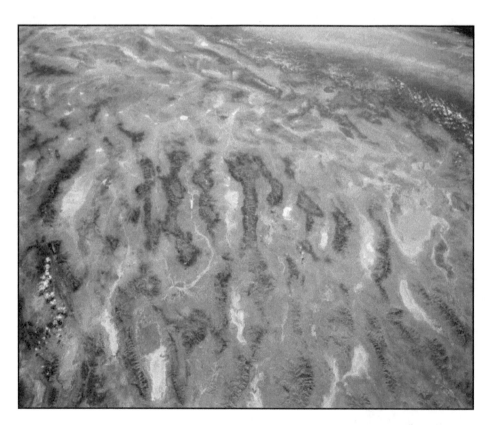

FIGURE 2.21 *Space station view looking southwest of the Basin and Range province stretching from the back side of the Sierra Nevada (top right running diagonally-note Mono Lake) across Nevada. (Courtesy NASA)*

CHAPTER 2   *Geological History and Landforms*   37

FIGURE 2.22   *Mountains and Joshua trees in the Mojave Desert. (Courtesy U.S. Geological Survey, http://pubs.usgs.gov/of/2004/1007/carbonate.html)*

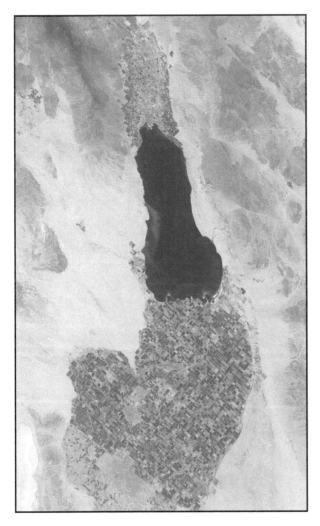

FIGURE 2.23   *Space station view of the Salton sea, Coachella Valley (north of the Salton Sea) and Imperial Valley (south of the Salton Sea) located within the Colorado Desert. (Courtesy NASA)*

Bibliography

Alt, D. and D.W. Hyndman (2000). *Roadside Geology of Northern and Central California*. Mountain Press Publishing Company, Missoula, MT.

Atwater, T. (1970). Implications of Plate Tectonics for the Cenozoic Evolution of Western North America. *Geological Society of America Bulletin* 81(12): 3513–3536.

Bailey, E.H. (1966). *Geology of Northern California*. California Division of Mines and Geology, Bulletin 190, San Francisco, CA.

Bateman, P.C. (1974). Model for the Origin of Sierran Granites. *California Geology* January: 3–5.

Blanc, R.P. and G.B. Cleveland (1981). Pleistocene Lakes of Southeastern California. *California Geology* April: 1–8.

Booth, S. (2008). *California Geography*. Course taught at Sierra College. [online] http://geography.sierra.cc.ca.us/booth/California/cal_index.htm.

Chesterman, C.W. (1971). Volcanism in California. *California Geolog*, August: 139–47.

Christopherson, R. (2008). *Geosystems*. Prentice Hall Press, Upper Saddle River, NJ.

Cox, C.J. (2008). *California Geography*. Course taught at Sierra College. [online] http://faculty.sierracollege.edu/ccox/california_geography/index.html.

Donley, M.W., Allan, S., Caro, P., and C.P. Patton (1979). *Atlas of California*. Pacific Book Center, Culver City, CA.

Durrenberger, R.W. and R.B. Johnson (1976). *California Patterns on the Land*. 5th edition, Mayfield Publishing Company, Mountain View, CA.

Ernst, W. G. (1979). California and Plate Tectonics. *California Geology* September: 187–198.

Guyton, B. (1998). *Glaciers of California*. California Natural History Guide 59, University of California Press.

Harden, D. (2003). *California Geology*. 2nd Edition, Prentice Hall, Upper Saddle River, NJ.

Hinds, N.E.A. (1952). *Evolution of the California Landscape*. Department of Natural Resources, Division of Mines, Bulletin 158, San Francisco, CA.

Holland, V.L. and D.J. Keil. (1995). Topography, Geology, and Soils. *California Vegetation*. Kendall/Hunt Publishing, Dubuque, Iowa.

Hornbeck, D. (1983). *California Patterns: A Geographical and Historical Atlas*. Mayfield Publishing Company, Mountain View, CA.

Jenkins, O.P. (1951). *Geologic Guidebook of the San Francisco Bay Counties*. Department of Natural Resources, Division of Mines, Bulletin 154, San Francisco, CA.

Lacopi, R. (1971). *Earthquake Country: How, Why and Where Earthquakes Strike in California*. Lane Magazine & Book Co., Menlo Park, CA.

Lantis, D.W., Steiner, R., and A.E. Karinen (1989). *California: The Pacific Connection*. Creekside Press, Chico, CA.

McPhee, J. (1994). *Assembling California*. Farrar, Straus and Giroux, New York, NY.

Michaelson, J. (2008). *Geography of California*. Course at UC Santa Barbara, Dept. of Geography. [online] http://www.geog.ucsb.edu/~joel/g148_f08/.

Miller, C.S. and Hyslop, R.S. (1983). California: *The Geography of Diversity*. Mayfield Publishing Company, Mountain View, CA.

Selby, W. (2006). *Rediscovering the Golden State: California Geography*. 2nd edition, John Wiley Press, New York, NY.

Selby, W. (2008). *Geography of California*. Course taught at Santa Monica College. [online]http://homepage.smc.edu/selby_william/california/chapter_1.html.

Sharp, R.P. (1993). *Geology Underfoot in Southern California*. Mountain Press Publishing Company, Missoula, MT.

U.S. Geological Survey (2006). *California Education Standards: USGS Online Resources for Science, Geography, History and Social Science*. [online] http://education.usgs.gov/california/

U.S. Geological Survey (2006). *Geology of Southern California*. [online] http://geomaps.wr.usgs.gov/socal/geology/index.html

U.S. Geological Survey (2006). *San Francisco Bay Region Geology and Geologic Hazards*. [online] http://geomaps.wr.usgs.gov/sfgeo/index.html

U.S. Geological Survey (2006). *Southern California's Major Faults*. [online] http://geomaps.wr.usgs.gov/socal/geology/inland_empire/socal_faults.html

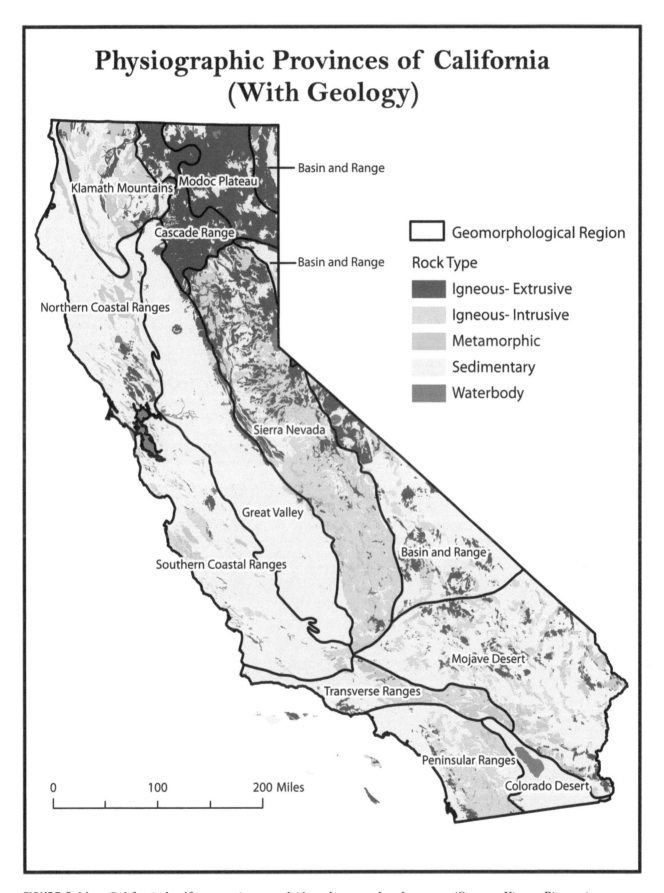

FIGURE 2.11 *California landform provinces overlaid on the general geology map. (Courtesy Kirstyn Pittman)*

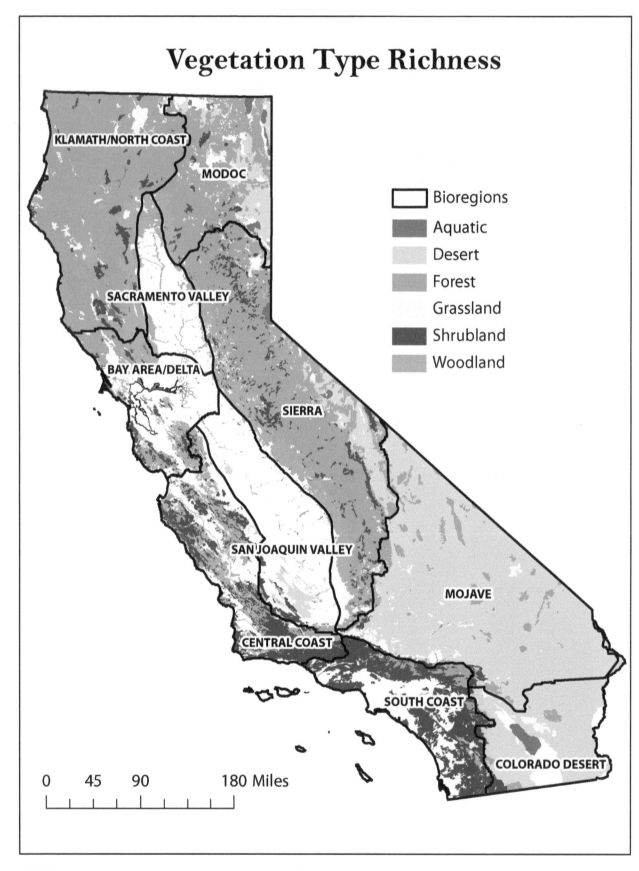

FIGURE 4.3 *General vegetation types of California. (Courtesy U.S. Geological Survey GAP analysis program, and Kirstyn Pittman)*

CHAPTER 3

California Climate
Scales and General Controls

Key terms

Air masses	Low pressure	Sea breezes
Aleutian low	Marine layer (coastal fog)	Solar radiation
Arizona monsoon	Maritime climate	Temperature inversion
California current	Mediterranean climate	Thermal inversion
Climate	Mountain breeze (katabatic)	Topography
Continental climate	North Pacific high	Trade winds
Delta breeze	Offshore breeze	Tule (valley fog)
El Niño/La Niña	Onshore breeze	Valley breeze (anabatic)
Environmental lapse rate	Orographic effect	Weather
Global-scale	Pineapple express	West coast position
Gyres	Polar jet stream	Westerly winds
High pressure	Rain shadow	Windward
Latitude	Regional-scale	
Leeward	Santa Ana winds	

Introduction

Climate is 'what to expect' as opposed to **weather**, which tells us what is happening—now. While a strong winter storm or an El Niño event can bring exciting weather to the state, what this chapter is interested in explaining is the general climate trends that occur in California. Climate includes the expected state of weather variables like precipitation or temperature, based on averages over time periods that are typically thirty years and longer.

California has greater climatic diversity than most of the U.S. and Canada, and its diversity of vegetation (see Chapter 4) reflects this varied climate. For example, in northwest California there are places that receive up to 250 cm (100 inches) of precipitation per year and that support temperate rainforests. On the opposite end of the spectrum down in the southeastern regions, places like Death Valley receive less than 5 cm (2 inches) per year and represent hot deserts with none to sparse

39

vegetation. In addition, the climate along California's coast is mild, seldom experiencing frost in winter, while in the interior on the mountains frost can occur any day of the year. The climate changes so rapidly that locations that are only a few kilometers apart may have significantly different climatic environments. For example, witness the climate complexity of the San Francisco Bay area where on any given summer day San Francisco can be 13° C (55° F), Berkeley across the bay can be 18° C (65° F), and then just a few kilometers inland over the Berkeley hills, Walnut Creek can be 35° C (95° F). This example covers just 40 km (25 miles).

In order to understand the general controls on California's climate and its influences on people it is best to scale the responses of various processes and variables to provide a greater understanding. The important elements of climate to consider include **global-scale** features, such as **latitude**, ocean **gyres**, **air masses** and **El Niño/La Niña**, and **regional-scale** features, which include **topography**. In addition, however, there are small-scale features we will address that relate to unique climate features.

3.1 Global-Scale Controls

The factors that have an effect on California's climate on a global scale relate to a spatial scale of hundreds to thousands of kilometers. At this scale they affect all of the state in similar ways by producing broad, gradual changes over large areas. In general, the importance of latitude and location relative to major land and ocean areas relates to an area's position on Earth's spherical surface.

3.1.1 Latitude

As a rule, latitude is a major factor that influences climates across the planet. Since California extends over a greater range of latitude than other states, except for Alaska, we can expect various climatic features to display themselves latitudinally from south to north. In fact, the predictable pattern for California is that precipitation increases and temperatures decrease from southern to northern California. We can link this to the effects of latitude on incoming **solar radiation**, since solar radiation variations affect temperatures, which then affect moisture regimes. A comparison between summer and winter latitudinal variations (Figure 3.1) reveals that, in the summer the day length increases northward (i.e., increasing solar radiation as the earth's axial tilt favors the northern hemisphere being tilted toward the sun), while in the winter day length decreases, moving northward (i.e., the northern hemisphere is tilted away from the sun). This phenomenon has to do with our planet being tilted 23° 30′ on its axis of rotation as it revolves around the sun and its spherical shape which affects the way parallel rays of radiation from the sun are intercepted by a curved surface. Higher latitudes receive less energy because the orientation of the Earth's surface relative to the sun's rays diminishes the intensity of solar radiation at high latitudes and the sun's rays must pass through more atmosphere at higher latitudes. Therefore, the Earth's tilt and its position as it revolves around the sun are responsible for the seasons by changes in day length and by the sun's altitude above the horizon over the course of the year (i.e., the sun is higher overhead as it traces a path overhead from sunrise to sunset in summer, while it traces a path at a lower angle above the horizon during the winter).

3.1.2 West Coast Position and Air Masses

California's latitudinal range of 32° 30′ N to 42° N places it between two major hemispheric circulation belts and their respective air masses that affect the region's climate. In the north Pacific Ocean, there is a subtropical high pressure belt (**North Pacific high**), representing warm and dry conditions, and a low pressure polar front zone to the north (**Aleutian low**). The polar front zone is the source of all of the region's mid-latitude winter storms, traveling west to east following the **Polar jet stream** and the mid-latitude Westerly winds. However, while this brings wet conditions in the winter months the amount of moisture received in the state can be highly variable.

An understanding of air masses is required to comprehend the unique **Mediterranean climate** in which California finds itself. An **air mass** is an area of air that

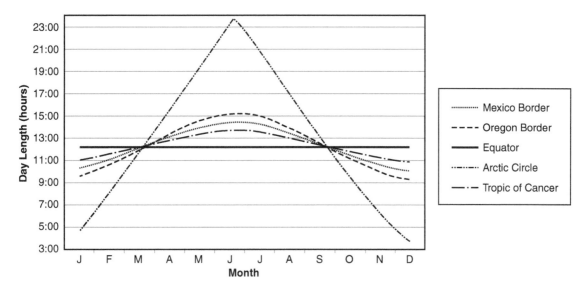

FIGURE 3.1 *Day length throughout the year as controlled by latitude for the Artic circle (66°), equator (0°), Tropic of Cancer (23° 30'), California's border with Mexico (32° 30'), and California's border with Oregon (42°). Higher latitudes receive less direct sunlight throughout the year, so the air there is colder.*

has similar properties of temperature and humidity. Air masses develop over areas of similar geographical character, like polor ice caps, hot deserts, warm or cold oceans. Two air masses influence California: a strong Pacific high pressure cell during summer and a strong Aleutian low during winter.

As warm air rises it expands—becomes less dense and cooler—and creates a **low pressure** region from which water vapor then condenses and forms clouds. In contrast, descending cool air heats, compresses and absorbs water vapor, causing evaporation from the environment that can lead to dry conditions; this creates a **high pressure** region. At the equator the extreme solar heating allows a dominant low pressure condition to occur with warm air rising cooling and condensing into clouds. This air then moves both north and south of the equator towards the poles (driving a strong planetary heat gradient because hot flows to cold). The air flow comes down and creates a high pressure center along the 30° latitude area, thus causing many of the desert conditions found (i.e., the Sahara, northern Mexico, etc.). During the summer the high pressure air mass in the North Pacific shifts farther north, so high pressure dominates California and creates warm and drought conditions (Figure 3.2a). In the winter, the Pacific high pressure belt shifts south, so wet conditions prevail, especially in northern California, because the **Polar jet stream** is able to pick up cool wet air masses from the Aleutian low (Gulf of Alaska), and its trajectory allows it to flow over northern California (Figure 3.2b). The **Westerly winds** catch the frequent winter storms, which travel across California from the northwest to the southeast. Sometimes, however, the Polar jet stream drops far enough south to pick-up a warm moisture laden air mass around Hawaii and brings it in a direct eastward movement to central or Southern California. This phenomenon is termed a **Pineapple express** or Hawaiian storm.

Position on the planet matters, and while latitude is important, having a **West Coast continental position** is very important for allowing the mid-latitude Westerlies to bring a strong marine influence to the coast in conjunction with a North Pacific cold water gyre. A **gyre** is a large oceanic current that moves clockwise in the northern hemisphere and counter-clockwise in the southern hemisphere (all via the Coriolis Effect). As water moves westward at the equator via drag by the eastern **Trade winds**, it picks up heat, turns right and moves up the eastern coast of the Asian continent and Japan. When the water current is forced to bend right again at the top of the Pacific (coastal Russia and Alaska), the water loses heat flowing through the high Arctic latitudes, which makes the water flowing south along the Western coast of North America cold (Figure 3.3). This southward flowing water is called the **California current**.

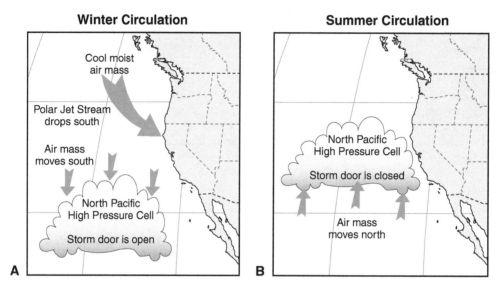

FIGURE 3.2 *(a) Winter circulation with the North Pacific high pressure cell dropped to allow a open 'storm door' for the Polar jet stream to bring moist air masses into the region; and (b) Summer circulation with the North Pacific high pressure cell moved up and thus closing the 'storm door' to the Polar jet stream, which is pushed farther north on its Westerly track now mostly through Washington and Canada.*

Therefore, there are moderate temperature variations (cool and mild) along the California coast as a result of the ocean serving as a giant heat reservoir. Any air masses passing over the California cold water current are moderated since during the summer the ocean absorbs heat and then releases the heat in the winter. There also tends to be high moisture content in storms at the coast, which decrease inland. Winter storms develop in the Aleutian low pressure area over the north Pacific (the Gulf of Alaska region) and flow southeastward, spinning counterclockwise and pulling more water off the ocean and onto the land.

On the coast, the climate is strictly a **maritime climate** as a result of the ocean's moderating effect (i.e. water heats up and cools slower than the land). Maritime climates occur on the lower seaward side of the coastal mountains. Therefore coastal places have a much narrower range of temperatures (summer to winter) than places in the interior (Figure 3.4). As one moves inland from the ocean's influence, the maritime climate gradually changes to a more **continental climate** from the interior coastal mountains, the Great Central Valley and into the interior mountain ranges. At higher elevations (1,800 m/6,000 ft), one finds montane climates. Continental and montane climates have distinct seasons and show much greater daily and seasonal temperature fluctuations than do maritime climates.

3.1.3 EL NIÑO/LA NINA EVENTS

El Niño—the 'Christ Child'—is a warming of the eastern Pacific that occurs at intervals between two and ten years and lasts for up to two years. The early Spanish explorers and colonists on the West coast of South America noticed changes in the climate during the Christmas holy days and named the phenomenon. Originally, El Niño referred to a warm current that appeared off the coast of Peru, but scientists now realize that this current is part of a much larger system. It has the ability to disrupt the world's climates.

During normal conditions in the Pacific Ocean, Walker circulation is present as an east-west circulation that occurs in low latitudes (Figure 3.5a). Near South America winds blow offshore, causing upwelling of the cold, nutrient rich waters. By contrast, warm surface water is pushed into the western Pacific by the Trade winds. Normally, sea surface temperatures in the western Pacific are over 28°C (83°F), causing an area of low pressure and producing high rainfall. By contrast, over coastal western South America sea surface temperatures are lower, high pressure exists and conditions are dry.

During El Niño episodes, the pattern is reversed. As warm water from the western Pacific flows into the eastern Pacific, water temperatures rise (Figure 3.5b).

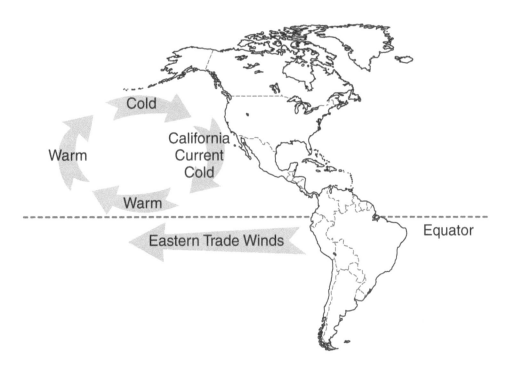

FIGURE 3.3 *North Pacific gyre and its effect on the West coast of North America by brining cold water southward from Alaska via the California current.*

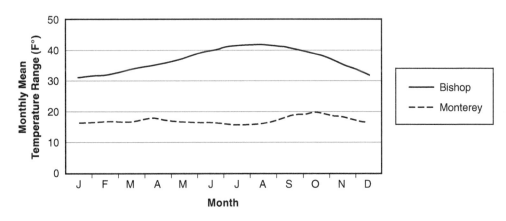

FIGURE 3.4 *Seasonal variations are much less along the coast as shown in this example of Monterey on the coast versus Bishop in the interior.*

This is due to the weakening of the eastern Trade winds, which normally blow west across the equatorial Pacific. As they nearly cease to blow, the warm waters piled up in the western Pacific (up to one-half meter difference in sea level height) rush back to the east, hitting equatorial South America and deflecting north and south of the equator. During El Niño events, sea surface temperatures of over 28°C (83°F) extend much further across to the eastern Pacific. Low pressure develops over the eastern Pacific and high pressure over the western Pacific. Consequently, heavy rainfall occurs over coastal South America and as far north as California, whereas the western Pacific island nations experience warm, dry conditions. Since the height of an El Niño occurs during the winter season in the northern hemisphere, California is already receiving its cooler mid-latitude storms generated in the Gulf of Alaska's low pressure region via the Polar jet stream,

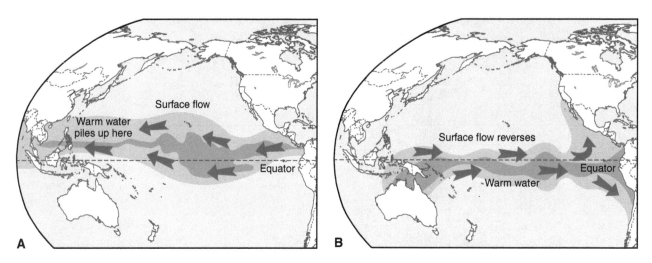

FIGURE 3.5 *(a) Normal conditions in the Pacific Ocean; and (b) El Niño conditions in the Pacific Ocean.*

but now warmer subtropical storms are also flowing northward generated by the warm waters off the coast (flowing against the cold California current). An El Niño leads to increased rainfall and flooding events in California. The 1982–83 event took 160 lives and caused $2 billion worth of damage in floods, mudslides, crop losses and coastal erosion. Again in 1997–98, seventeen lives were lost, 42 of 58 counties proclaimed a state of emergency and the damage estimates were $1.1 billion. Differences in storm intensity and duration accounted for some of the reduced costs in the later El Niño, but the better forecasting models and hazard preparation that existed six months before the start of the 1997–98 event had the state prepared for the worst.

The exact opposite of an El Niño condition is the **La Niña** condition, which is a resumption of the eastern Trade winds pushing the surface flow of warm waters back to the western Pacific. However, it is much stronger than during normal conditions. This leads to a very strong upwelling of cold waters off of South America (cooler than usual ocean temperatures occur on the equator between South America and the International Date Line, 180°) and a much larger high pressure air mass region in the subtropical latitudes of the eastern Pacific. Consequently, a La Niña event leads to a shift of the Polar jet stream farther north, bringing warmer and drier winters to southern California and therefore increased drought conditions, but leaving extra-wet conditions in the Pacific Northwest where the Polar jet stream is largely contained. California had been experiencing a La Niña condition from 2006–2010, which coincided with an official state-declared drought in 2008, the first such statewide water emergency since 1991.

3.2 Regional Scale Controls—Influence of Topography

While most Californians experience the general seasonal climate shifts of winter rainfall to summer drought that define a Mediterranean climate type, it is the diversity of climates within the state that also makes it exceptional. Because plate tectonic activity has developed complex **topography**, precipitation and temperature patterns also vary greatly with respect to elevation and major topographic features. In general, topography forms barriers to storms, forcing them to rise, and higher elevation leads to lower temperatures. Precipitation increases with elevation due to the **orographic effect** (Figure 3.6). The mid-latitude Westerly winds carry moist air masses coming from the Pacific over the western slopes of the California mountain ranges. As this air rises over the mountains, it cools, reducing the air's ability to hold water vapor, which condenses and falls as precipitation or snow on the **windward** slopes. Precipitation totals increase with ele-

FIGURE 3.6 *The orographic effect. The major driver for California's precipitation and snow fall.*

vation until the air mass is pushed over the other side of the mountains and forced downslope on the **leeward** side. The compression of air causes it to heat, evaporating the condensed water to water vapor, therefore decreasing the amount of rainfall on the leeward side and causing a **rain shadow** effect that produces local dry high pressure conditions. The orographic effect is cumulative, as storms move across California they lose more moisture with each mountain range they are forced rise over. Because air cools roughly the same amount in 100 m (328 ft) of altitude as it does in a 482 km (300 miles) shift northward, a mountain is a microcosm of climate varieties and a regional surrogate for latitude.

As for topography and temperature, usually temperature decreases with increasing elevation based on the generalized trend. Scientists call this the **environmental lapse rate**. The temperature decreases at an average rate of 0.6° C per 100 m increase in elevation (3.5° F per 1,000 ft).

Topography also determines the inland penetration of marine influence (onshore flow). **Sea breezes** can cool inland areas during the summer. Mountain barriers block much of the coastal sea breezes; however, valleys and gaps in the mountain ranges can channel and intensify sea breezes. You can see this marine flow as it moves through San Francisco Bay's Carquinez Straits into the Sacramento-San Joaquin Delta region (known locally as the **Delta breezes**).

3.3 Small Scale Issues

Finally, California lies within the zone of prevailing Westerly winds. Winds are the horizontal movement of air. Surface winds flow as rivers of air around the globe. The basic flow in the free air above the state is from the west or northwest during most of the year. The coastal mountain chains are responsible for deflecting these winds southward parallel with the coastline, and wind direction in the state is likely to be more a product of local terrain than of prevailing circulation. Thus, at the local level we can classify wind movement based on the effects of continentality and topography.

Onshore and **offshore breezes** represent the fact that water heats and cools more slowly than does land (Figure 3.7). In the case of California, the offshore large scale current is a cold water one running from the north to the south. Due to the effects of solar heating via latitude and the Earth's axial tilt, the current is colder in the north near, for example, Eureka, and warmer near San Diego in the south. The basic process for an onshore flowing sea breeze relies on land heating faster than water in the morning. For an offshore flowing land breeze to occur, the land cools faster than water in the

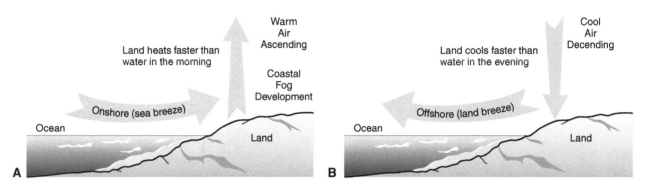

FIGURE 3.7 *(a) Onshore breezes (sea breeze) in the morning hours; and (b) Offshore breeze (land breeze) in the afternoon hours.*

afternoon and into the evening. Depending on the latitude along the California coast, the breezes can have differing strengths, i.e., they are usually stronger in the north and weaker in the south.

Recalling that temperature drops with increasing altitude, the other important local winds are the **valley** (anabatic) and **mountain** (katabatic) **breezes** (Figure 3.8). These breezes are important for much of the inland areas away from the coast. Starting in the morning and continuing during the day, the valley floors warm more slowly than the mountains tops which starts a flow of wind up the slopes of the mountains, causing a valley breeze. These breezes are commonly experienced in river canyons. At night the flow reverses itself as the mountains cool faster than the valleys and thus the cooler mountain air slides down (e.g., cold air drainage) the mountain slopes to fill the valleys and cause a mountain breeze. One way we can experience mountain breezes is during mountain forest fire events. During the day a valley's air can be kept clear, but at night the smoke flows down into the valley, making for bad air quality in the morning. The problem, however, is that the smoke can become trapped and not clear because of **thermal inversions**. The next section will discuss this situation.

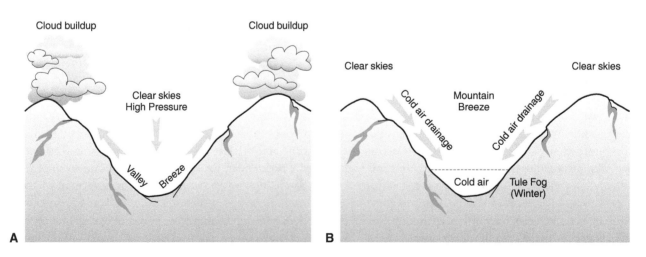

FIGURE 3.8 *(a) Valley breezes (anabatic) by day; and (b) Mountain breezes (katabatic) by night.*

3.4 Precipitation and Temperature Spatial and Seasonal Patterns

As discussed earlier, precipitation and temperature in California vary in complex ways in response to latitude, seasonality, West coast position, El Niño/La Niña conditions and especially topographic factors. This section will discuss more specific mechanisms and patterns of precipitation and temperature, especially in relation to the seasons, and descriptively for each landform region.

3.4.1 Precipitation Controls

According to California's state climatologist (a scientist who studies the long term averages of weather), annual precipitation totals in excess of 127 cm (50 inches) per year are characteristic of the west slope of the Sierra Nevada north of Stockton, the west slope of the Coast Range from Monterey County northward and parts of the Cascades (the taller volcanic peaks, i.e., Mt. Shasta and Mt. Lassen). More specifically, however, there are more local details that make for a complex precipitation regime, both latitudinally and longitudinally (see page 6 in *California Atlas*).

Mediterranean climates are unique in the world, and within California there are a large range of Mediterranean climate variations. Cool, wet winters and hot, dry drought condition summers characterize the climate. For most of California, the arrival of precipitation marks the seasons, not temperature.

3.4.1.1 Seasonal Precipitation Cycles

During the summer drought a strong North Pacific high pressure cell (also known as the Hawaiian high) sits over the marine layer and makes rainfall unlikely by pushing the mid-latitude storm track brought in by the Polar jet stream far to the north. Only the odd thunderstorm may occur near the coast and over the Great Central Valley at any time of the year, but they are usually light and very infrequent.

With the onset of the weakening and the shifting southwards of the North Pacific high in fall the mid-latitude westerly storm track strengthens and moves south with the Polar jet stream. It picks up storms in the Aleutian low cell (Gulf of Alaska) and moves them into the lower latitudes of the Mediterranean zone, bringing rain, with snow at higher elevations. As a result, California receives almost all of its yearly precipitation during the winter season and may go anywhere from six to eight months during the late spring-summer to fall without having any significant precipitation. Because the storms start in California's northwest and trend toward the southeast of the state, the northern half of California gets two-thirds of the state's precipitation. This is because of topography that will be explained further on. In addition, occasional tropical storms reach Southern California either via El Niño events or Pineapple Expresses (as discussed earlier); these usually bring very heavy rainfall and intense local flooding.

Finally, the transmontane regions of California, which you should recall from Chapter 2, contain areas east of the main mountain ranges (i.e., the Cascades, Sierra Nevada, San Gabriel, San Bernardino and Peninsular ranges) that have summer rainfall events. In the Mojave and Colorado Desert, the Salton Trough and northwards into the Basin and Range provinces summer rainfall moves through the far eastern border regions in a northward direction. These storms are thunderstorms originating from the **Arizona monsoons** that penetrate the southwestern U.S. due to moisture-laden airflow coming northwards from the Gulf of California and the Gulf of Mexico then up through Northern Mexico into the U.S. This then interacts with rising air coming from the intense surface heating of the deserts. The convective thunderstorms break the typical Mediterranean precipitation patterns for the eastern Californian regions by bringing short but intense rainfall events to the eastern deserts and the Basin and Range region.

3.4.1.2 Controls on Precipitation Spatial Patterns

Winter rainfall increases northward with latitude because the mid-latitude Westerly storms track from the northwest towards the southeast. For example, on the coast from Chula Vista (San Diego County) to Crescent City (Del Norte County) rainfall increases eightfold south to north (Figure 3.9a). In contrast, summer rainfall in the transmontane regions increases longitudinally to the east and south of the state.

As discussed earlier, topography has the strongest effect on the rainfall distribution in California due to the nature of the marine frontal storm systems acted upon by the orographic effect. Rainfall totals generally increase with higher elevations, eventually turning to high snow levels at the highest elevations. Recall that windward slopes are wetter than leeward slopes (rain shadows), Thus, the conditions in California are for windward to be on west-facing slopes in winter and to reverse to east slopes in summer for the Sierra Nevada and Cascades due to the Arizona monsoons. Because winter storms move in a northwest to east and south-east manner across the state from the ocean and perpendicular to the north-south trending mountains, rain shadows occur to the east of the mountains. Because there are so many mountain ranges there is a progressively stronger rain shadow moving inland from the coast. The Great Central Valley is in a rain shadow from the relatively low elevation North and Central Coast Ranges (Figure 3.9b). However, because the storms move from the northwest to the southeast areas, the south-western San Joaquin Valley can experience near desert-like conditions from very low rainfall. The deserts of the state are all on the leeward side of one or more mountain ranges. In addition, consider the impact of the orographic effect on Nevada, Utah and Arizona where the taller mountains in California tend to cast a very effective rain shadow on them. The rain shadow of the Cascade-Sierra Nevada axis blocks moisture from about a third of the U.S. Nevada, Utah, and Arizona are large inland high altitude deserts thanks

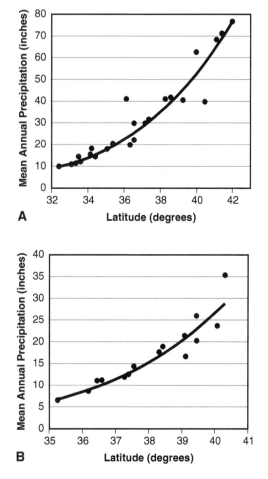
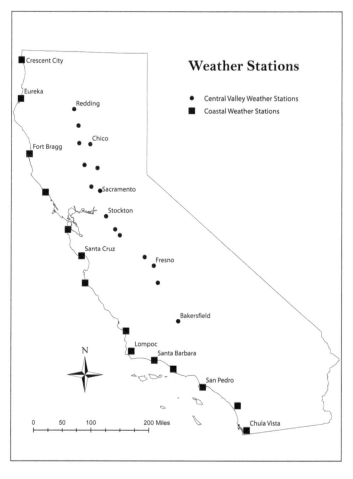

FIGURE 3.9 *(a) Rainfall increase from the southern to northern borders along the coast; and (b) Rainfall increase from the southern end of the San Joaquin Valley to the northern end of the Sacramento Valley. (Courtesy Kirstyn Pittman)*

to the combination of California's plate tectonically derived mountain ranges blocking the winter frontal storm activity. For example, the average annual rainfall of Truckee, California on the Interstate-80 freeway is 81 cm (32 inches), while Reno, Nevada, just thirty miles down the freeway, receives 19 cm (7.5 inches). One can also see the cumulative effects of orographic lifting and the reduction of precipitation while proceeding south and inland. Death Valley, which receives 5 cm (2 inches) per year, is in the triple rain shadow of the Central Coast Ranges, the Sierra Nevada and the Panamint Range.

3.4.2 TEMPERATURE CONTROLS

Areas of moderate temperatures and other places where temperatures reach extreme values of either heat or cold exist in California. Along the coast the small range in temperature from day to night and from winter to summer produces an unusually moderate climate regime. As one leaves the coast and continues inland, temperature ranges become wider. However, this depends upon the amount of maritime influence experienced inland. Finally, there is also a large temperature variation within the state's higher mountain elevations (see page 7 in *California Atlas*).

3.4.2.1 Seasonal Temperature Cycles

In the following discussion keep in mind that California receives the cold water California current as part of the large clockwise rotating north Pacific oceanic gyre. Based on the fact that solar incoming radiation increases with a decrease in latitude, the California current will be at various stages of heating depending on the latitude and the season of the year. During the summer months the northern coastal areas are colder because the ocean current is colder at the higher latitudes, though warmer than it is during the winter months (Figure 3.10). In summer, the longer day length offsets the lower sun angles at the higher latitudes. In the interior there is little difference in maximum or minimum summer temperatures whether they are in northern California (i.e., Yreka, Siskiyou County, 22°C [72°F]) or in southern California (i.e., Campo, San Diego County, 22.7°C [72.9°F]).

The onset of winter changes the temperature range situation largely for the interior, but the coastal areas remain moderately cool to warm due to the differential heating and cooling of water versus the land (Figure 3.10). During the summer the ocean waters are warm and hold onto the heat longer into the winter because water has a higher specific heat than land. In the northern coastal areas it will be generally colder than in the southern coastal areas because of the cold water current and the lower sun angle in combination with shorter day lengths due to the tilt of the northern hemisphere away from the sun (Figure 3.11). During the winter, coastal temperatures follow an average decrease of 0.83°C–1.1°C (1.5°F–2.0°F)/degree of latitude. In the interior there is a greater difference in maximum or minimum winter temperatures between sites in northern (i.e., Yreka, 0.9°C (33.6°F) versus southern California (i.e., Campo, 4.7°C [47.7°F]).

3.4.2.2 Controls on Temperature Spatial Patterns

Topography is the main control on temperature patterns in California, and since plate tectonics has created such an immense complexity of forms—hills, mountains, orientations, gaps, valleys, etc.—it is the main issue that explains many temperature phenomena used throughout the following discussion.

The autumn to winter patterns in temperature and pressure change dramatically across the state as the incoming intensity of solar radiation shifts southward and the Pacific high also weakens to shift south. In general, during winter the state experiences the most minimum temperatures, whether they are coastal or inland and valley or mountain. However, since the ocean cools more slowly, the land areas with a strong marine influence will be not as cold as the non-marine influenced inland regions. In addition, we can expect that generally temperatures will be colder at higher elevations as temperature already follows the environmental lapse rate in addition to the decrease in solar radiation angle and the day length (see Figure 3.1). The topographic aspect also plays a role combined with the sun's southern track across the sky by having warmer slopes on south-facing aspects and cooler (in shadow) slopes on north-facing aspects.

The changes in latitudinal heating in autumn leads to one of the first major issues related to temperature and pressures of the season—the **Santa Ana winds** (aka "Devil Winds"). These winds are drainage winds that start when a high pressure (descending) air mass builds on the cooling landscapes of the Great Basin

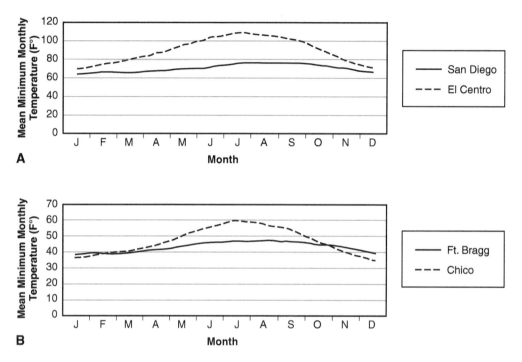

FIGURE 3.10 *(a) Exposure to marine influence during the summer usually has a greater impact as the ocean is cooler than the land. Here is an example of two sites on the same latitude: San Diego on the coast versus El Centro (Imperial Valley) in the interior; and (b) Exposure to marine influence during the winter usually has a minimal impact as the ocean is as warm as the land. Here is an example two sites on the same latitude: Ft. Bragg on the coast versus Chico in the interior.*

over Nevada and Utah. Since the ocean temperatures take longer to cool, the coastal waters are warmer (after having warmed the whole summer) and thus represent a low pressure (ascending) situation. Because high pressure flows to low pressure, a regionally large offshore (land) breeze develops across largely southern California (Figure 3.12). As the winds come down from the interior with its high elevation to the coast, they heat up via adiabatic heating (compression). Because they already start with a low humidity as cool and dry desert air, they dry further to below 10 percent humidity. Therefore, the coastal areas experience some of the warmest and driest conditions of the year, even warmer than the interior deserts, and the pollution and smog in the coastal basins blow out to sea. Of course, this "extended" summer into autumn also has its dangers, as the combination of wind, heat and dryness turns California's infamous chaparral vegetation into explosive fuel that feeds the wildfires for which the region is known. Wildfires fanned by Santa Ana winds burned 721,791 acres in southern California during October 2003 (Figure 3.12), and some 426,000 acres again in October 2007.

During the height of winter the colder temperatures in the inland valleys and surrounding mountains cause a condition called a **temperature inversion** (Figure 3.13). A temperature inversion results when warmer air overlays colder denser air, resulting in a stable condition as the colder air resists rising. In the inland valleys of California, the flow of cold denser air downslope from the surrounding mountains—cold air drainage—that the wind does not move results in the formation of **tule** or valley fog (see Figure 3.8). This wintertime fog (November 1 to March 31) forms on clear winter days when heavy cold air settles into the mountain valleys, especially the Great Central Valley (from Red Bluff to Bakersfield) while warmer air moves over the surrounding mountains and caps the cold air. At night the already frigid ground further cools the air immediately above it, bringing the water vapor in the air to dew point and thus condensing the water so that it forms very thick ground level fog. This fog can last for days and leads to extremely hazardous driving conditions on the Interstate-5 freeway and on Highway 99, usually involving multiple vehicle accidents at a time (more than a hundred vehicles in November 2007).

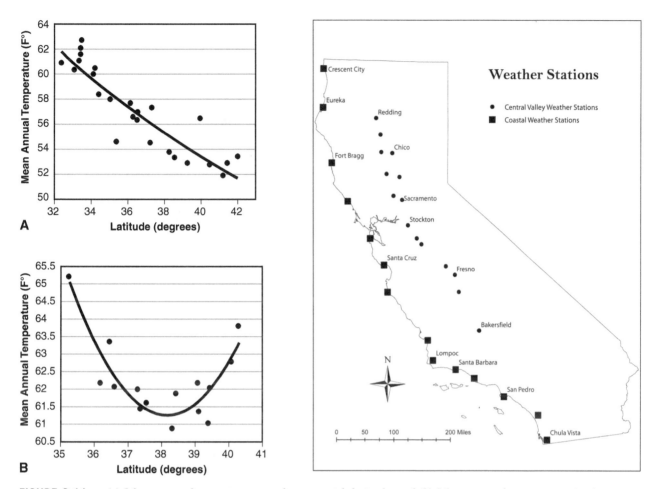

FIGURE 3.11 *(a) Mean annual temperature on the coast with latitude; and (b) Mean annual temperature in the Central Valley with latitude. (Courtesy Kirstyn Pittman)*

The temperature and pressure patterns from spring to summer are the opposite of the autumn to winter activities. The Pacific high pressure is reforming with the increasing solar radiation from the northern hemisphere's tilt toward the sun, and over the winter the coastal waters have lost their heat, but now their colder temperatures help to moderate the warming coastal lands with a stronger marine onshore flow. The interior valleys and deserts of California are heating creating low pressure (ascending) air masses that start the horizontal (advection) onshore flow of cooler denser air from the oceans. The thick **marine layer** of **coastal fog** develops far into coastal valleys from April to July and keeps these areas cool, especially in the morning hours (see Figure 3.7). For Californians along the Central Coast and in the south this is the season of "June gloom." The sea breezes that develop can be strong enough to penetrate any gaps in the coastal ranges to bring cool relief to the inland valleys, such as the Santa Clara Valley and Great Central Valley. This marine influenced cooling occurs as the Delta breezes pass through the Carquinez Straits, which then keep Sacramento and Stockton cooler than normal for inland valley locations (Figure 3.11b).

Lastly, the summer temperatures bring another **thermal inversion** to the Los Angeles Basin, including the San Fernando, San Gabriel and Pomona Valleys, and into the inland empire valleys of San Bernardino and Riverside, as well as the Great Central Valley. The surrounding mountains cause the high pressure air over California that is sinking and therefore compressing and heating to form a warm, dry lid of air above the valleys and basins. On the coast the cooler, denser marine layer drawn in from the ocean produces a stable air mass under the warmer air. In addition, the normal adiabatic cooling rate becomes reversed for any rising pollution leaving the landscape when it hits the warm air lid and thus becomes trapped (Figure 3.13). This helps develop the chronic smog conditions found in the areas mentioned above.

52 CHAPTER 3 California Climate

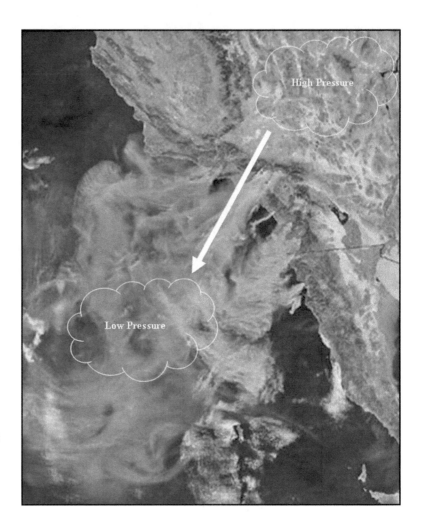

FIGURE 3.12 *Satellite image of Southern California 2003 fires during the fall Santa Ana wind period with labels showing pressure regions and wind direction. (Courtesy NASA)*

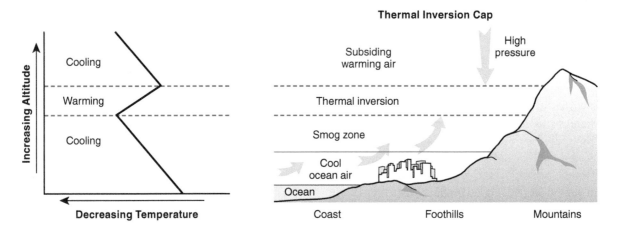

FIGURE 3.13 *The process of a summer thermal inversion cap caused by topography, where the normal environmental lapse rate process is blocked by a cap of warmer air. (Adapted from Durrenberger)*

California In Focus

SURFING CULTURE

Dude! Wat's up brah? One could not talk about California without mentioning surfing and surf culture. Surfing is derived from the ancient royal Hawaiian sport of riding a coastal wave by standing up on a flat board. Surfing in California has its humble beginnings in the early 1900s when Americans living in Hawaii and a famous Hawaiian surfer, Duke Kahanamoku, brought the sport to southern California and demonstrated it near San Diego at Redondo Beach and Ocean Beach. By the 1930s surfing had become popular, but it was the 1950s when Hollywood movies and Southern California surf culture created an explosion in popularity with *Gidget* and, later in the mid-1960s, *Endless Summer*. It was not until the mid-1980s, however, that California became a destination for the World Pro Surfing competition tour.

A side development that Southern California surf culture spawned in the 1950s was skateboarding. If the waves were blown out, especially in the afternoon with the offshore breezes, then surfers came up with the idea that the streets could be surfed instead. It was the mid-1970s when skateboarding really took off at the Del Mar Ocean Festival. The Zephyr team (Z-Boys) from the Dogtown area of Venice Beach used surfing moves to make skateboarding exciting. They were also very creative in their use of skateboarding in empty private pools found around Santa Monica and Venice Beach during a major drought that occurred in California at the same time. The pool skateboarding that they perfected became the catalyst for skateboard park design throughout the U.S. from the 1980s until the present.

Mavericks Surf Contest 2010 in Half Moon Bay, California.
© *Rick Whiteacre, 2012. Shutterstock*

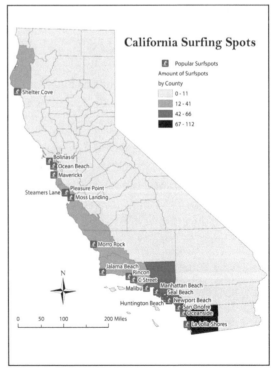

Courtesy Kirstyn Pittman

3.5 CLIMATIC VARIATION BY LANDFORM REGION

3.5.1 SIERRA NEVADA

The climatic focus for this region is its precipitation patterns that have the following characteristics:

- This long mountain range represents the major source of water for California, as it poses a strong orographic effect for one half of the state (see pages 5 and 6 in *California Atlas*).

- The large snowpack allows for a delay in river runoff peak for several months (mid-summer) after its precipitation peak in early spring, which helps fill many of the state's reservoirs and allows for generation of hydroelectric power (see page 23 in *California Atlas*).

- Spatially, the northern Sierra Nevada receives the majority of the precipitation. The southern Sierra Nevada region is considerably drier, but the southern area's snowfall totals are still impressive at higher elevations.

- The temperature patterns are strongly tied to the elevation gradient with limited latitudinal differentiation (see page 7 in *California Atlas*).

3.5.2 KLAMATH RANGES

The topographic complexities of this block of mountains in the northwestern corner of California have the following climatic characteristics:

- The precipitation regime ranges from very wet conditions (temperate rainforest-like, i.e., redwood forests) on the coastal (windward) side to moderately dry in the interior to leeward side facing Mt. Shasta and the northwestern Sacramento Valley (see pages 5 and 6 in *California Atlas*).

- Cool wet conditions at the coast for the winter lead to cool, foggy coasts during the summer; however, summer temperatures rise rapidly and trend inland (see page 7 in *California Atlas*).

- The winter temperatures are mild at lower elevations near the coast, but they become dramatically colder inland (the continentality effect) and at high elevations (see page 7 in *California Atlas*).

3.5.3 PENINSULAR RANGES

The climatic variations of this region of parallel trending southern coastal mountains are fairly easily defined:

- As this mountain range represents the most southern in California, it offers the last chance for any meaningful yet very limited orographic precipitation that increases rapidly with height on the windward slopes in San Diego County (see pages 5 and 6 in *California Atlas*).

- The ocean, which is at its warmest for California near the border with Mexico, moderates the seasonal temperatures. The interior weather stations, however, show wider variations in the summer and winter from the moderate elevations and parallel trending valleys (see page 7 in *California Atlas*).

3.5.4 SOUTHERN CASCADES AND THE MODOC PLATEAU

These two landform regions are grouped together because they represent a typical climate of the interior West:

- There is significant precipitation and snowfall on the higher volcanic peaks and low to moderate precipitation elsewhere, especially in the Modoc Plateau, due to the rain shadow posed by Mt. Shasta (see pages 5 and 6 in *California Atlas*).

- The interior continentality of these regions leads to warm to hot summers and to cold to very cold winters. The region is considered semiarid-arid except for the areas that surround the volcanic peaks, which drive a strong orographic effect (see page 7 in *California Atlas*).

3.5.5 GREAT CENTRAL VALLEY

The climatic patterns within the Central Valley are largely driven by the coastal ranges and San Francisco Bay's Carquinez Straits on the western side and by the Cascade-Sierra Nevada Ranges on the eastern border:

- The pattern of precipitation generally decreases from north to south and increases from west to east (see pages 5 and 6 in *California Atlas*). Therefore the

Western side of the San Joaquin Valley is considered a desert climate—very low rainfall and cool to hot winter temperatures.

- The temperature regime of this region has a wide range with cold winter temperatures and its affected by tule fog, while summer is generally hot throughout the entire valley. The Carquinez Straits, however, allow the cooler delta breezes to penetrate. Therefore, the middle of the valley near Sacramento and Stockton is cooler. The warmer valley air then becomes pushed to either end of the Central Valley, concentrating around Redding in the Sacramento Valley and Bakersfield in the San Joaquin Valleys (see Figure 3.11b). These opposite ends are typically 4–5° C warmer than around Sacramento and Stockton (see page 7 in *California Atlas*).

3.5.6 NORTH AND CENTRAL COAST RANGES

These sets of several parallel trending low mountain ranges with long valleys are broken up by the San Francisco Bay area. They have a wide variety of climates that have found favor with California's highly rated viticultural industry:

- In short, the precipitation generally increases northward (north coastal areas are quite wet, for instance, redwoods country), but it depends on topography as the Big Sur region in the central coast has very high rainfalls from its strong orographic effect next to the ocean. In general, though, the many north-south trending mountain ranges show progressive decreases in rainfall eastward and southward (see pages 5 and 6 in *California Atlas*).

- The temperatures for this region show large variations especially in the summer over very short distances. The coastal areas tend to be cool and foggy with small seasonal and daily temperature ranges, whereas the air temperature increases the farther one goes from the coast. The interior valleys can be hot. A penetrating marine layer that helps moderate temperatures, at least in the early summer, influences many of the valleys in the coastal ranges to some extent (see page 7 in *California Atlas*).

3.5.7 TRANSVERSE RANGES

Since this region represents California's only east-west trending set of mountain ranges, the element of the windward slope for precipitation is very different than for the rest of the state's large north-south trending mountain ranges:

- Storms that come down the coast are affected by Pt. Arguello and Pt. Conception (California's Cape Horn found in Santa Barbara County). This significant landform feature forces the storms south back out to sea even though the California coast turns to the east at this point. The storms then spin in a counterclockwise spiral to make the south-facing slopes of the Transverse ranges the dominant windward slope. Precipitation then increases rapidly with height via the orographic effect (see pages 5 and 6 in *California Atlas*).

- The temperature regimes are a typical pattern of cooler coastal areas moderated by the ocean with interior weather stations showing wider variations in summer and winter (see page 7 in *California Atlas*).

3.5.8 BASIN AND RANGE

Mary Austin wrote in 1903 the book *Land with Little Rain* to describe the Basin and Range environment and especially Owens Valley in particular. The Sierra Nevada overpowers this region through its influence on many of its climatic characteristics:

- The entire region represents a very strong rain shadow from the Sierra Nevada with arid valleys and semi-arid mountains running parallel north and south across the region all the way to Utah. There are some slight differences in moisture regime as the southern ranges get some summer rain from the northward moving Arizona monsoons, while the northern regions get most of their precipitation as snow (see pages 5 and 6 in *California Atlas*).

- The temperatures depend on elevation, with the valley floors being hot to very hot in the summer and cool to mild in the winters. The mountain ranges are mild to cool in the summer to very cold in the winters. This region represents California's strongest continentality effect (see page 7 in *California Atlas*).

3.5.9 MOJAVE

The Mojave is a high elevation desert that divides climatically into three regions with their own characteristics (see pages 5, 6 and 7 in *California Atlas*):

- The western Mojave receives limited precipitation entirely in the winter, and snow is not uncommon. The winter temperatures are cool, and the summer temperatures are hot.

- The central Mojave represents the driest portion of the desert even though it receives some rain in the winter and the summer via the Arizona monsoons. The winter temperatures tend to be mild, but the summer temperatures are hot, with temperatures above 120°F) in the lower elevations' playa lake beds.

- The eastern Mojave receives limited rainfall that is evenly split between winter and summer, with Arizona monsoonal rain probably being more important for its overall moisture regime totals. This part of the desert is generally higher in elevation, which leads to cool winter temperatures and warm summer temperatures. However, it becomes very hot at the lower elevations, especially near the Colorado River basin.

3.5.10 SALTON TROUGH/COLORADO DESERT

The Salton Trough/Colorado Desert represents the only extension of a warm desert into the state. One of the characteristics of this region is that winter nighttime temperatures never drop below freezing; therefore, it is California's only area where cactus can survive.

- An apt description for this region would be dry and drier. The combined rain shadow from the Peninsular ranges and its position in the subtropical latitudes near 30° N where there is significant high pressure subsidence leading to warm dry air make for bleak rainfall conditions (see pages 5 and 6 in *California Atlas*). However, there is a slight Arizona monsoon along the east side of the Salton Trough to the Colorado River.

- Another apt description for this region would be hot and hotter. This region is known for its record hot temperatures. Summer temperatures are well over 100°F daily, with 120°F not uncommon on many days during high summer. Winter temperatures are mild, rarely below freezing (see page 7 in *California Atlas*).

Bibliography

Baldwin, John L. *Climates of the United States.* U.S. Department of Commerce (NOAA), 1973.

Booth, S. (2008). *California Geography.* Course taught at Sierra College. [online] http://geography.sierra.cc.ca.us/booth/California/cal_index.htm.

Carle, D. (2006). *Introduction to Air in California.* University of California Press, Berkeley, CA.

Christopherson, R. (2008). *Geosystems.* Prentice Hall Press, Upper Saddle River, NJ.

Cox, C.J. (2008). *California Geography.* Course taught at Sierra College. [online] http://faculty.sierracollege.edu/ccox/california_geography/index.html.

Donley, M.W., Allan, S., Caro, P., and C.P. Patton (1979). *Atlas of California.* Pacific Book Center, Culver City, CA.

Durrenberger, R.W. and R.B. Johnson (1976). *California Patterns on the Land.* 5th edition, Mayfield Publishing Company, Mountain View, CA.

Felton, E.L. (1965). *California's Many Climates.* Pacific Book Publishers, Palo Alto, CA.

Holland, V.L. and D.J. Keil. (1995) California Climate. *California Vegetation.* Kendall/Hunt, Dubuque, Iowa.

Hornbeck, D. (1983). *California Patterns: A Geographical and Historical Atlas.* Mayfield Publishing Company, Mountain View, CA.

Michaelson, J. (2008). *Geography of California.* Course at UC Santa Barbara, Dept. of Geography. [online] http://www.geog.ucsb.edu/~joel/g148_f08/.

Miller, C.S. and Hyslop, R.S. (1983). *California: The Geography of Diversity.* Mayfield Publishing Company, Mountain View, CA.

Russell, R.J. (1953). *Climates of California.* University of California Press, Berkeley, CA.

Selby, W. (2006). *Rediscovering the Golden State: California Geography.* Second edition, John Wiley Press, New York, NY.

Selby, W. (2008). *Geography of California.* Course taught at Santa Monica College. [online]http://homepage.smc.edu/selby_william/california/chapter_1.html.

State Climatologist Report (2005). *Climate of California.* [online] http://www.wrcc.dri.edu/narratives/california.htm.

PRISM Group. (2008). *Northern American Climate Mapping.* Oregon State University, Corvallis, OR. [online] http://www.ocs.orst.edu/prism.

CHAPTER 4

California Flora and Fauna
Diversity, Hazards and Conservation

Key terms

- Annual grasses
- Arcto-Tertiary
- Bay/Delta
- Bioregions
- Central Coast
- Chaparral
- Cistmontane
- Closed cone pines
- Coastal sage scrub
- Colorado Desert
- Decidousness
- Degradation
- Ecosystem
- Endangered species
- Endemics
- Fire
- Habitat removal
- Habitats
- Klamath/North coast
- Madro-Tertiary
- Mediterranean-type
- Modoc
- Mojave
- Neotropical-Tertiary
- Oak woodland
- Perennial bunch grasses
- Plant diversity
- Ponderosa pine
- Redwood
- Sacramento Valley
- San Joaquin Valley
- Sclerophyllous
- Serpentine
- Sierra
- South Coast
- Species diversity
- Transmontane
- Vegetation types

Introduction

Scientists describe California as having a **Mediterranean-type** ecosystem. This stems from the Mediterranean-type climate that characterizes two-thirds of the state, excluding the deserts. This type of **ecosystem** is related to unique floral and faunal characteristics and histories, evolving dynamic landscapes and problems with expanding human impact and subsequent conservation efforts.

Compared to the world's tropical zones the Mediterranean-climate zones are second in their **species diversity** of vascular plants (the number of different species) and in the number of **endemics** represented globally (e.g., California is ranked number 17 for global biodiversity richness). Endemics represent species that are only found in a particular locality. California is a primary example of a uniquely Mediterranean type of flora with a nearly equally distinct set of fauna that operate under a specific set of **habitats**. This region, however, also mimics the other Mediterranean-climate zones for being affected by the

57

propensity of its vegetation to require **fire** for continued survival. Humans also like Mediterranean climate zones; thus past and present pressures from urban, agricultural, logging and recreation expansion have caused severe impact on the biotic legacy.

4.1 Bioregions of California

In contrast to the natural landform regions outlined in Chapter 2, in general California is divided ecologically into **bioregions** for biodiversity assessment activities (Figure 4.1). While the ten bioregions were initially based on the natural landform regions found in the state, there are differences that tended to split and group the bioregions based on ecological management activities and coordination among state and federal agencies. This chapter will use these bioregions to describe the state's unique biodiversity. The following brief descriptions of them are from the California Biodiversity Council Website (http://biodiversity.ca.gov/):

■ **Klamath/North Coast**
Includes the Klamath and North Coast Ranges, but also the western Sacramento Valley foothills.

■ **Modoc**
Includes the Modoc Plateau and Southern Cascades, but also including the northern half of the eastern foothills of the Sacramento Valley.

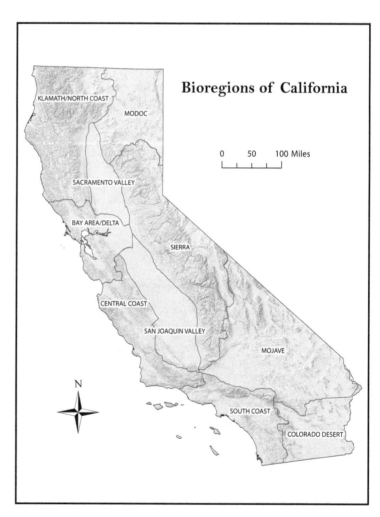

FIGURE 4.1 *These are the defined bioregions of California. (Courtesy California Department of Fish and Game, and Kirstyn Pittman)*

- **Sacramento Valley**
  Represents the flat valley floor north of the Sacramento-San Joaquin Delta.
- **Bay/Delta**
  Essentially the immediate watershed of the San Francisco Bay Area and the Sacramento-San Joaquin Delta, not including the major rivers that flow into the Delta.
- **Sierra**
  Includes the Sierra Nevada Range, Tehachapi Mountains and the Basin and Range sections of California east of the Sierras.
- **San Joaquin Valley**
  Represents the flat valley floor south of the Sacramento-San Joaquin Delta, but also the Carrizo Plain and the Bureau of Land Management's Caliente Resource Area in eastern San Luis Obispo County.
- **Central Coast**
  Coastal mountains and inland valleys from Santa Cruz to Ventura Counties.
- **Mojave**
  Nearly same boundary as presented for the Mojave Desert landform region.
- **South Coast**
  Mountains, valleys and coastal plains/basins from Ventura County to the Mexico border.
- **Colorado Desert**
  Nearly same boundary as presented for the Colorado Desert/Salton Trough landform region.

4.2 Determinates of Vegetation Patterns

Multiple competing factors that combine to create some fascinating landscapes largely determine vegetation patterns. In California, there are over five hundred vegetation communities defined by vegetation ecologists that they can map. The basis for this incredible amount of vegetation diversity, which leads to habitat diversity for animals, is largely a set of parameters that have changed over both geological and recent time scales.

4.2.1 Climate (Past and Present)

As discussed in Chapter 3, climate is a strong characteristic affecting California's landscapes, and it represents the largest contributor to vegetation patterns. In broad terms, a very dry, warmer climate will tend to lead to vegetation communities having a lack of trees that are more likely to consist of shrubs and annuals and, depending on the winter temperatures, even succulent species. This would characterize desert vegetation communities, for example, in the Mojave and Colorado Deserts. In contrast are the very wet, cooler climates that tend to have dominant tree cover, such as the Pacific Northwest temperate rainforests found in Humboldt and Del Norte counties in the region of the Klamath Mountains. Between these two broad types of moisture and temperature regimes various climatic patterns can come together to help determine many different general vegetation categories, i.e., moist but very cold would lead to alpine dwarf shrub, as found in the high Sierra Nevada. In addition, climates are a scaled phenomenon. Thus, while California has a host of regional climates (Mediterranean cool summer, warm summer, highland; hot desert, cold desert, steppe, and alpine), which can be broadly linked to major vegetation formations, there are also local climates that are the result of local variations in temperature, moisture, wind, humidity, etc. within a regional climate. For example, the north versus south-facing slopes within a canyon will tend to play out to differing vegetation communities, i.e., cooler-moisture requiring communities found on the north- facing slopes versus warmer-drier communities found on the south-facing slopes.

What is important to note here is that climates are not static, but instead they are very dynamic over long time scales. This means that the general climate of California currently known as a Mediterranean-type did not exist in this region several million years ago, and, in fact, scientists can really only account for it within the last 20,000 to 30,000 years. Climates change from glacial periods to interglacial periods (with a periodicity of about 100,000 years), like the one we are in now. Future climates will be different, as currently posed by global climate change research that links the rapid climate changes we are experiencing to human industrial development. Under all the various climatic changes, cooler to warmer and wetter to drier, the vegetation types and patterns found within the state have changed and will change as well.

4.2.2 TOPOGRAPHY

As you explore the affects of topography on vegetation pattern, it is important to make sure that you thoroughly covered Chapters 2 and 3 and understand that they hold the keys to the ways topography dominates the state's biological legacy after climate. In particular, the two terms **cistmontane** and **transmontane** are germane to this discussion (as they were in Chapter 2). To reiterate, cistmontane refers to that portion of California to the west of the crest of the Cascade Range, the Sierra Nevada, the high peaks of the Transverse Range and the Peninsular ranges. This accounts for 70 percent of the state's land area. Geologists refer to the remainder as the transmontane, or the area east of the mountain crests, which includes the Basin and Range, Mojave Desert and the Colorado Desert/Salton Trough regions (see page 5 in *California Atlas*).

Since topography plays such a strong role in the regional climate of California (i.e., the interaction of elevation and temperature, as well as the patterns of the orographic effect that create moister windward slopes versus drier leeward or rain shadow slopes), broad vegetation patterns strongly relate to the broader cistmontane (i.e., moister vegetation types) versus transmontane (i.e., drier vegetation types) divisions, as well as to the regional patterns within these two descriptive divisions.

More importantly, at the local level, topographic variation greatly affects the amount of solar radiation received at a particular site, depending on the slope and aspect. Therefore, while increasing elevation should come with a corresponding decrease in air temperature, different slopes located at the same elevation will have very different temperatures (i.e., north-facing slopes—cooler versus south-facing slopes—warmer). This situation in turn will affect the local soil moisture balance and thus affect the type of vegetation a site is able to accommodate.

4.2.3 GEOLOGY AND SOILS

The geological diversity in California and its dynamic climates have created some of the world's most complex soil types. They range from nutrient poor to nutrient rich (based on available nitrogen and phosphorus). This is a typical trend for the other Mediterranean-type ecosystems in the world. In California, the soils tend to be more nitrogen limiting than phosphorus. Soils as a rule support and often determine the patterns of the state's vegetation communities.

The soils that represent the least productive communities exist on the steep, rocky slopes in the high mountain and desert environments. In contrast, the rich deposits found in the coastal valleys and interior cistmontane valleys and floodplains are usually the most nutrient rich. In addition, California's geological processes have allowed for rare soil types to develop, some based on metamorphic rock geology. One of these interesting soil types, **serpentine** soils (serpentine is the state's official rock type), is the home to the rarest and most endangered plants found in the state. They exist largely in the Klamath and coastal ranges and in the Sierra Nevada foothills. These sites also represent the most nutrient poor soil types available to plants in California.

To summarize, California's soils and geology derive from the actions of earthquakes, volcanoes and glaciers that occurred over the last 1.8 billion years. Granitic, volcanic and marine sedimentary rocks (e.g., Franciscan mélange) and associated soils cover much of the state. However, scattered among this matrix are serpentine, limestone and highly acidic soils that support specially adapted plant communities. The story of California's soils is far more complex than space allows in this textbook; just remember that they are largely unique and have led to major diversification in California's floral inventory.

4.2.4 NATURAL DISTURBANCES—FIRE

California's human residents fear it, but our ecosystems need it—fire is an integrated component that has evolved with California's vegetation types. Fire has affected almost the entire floral inventory of California over its evolution, and accordingly plants have adapted in several ways.

One true characteristic representative of all Mediterranean-type climates and ecosystems is that the dry heat at the end of the summer and in early autumn creates conditions for wildfires. These wildfires were frequent even before humans came to California 20,000 years ago, and as such California's plants have evolved to be typically fire-adapted. Wildfire risk is much greater in cistmontane California (with the exception of the Pacific Northwest temperate rainforest regions to the north) than in adjoining transmontane.

Many of the major plants that dominate the California landscapes require fire to continue their life cycles. In many respects, fire is a fundamental part of the California ecosystem experience, an aspect that the California public has a hard time coming to terms with, especially since many housing developers have started to encroach on these fire-prone vegetation communities on the steep, rocky slopes in the Southern California mountain ranges (Figure 4.2). The seeds of many of these plants require a fire to pass over the soil beds to help start sprouting or the cones on older plants hold the seeds tightly (i.e., the closed pine cone community). The cones finally pop open only when a fire hot enough scorches them. Many chaparral species crown sprout (regrow from root crowns) after fires. Many of the herbaceous perennials, annuals and bulbs require fire to clear out areas so that they can compete for all the newly available light and moisture resources. This occurs in our conifer forest vegetation communities in the mountainous regions of the state with slow moving ground fires. A re-ordering of the ecosystem occurs with a release of nutrients back into the soil via ashes from burned plants and leaf litter. It also removes the older moribund vegetation, allowing for the next generation of California's fascinating Mediterranean-type plants to utilize the light, water and soil nutrients.

4.2.5 HUMAN ACTIVITY

Finally, the last 20,000 years of human occupation and co-evolution with the various landscapes found in

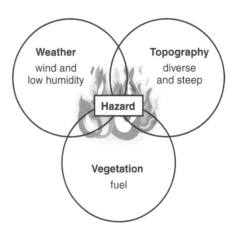

FIGURE 4.2 *The ingredients for a fire prone landscape.*

California have determined some of the vegetation patterns that exist around the state. This is especially true for grasslands, which have both seen an expansion in their aerial coverage as well as a complete change in their species makeup. The expansion came through the Native American use of fire as a tool to open up shrublands for hunting and to allow for desired annual food plants to dominate a landscape. Unfortunately, since the European and American occupation of California in the last 240 years, the state has lost its original perennial bunch grasses. Annual grasses brought in from the Mediterranean basin by the Spanish explorers and missionaries replaced them.

The final piece of this story of increased human domination of the landscapes of California really is one of natural vegetation removal for the development of agriculture, forestry and settlement. As long as California's population increases, pressure will continue on the natural vegetated landscapes of the state for housing, food and fiber production.

4.3 PLANT COMMUNITIES—GENERAL SUMMARY

California's Mediterranean climate produces four major plant community groupings with desert being a fifth major category to note in the transmontane regions of the state. While the groupings of forest, woodland, shrubland and grassland are very broad, each includes many different and distinctive plant communities that make up the over five hundred types identified by the state's resource agencies. The following description of California's general vegetation pattern relates to Figure 4.3.

- Forest—30% of the land area
  - This consists of closely spaced trees in a continuous tree cover pattern
  - Notable examples include coniferous forests in the Cascade/Sierra Nevada Mountains dominated by **ponderosa pine** and northern coast, mixed oak forests in canyons and **redwood** forests located along the coast from Big Sur north to Del Norte County.

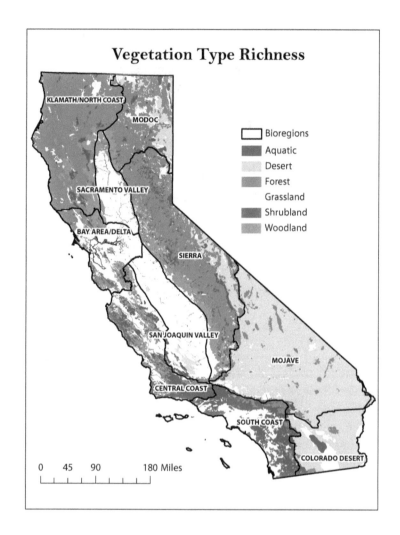

FIGURE 4.3 *General vegetation types of California. (Courtesy U.S. Geological Survey GAP analysis program, and Kirstyn Pittman)*

- Woodland—10% of land area
  - Woodland consists of trees that are in discontinuous cover, i.e., they are spaced far enough apart to leave room for a variety of shrubs, herbaceous plants and grasses in their understory.
  - The composition of trees represent mainly evergreen and deciduous oaks in drier inland valleys of California, typically termed as **oak woodland** (or foothill oak woodland)
  - Woodland generally occurs in areas suitable for agricultural use, so that much of the vegetation has been converted to pasture or cleared for growing various crops, especially wine grapes in central California.
- Shrubland (**chaparral**)—15% of land area
  - This area consists mainly of evergreen shrubs with **sclerophyllous** leaves (i.e., leather-like and thick for drought tolerance); few trees.
  - Called chaparral in California, it usually consists of a dense, single layer of tall shrubs with relatively few plants in the understory, except after a fire—which they require for continuation. Usually found on steep, rocky slopes with thin soils, but good drainage.
  - On the coast, it is called **coastal sage scrub** and includes low shrubs adapted to wind and salt air off the ocean. In California, it is divided into northern coastal scrub, central coastal sage scrub and south coastal sage scrub.
  - Chaparral and costal scrub communities dominate in cistmontane southern California.
- Grassland—12% of the land area
  - Grassland consists of few trees or shrubs, annual grasses and interspersed dominant herbaceous plants.

- Generally found in dry coastal foothills and interior valleys.
- Exotic European Mediterranean basin **annual grasses** and weeds have now completely replaced California's native grasses, formerly **perennial bunch grasses**. The annual grasses have a short lifecycle that leaves the landscapes in a "golden-brown" state from late spring until the rain comes in late autumn. When the California grasslands were perennial bunch grasses the landscapes were greener for a greater length of time.

- Desert—33% of the land area
  - Desert generally ranges from barren salt scrub to short largely spaced shrubs to open woodlands consisting of pinyon pines and Joshua trees.
  - These varying arid to semi-arid regions represent most of transmontane California.
  - The winters get too cold in the Mojave Desert region to support cactus, but
  - The Colorado Desert/Salton Trough region, which is California's only true warm desert, supports cactus species. In addition this warm desert supports unique native palm oases.

4.4 Plant Diversity

The size and complexity of California's flora, among the richest in North America (Table 4.1), add to its fascinating landscape diversities. Of the approximately 15,890 species of vascular plant species in the continental United States, 6,272 or about 32% occur somewhere in the state. Among the total species known to occur in California, 58% are native, of which 33% are endemic and 17% are naturalized aliens (i.e., brought into California and able to establish naturally). The size of California (414,400 km^2), its range of elevations (–85m to 4420m) and its large latitudinal range (32° 30′ to 42° N) allow for steep environmental gradients that make the variety of available habitats quite large. Much of the **species diversity** derives from these factors.

The following summary outlines the patterns of vegetation and plant diversity found within the state:

- **Vegetation types**—A recent assessment by the California Department of Fish and Game suggests that California may have over 2,000 distinct types of plant associations (Figure 4.4). An association is a distinctive set of plants that have ecologically similar requirements and is defined by the dominant species. This is almost half of the currently identified number in the entire U.S.
  - High diversity among vegetation types occurs in the Klamath/North Coast region, the San Francisco Bay Area/Delta area, the Central Coast, the South Coast and the northern Sierra.

TABLE 4.1 A Comparative Example of Native Flora in Various Areas of the World.

Region	Area (km^2)	Latitudinal Range (°)	Climate Classification	# of Species
California	414,000	32 30′–42 N	Mediterranean	6272
Alaska	1,479,000	55–70 N	Subarctic	1366
Texas	751,000	25–34 N	Humid Subtropical	4196
Sonoran Desert	310,000	25–32 N	Arid Desert	2441
North Eastern U.S.	3,238,000	40–45 N	Humid Continental	4425
Guatemala	109,000	16–18 N	Tropical Rain Forest	7817

- The lowest diversity occurs in the Central Valley (largely due to previous removal of natural vegetation for agricultural production), the Modoc plateau, the eastern Sierra Nevada and the Mojave and Colorado Deserts.
- **Plant diversity**—This diversity stems from the unique combination of climate and topographic, geologic and soil diversity that has allowed for incredible speciation. In addition, many ancient relic species survive here because of the mild climate.
  - Species diversity closely follows shifts in moisture and temperature produced by topography and climate (see pages 5, 6, and 7 in *California Atlas* for comparison).
  - The Sierra Nevada shows the most extreme topographic and moisture gradients in the state, and thus the most plant diversity, followed by the Klamath, the outer North Coast ranges, the Cascades and the San Bernardino Mountains (Figure 4.5).
  - The lowest diversity of plant species occurs in the deserts and Central Valley.

California supports a tremendous diversity of natural communities, including many unique ones that occur only within the state. The approximately five hundred natural communities covering both terrestrial and aquatic communities combine at the ecosystem level into diverse and complex living landscapes. These landscapes not only provide for a high diversity of plant and animal species in most remaining areas of the state, but they also provide much of the aesthetic beauty that Californians prize.

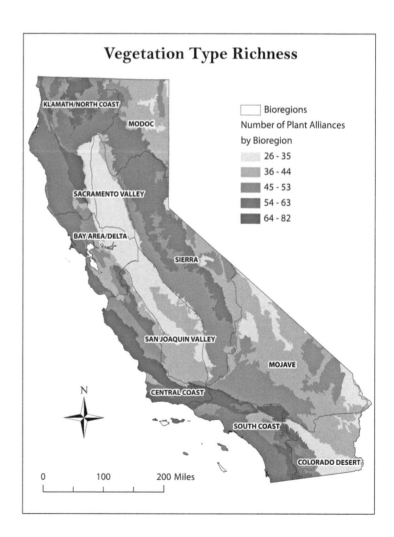

FIGURE 4.4 *Diversity of vegetation types in California. (Courtesy Kirstyn Pittman)*

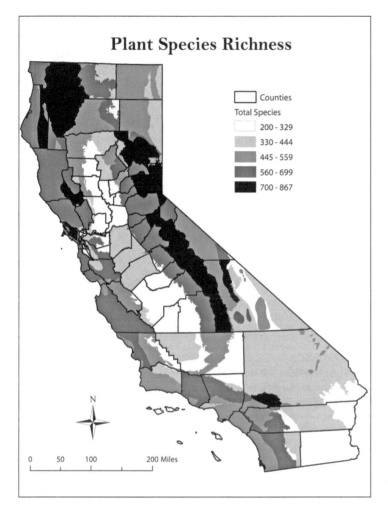

FIGURE 4.5 *Plant diversity in California. (Courtesy Kirstyn Pittman)*

4.5 FLORAL HISTORY

The land we now call California has changed dynamically throughout its approximately 150 million year existence. The interaction of long-term climatic changes and geologic processes coupled with evolutionary mechanisms has created its biological wealth. This section, while not going into a detailed survey of the natural history of California, will give the brief overview necessary to understand the present distributions of the major vegetation communities.

The basis of the discussion that follows is the Cenozoic Era in geologic history, approximately 65 million years b.p. to the present (Table 4.2). A very great amount of climatic, geologic and vegetational changes occurred during this time period. The events of this Era primarily influenced present conditions.

The history of the plant life of California is closely entwined with the history of the state's climate and topographical features. Fossil records have shown the climate changing from wet and tropical, with high precipitation year round, to being strongly seasonal with cool, damp winters and warm, dry summers. Over millions of years strong tectonic forces have modified California's topography. Many of the classical examples of topographical features (e.g., volcanoes, folded mountains, glacial U-shaped valleys, etc.) have been created in California. Their remnants, and newer examples of volcanoes, mountain ranges and plateaus, are still exerting an environmental force on the terrain today. Throughout these numerous changes the plant life has changed in tandem with the dynamic landscape.

TABLE 4.2 Geologic Timeline with Associated Environmental Events Affecting California's Flora.

Era	Period	Epoch	Event	Millions Years BP
Cenozoic	Quaternary	Holocene	Major human influence on landscapes	0.1
Cenozoic	Quaternary	Pleistocene	Great climatic fluctuation: cool-moist glacial periods alternating with warm-dry interglacial periods. Final rapid elevation of the Sierra Nevada. Last 10,000 years of Epoch saw the arrival of human groups into California.	2
Cenozoic	Tertiary	Pliocene	Spread of arid climates. Northward migration of Madro-Tertiary flora.	5
Cenozoic	Tertiary	Miocene	Separation of Eurasian and American Floras.	15
Cenozoic	Tertiary	Oligocene	Southward migration of Arcto-Tertiary flora.	40
Cenozoic	Tertiary	Eocene	Circumpolar angiosperm floras.	50
Cenozoic	Tertiary	Paleocene	Warm-moist climates prevalent.	65

During the early Tertiary period California looked vastly different from its present conditions. Many of the mountain ranges (i.e., the Klamath, the Coast Ranges and the Sierra Nevada) were more likely low in relief, or nonexistent in some cases. As outlined in Chapter 2, the Pacific Ocean covered much of what we know as the western half of the state and made shores along the present-day Sierra Nevada and Cascades foothills. Scientists have reconstructed California's climate as moist (summers) and warm (winters). Evidence from fossils indicates that the vegetation of the state was tropical, with rainforests to the north and tropical savannas extending well to the south along the Pacific Coast. What we know as summer drought was nonexistent, and year-round rainfall was common. At this time the deserts of present day California had not yet formed, and the rain shadow areas east of the Sierra Nevada were actually quite moist and forested. Plant communities included many whose present day relatives exist in the frost-free tropics. These ancient tropical plants are grouped together and called the **Neotropical-Tertiary** geoflora.

The climate and the vegetation of the state remained largely tropical with only minor fluctuations until the end of the Eocene Epoch. Near the end of the Ecocene the climate became more strongly seasonal and a gradual cooling and drying took place, accompanied by shifts in vegetation patterns. Tropical elements of the flora shifted from interior locations to southern coastal areas. Species by species, the Neotropical-Tertiary flora became eliminated from the state's floral inventory. Northern forests moved southwards and began replacing the tropical vegetation. Scientists recognize the species composition of these forests as the **Arcto-tertiary** geoflora. Only some modern-day California plants directly descended from Arcto-tertiary ancestors (e.g., maple, spruce, sequoia). These species exist in the forested areas of the state (e.g., primarily the Sierra Nevada, the Cascades, the Klamath Mountains and the northern Coastal Ranges). The vast majority do not exist in California anymore (e.g., beech, chestnut, elm, sweet gum), while others are extinct. Those that have survived grow today in the Eastern U.S., Europe, and East Asia.

During the mid-Tertiary (Miocene), California underwent many changes. These included volcanism that created the Cascade Range, the creation of the Coastal Ranges through folding/faulting and the creation of the Baja California peninsula by tectonic lifting and rifting. In tandem with the tectonic activity came climatic shifts of increasing seasonality with a subsequent loss of the state's tropical character. Mountain uplift created rain shadows, and the climate became cooler, changing the composition and structure of the vegetation. This cooling trend that began in the Tertiary became the great ice age of the Pleistocene Epoch. This period created both latitudinal and altitudinal shifts in vegetation. Species adapted evolutionarily to the new environmental conditions, migrated or became extinct. The drying of California caused many moisture-loving species to move to the moister mountain environments or to stay close to the milder coast. The fossil record indicates that at this time a semi-recognizable Mediterranean climate of dry-warm summers and moist-cool winters began. Many species of the state's deciduous hardwood forests and a few conifer species became regionally extinct during this period.

A new assemblage of species with warm-temperate or dry sub-tropical affinities began to expand into the southwest region of the U.S. from Mexico by the late Miocene-early Pliocene. This group of species known as the **Madro-Tertiary** geoflora was unique for its small leaved evergreen **sclerophyllous** foliage (i.e., thick and leather like) and some ability for drought **deciduousness** (i.e., dropping leaves with onset of water stress). These plants became concentrated in the southern part of California. Today, though, what remains of this geoflora resides in Mexico in the form of derivates in sub-tropical thorn woodlands. Only the relatives of the Madro-Tertiary geoflora (e.g., acacia, mesquite, ocotillo) are prominent in California's modern southern desert flora (i.e., the Colorado Desert/Salton Trough). Characteristic elements of the geoflora, however, exist in the **chaparral**, **coastal sage scrub** (Figure 4.6), foothill **oak woodland** and grassland communities throughout the state. With a pronounced severe summer drought that characterized the end of the Tertiary and the beginning of the Quaternary Periods, species requiring summer moisture became restricted to localized moist sites or eliminated from the state's flora.

Although the end of the Pliocene Epoch saw many components of modern flora and vegetation in place, changes during the Quaternary have had the most impact on California's floral composition and distribution patterns. Around 1.5 million years ago in the Pleistocene, a series of cold episodes began to create continental glaciers in the north and mountain glaciers

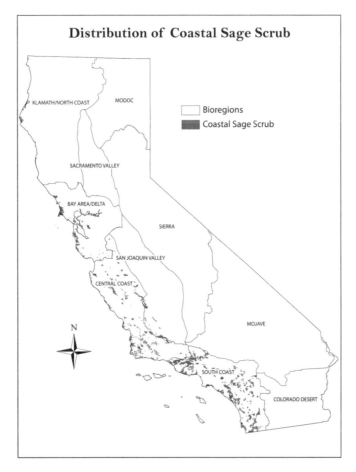

FIGURE 4.6 *Distribution of coastal sage scrub. (Courtesy Kirstyn Pittman)*

throughout the newly developed western ranges. This shifting of the climate changed both latitudinal and elevational vegetation zones. Ice sheets came and went repeatedly, cooling California for many thousands of years. The extreme changes in temperature forced cold-intolerant species to either adapt, shift their ranges latitudinally or altitudinally or become extinct. Many forest communities spread down from the mountains into valleys or found refuge along the much milder coast. Today only fragments of these forests exist along Califonia's central and southern coast (e.g., **closed cone pine** communities with Torrey, Bishop, Monterey and Knobcone pines).

Interspersed with these cold spells were warming periods of shorter duration. These short climatic warming periods caused the vegetation to shift back towards their places of origin. This cold-warm cycle has continued into the present (Holocene Epoch), marked by the retreat of the last ice sheets 12,000 years ago.

Following the retreat of the glaciers was a period known as the Xerothermic that lasted approximately 8,000 to 4,000 years ago. This section of time was a period marked by a warmer and drier climate than that of present day California. This prolonged drought condition caused many distributional changes to take place. The large inland valley pluvial lakes (see Chapter 2) dried out, desert communities expanded into rain shadow areas, forest communities retreated up mountain slopes, and there was increased forest fragmentation along the coast. Drought tolerant communities like chaparral, coastal sage scrub, grasslands and oak woodlands increased their range by moving into the newly vacated areas where coniferous forests once stood.

Since this period, the climate of California has moderated somewhat into what we know as the Mediterranean-type climate. Humans, as an element of change, have helped in large part to keep the scrubby drought-tolerant and fire-tolerant communities in place and even to expand their range. They have also introduced many plants from the old world (i.e., Eurasia, Africa, as well as Australia and South America) that have out-competed many native Californian species, forever changing communities like grasslands. Humans have also done their share of converting vast tracts of land into areas of urban and agricultural usefulness. The current human induced global climate warming from increasing greenhouse gases will provide pressure on many plant species to move up in elevation or north in latitude to survive the temperature increases and potential changes in precipitation.

4.6 California Animal Diversity

The variety of native vertebrates and invertebrates found in California because of its unique geography and geologic history is truly astounding. Outside of Hawaii, California is a state with high biodiversity. Of the 1,002 native vertebrates found within its borders the breakdown among various groups is as follows: 602 birds, 198 mammals, 84 reptiles, 51 amphibians and 67 freshwater fish. Of these, 8 percent only exist in California and nowhere else on the planet. California has the highest total number of animal species of any state (i.e., continental, not including Hawaii) and the highest number of endemic species. Scientists estimate that California has 30,000 invertebrate species. In addition, combined with plant diversity, they recognize it as home to several of the nation's biological "hotspots" and identify it as one the twenty-five biodiversity "hotspots" worldwide (i.e., ranked 17th).

Areas of high species diversity, rarity and endemism in animals occur in every region of California because of climate, resulting vegetation, rare local geology or geographic isolation. The following is a summary of the patterns in the vertebrate groupings:

- **Birds**—use every available habitat found in California (Figure 4.7).
  - The state's large size, varied topography, mild climate and habitat diversity are largely responsible for its unique and rich bird life.
  - Found nowhere else in the world are the Channel Island Scrub-Jay and Yellow-billed Magpies.
  - Virtually confined to California: Ashy Storm Petrel, Nuttalls' Woodpecker, Oak Titmouse, Wrentit, California Thrasher, Lawrence's Goldfinch, Tricolored Blackbird, and California Condor.
  - When categorized based on whether or not they migrate: 25% are year round residents, 33–50%

migrate (summer breeders, over winter, or spring/fall migratory), and 33% are vagrants (neither breeding or wintering or migratory routes in state) (see page 20 in *California Atlas*).

- Summer bird diversity—The greatest numbers live in woody vegetation of the coastal regions, foothills and mountains and valleys of northeastern California.
- Winter bird diversity—The greatest numbers live in the coastal regions, the Central Valley, the foothills and mountains of Southern California, along the Colorado River and around the Salton Sea.

■ **Mammals**—While mule deer and mountain lions are the largest and best known, the majority of mammal species in California are relatively small (Figure 4.8).

- The greatest species diversity exists in the squirrel and chipmunk family and in the woodrat and vole family.
- The highest mammal endemism of any state in the country: Mount Lyell shrew, Alpine chipmunk, Sonoma chipmunk, Yellow-cheeked chipmunk, San Joaquin antelope squirrel, Mohave ground squirrel, San Joaquin pocket mouse, the white-eared pocket mouse, the narrow-faced kangaroo rat, the Pacific kangaroo rat, Heerman's kangaroo rat, the Giant kangaroo rat, Stephen's kangaroo rat, the Salt-marsh harvest mouse, the California red tree vole and the Island gray fox.
- High species diversity is largely confined to these forested habitats: Klamath/North Coast, Modoc, and Sierra regions.
- Pocket mice and kangaroo rats account for much

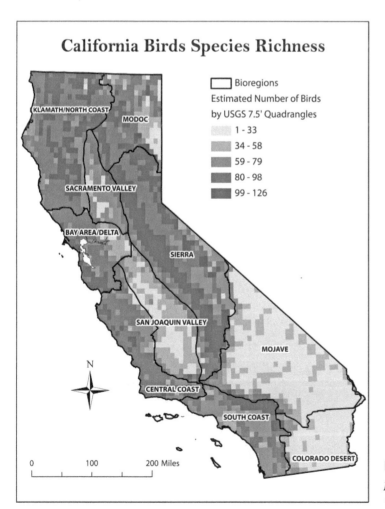

FIGURE 4.7 *Species richness (number) of birds per U.S. Geological Survey 1:24,000 map sheet. (Courtesy Kirstyn Pittman)*

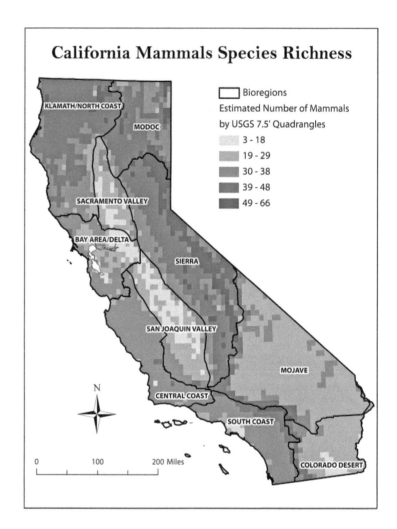

FIGURE 4.8 *Species richness (number) of mammals per U.S. Geological Survey 1:24,000 map sheet. (Courtesy Kirstyn Pittman)*

of the species diversity on the eastern side of the Modoc and Sierra regions and in the Colorado Desert.

- **Reptiles**—The highest diversities are largely confined to the desert regions, as reptiles' body temperatures rely on the surrounding environment to keep warm (Figure 4.9).
  - While there are reptiles in all parts of California, the deserts have the species that are most adaptable to such extreme environments.
  - The Sierra Nevada foothills have the highest species diversity of snakes due to the variety of rocky habitats and a reduction in flooding possibilities that California's valleys can succumb to each winter.
  - A high number of lizards and snakes are range restricted to the South Coast region and have become rare due to high development pressure.

- **Amphibians**—These represent a group of species that include frogs, toads and salamanders, which are the most environmentally sensitive of the vertebrates (Figure 4.10).
  - Since these species prefer wet places, the species diversity is highest in those parts of California with the highest precipitation, including the Klamath/ North Coast and the San Francisco Bay/Delta. Other high precipitation areas include areas below the snowfall line in the Sierra Nevada foothills, the Central Coast and the South Coast.
  - The Sierra region contains several rare frogs and

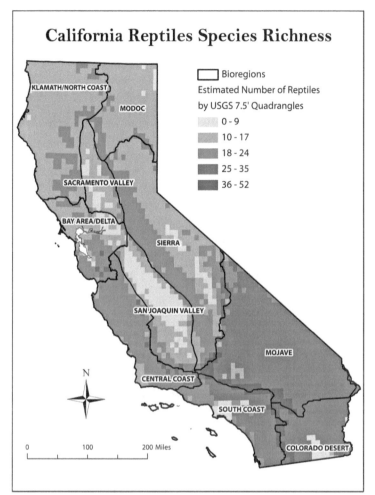

FIGURE 4.9 *Species richness (number) of reptiles per U.S. Geological Survey 1:24,000 map sheet. (Courtesy Kirstyn Pittman)*

toads. A number of salamanders with very restricted distributions also exist in the southern Sierra.

- **Freshwater fishes**—In a dry state, diversity is confined to the state's perennial streams and rivers in the northern half of the state and largely tied to the Delta region (see pages 5 and 22 in *California Atlas*).
  - The highest diversity is in the low elevation rivers and lakes, with the Sacramento valley and San Francisco Bay/Delta regions being the highest because of a very productive aquatic habitat.
  - The desert regions are the lowest in diversity, but they are home to some of the rarest species, such as the pupfishes found in springs near Death Valley and the Owens Valley.
- **Invertebrates**—Scientists estimate there are well over 30,000 species found in the state, with butterflies accounting for 636 species.
  - This group includes such groups as worms, mollusks, crustaceans, spiders and insects.
  - Many invertebrate species that were formerly common are now rare and endangered. This has occurred largely in the San Francisco Bay/Delta and South Coast regions where urbanization pressure has led to severely restricted habitats. Species include the Bay Checkerspot butterfly, the Myrtles Silverspot butterfly, the El Segundo blue butterfly and Lange's El Segundo dune weevil.

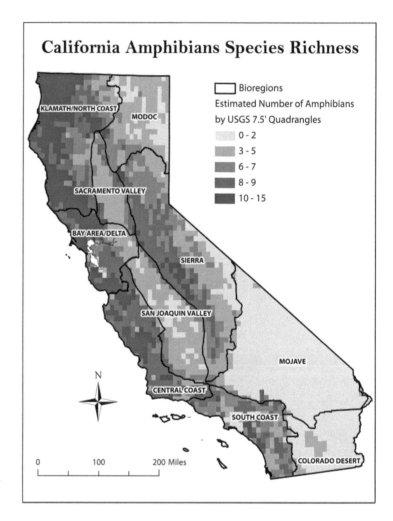

FIGURE 4.10 *Species richness (number) of amphibians per U.S. Geological Survey 1:24,000 map sheet. (Courtesy Kirstyn Pittman)*

4.7 ENDANGERED SPECIES AND CONSERVATION

At last count in 2012, California had 303 species on the Federal **Endangered Species** List and 20 on the pending list, and it can lay claim to at least 13 recorded extinctions due to human "progress." California's citizens see one of these species, which has been extinct locally since 1913, everyday on the state flag: The grizzly bear is gone but not forgotten. Any successful effort to maintain biological diversity in California depends on the support of its citizens. Ultimately, support for the conservation of biodiversity in California requires that the public understand the severity of the problem and support the needed planning and management actions. Accordingly, there are two issues that affect biological diversity: **degradation** and **habitat removal**.

The adverse modification of habitat from its original condition causes the degradation of biological diversity. Degradation of biological resources may be more subtle than the more apparent complete elimination of a vegetation community (and its faunal associates) or ecosystem. Some examples of major current habitat degradation include: the reduction of stream flows from water diversion; the reduction in tree species diversity from timber production and intensive fire suppression efforts; the long-term reduction in tree cover as a result of air pollution in the mountains of southern California; the disruption of natural ecosystems by aggressive non-native alien plants; and human disturbance in recreational developments. Of these, the impact of fire suppression has been the most damaging. While fire prevention and containment has been a well-intentioned form of human management intervention, it has substantially altered the vegetation communities. Natural fire once maintained chaparral and forest communities in a mosaic of stands of different ages, and it created open, park-like stands and encouraged the growth of many plants and tree diversity in conifer forests. Unfortunately, aggressive fire prevention and suppression programs have led chaparral and forest communities to build up huge fuel loads and to become dangerous to other dependent plants and animals, and now human settlements.

Habitat fragmentation and removal is the not-so-subtle effect on California's landscapes. Human settlement expansion and agricultural activities reduce the habitat for plants and animals to survive, and the smaller patches are no match for larger pieces of landscape to maintain biological diversity. Examples of species affected by habitat removal include: the California condor, California spotted owls, the California gnatcatcher, mountain lions, mule deer, the San Joaquin kit fox, the San Joaquin riverine rabbit and the kangaroo mouse. Ultimately, nearly all current and future threats to California's biological diversity from habitat loss and degradation are caused by the expanding human population, conflicting public attitudes toward biological resources and the poor decision-making processes that weakly try to balance economic development with the environment in land use planning.

At the end of the day, increased support for biodiversity conservation must come from expanded efforts to educate all segments of California's population about the state's biological uniqueness to the U.S. and the world, and to demonstrate the benefits of resolving environmental and economic development conflicts.

Bibliography

Barbour, M., Keeler-Wolf, T., and A.A. Schoenherr (2007). *Terrestrial Vegetation of California*. University of California Press, Berkeley, CA.

Booth, S. (2008). *California Geography*. Course taught at Sierra College. [online] http://geography.sierra.cc.ca.us/booth/California/cal_index.htm.

California Department of Fish and Game (2003). *Atlas of the Biodiversity of California*. Sacramento, CA.

Cox, C.J. (2008). *California Geography*. Course taught at Sierra College. [online] http://faculty.sierracollege.edu/ccox/california_geography/index.html.

Davis, F.W., D.M. Stoms, A.D. Hollander, K.A. Thomas, P.A. Stine, D. Odion, M.I. Borchert, J.H. Thorne, M.V. Gray, R.E. Walker, K. Warner, and J. Graae. (1998). *The California Gap Analysis Project—Final Report*. University of California, Santa Barbara, CA. [onlinr] http://www.biogeog.ucsb.edu/projects/gap/gap_rep.html.

Holland, V.L. and D.J. Keil (1995). Vegetation history of California. *California Vegetation*. Kendall/Hunt Publishing, Dubuque, Iowa.

Johnson, S.G. (1998). *Oaks at the Edge: Land Use Change in the Foothill Woodlands of the Central Sierra Nevada, California*. UC Berkeley, Dept. of Geography, PhD Dissertation.

Major, J. and M.G. Barbour (1977). *Terrestrial Vegetation of California*. John Wiley, New York, NY.

Mensing, S. and R. Byrne 1999. Invasions of Mediterranean weeds into California before 1769. *Fremontia*, 27: 6–9.

Michaelson, J. (2008). Geography of California. Course at UC Santa Barbara, Dept. of Geography. [online] http://www.geog.ucsb.edu/~joel/g148_f08/.

Quinn, R.D., Keeley, S.C., and M.D. Wallace (2006). *Introduction to California Chaparral*. University of California Press, Berkeley, CA.

Sawyer, J.O. and T. Keeler-Wolf (1995). *Manual of California Vegetation*. California Native Plant Society, Sacramento, CA.

Schoenherr, A.A. (1995). *A Natural History of California*. University of California Press, Berkeley, CA.

Selby, W. (2006). *Rediscovering the Golden State: California Geography*. 2nd edition, John Wiley Press, New York, NY.

Selby, W. (2008). *Geography of California*. Course taught at Santa Monica College. [online] http://homepage.smc.edu/selby_william/california/chapter_1.html.

Sugihara, N.G., van Wagtendonk, J.W., Shaffer, K.E. and J. Fites-Kaufman (2006). *Fire in California's Ecosystems*. University of California Press, Brekeley, CA.

CHAPTER 5

California Native American Landscapes

Key terms

Algonquian	Language stocks	Proto-agriculturist
Athapascan	Linguistics	Sedentary
Cultural area	Localism	Spanish
Determinism	Mission-presidio-pueblo system	Trading trails
Exotic species	Native Americans	Transhumance
Fire	New world	Tribe
Hokan	Nomadic	Tribelet
Hunter-gatherer	Old world	Uto-Aztecan
Isolation	Penutian	Worldviews
Language families	Pest and pathogens	Yukian

Introduction

California may hold the key to New World prehistory. Evidence gathered on the Santa Barbara Channel Islands indicates that the first Americans arrived by watercraft along a coastal "kelp highway" up to 14,000 years ago. Unfortunately, since the end of the last ice age 18,000 years ago, the ocean levels have risen around 300 feet or the hypothesis of a North American extraterrestrial impact at 12,900 b.p. have disrupted the archaeological record. In any case, much of the evidence of these coastal migrations is lost to us that may have provided evidence to push the arrival of the first "Californians" back even further. Just these statements in themselves reveal the importance of our need to realize the significance of California in understanding the peopling of North and South America. We especially need to understand how these first inhabitants were ancestors of the approximately sixty to eighty California "tribes" (covering six **language stocks**, approximately twenty-one **language families**, and around three hundred mutually exclusive language dialects), who later developed into some of the world's most complex hunter-gatherer societies.

It is unfortunate to note, however, that California's Native populations were described derogatorily as "digger Indians" who in the view of their European-based conquerors had failed to grasp the minor components of civilization, such as classical agricultural production. The climate of California and its biological heritage that the early **Native Americans** encountered and worked with can explain many of these issues. One theme is true of California's first human settlers: The coming of European explorers and settlers led to their cultural destruction and population demise. By the latter part of the 1800s California's indigenous populations had succumbed to holocaust-like conditions, becoming enslaved often times killed and eventually placed on small reservations. By the end of the era California had become known for the worst slaughter of Native American peoples in U.S. history.

5.1 Pre-Contact California: California's First Peoples

Prior to contact with European explorers the native population of California was considered relatively dense with an estimated population of approximately 350,000. The California region probably represented the largest, densest set of Native Americans north of the Valley of Mexico (the Aztec cultural area) and contained at least 20 percent of the total Native American population in the U.S. and Canada. More to the point, there were more Native Americans in California than in any other part of the U.S.

Despite the state's isolated nature, descendants of the California tribes arrived from Asia via the last Ice Age's land bridge at the Bering Strait, which formed during the lowered sea levels between Russia and Alaska. Eventually large numbers migrated into the region from the north via overland and coastal waterways and later on from the U.S. southwest and from northern Mexico. They illustrate a long and complex history involving hundreds of tribes that contributed a minimum of twenty-one minor language families represented from six major language stocks.

5.1.1 Native Languages: A Tower of Babel Allegory

Before European contact the California region contained over three hundred language variants (this was more **linguistic** variety than found in Europe). This suggests at least one reason for the lack of early Native American unity against intruding European colonists. The highly diverse linguistic landscape of early California is tied to its geographical **isolation**, landforms and environmental complexity (see pages 5 and 9 in *California Atlas*). Researchers consider the following an outline of the major language stocks and their cultural areas listed by oldest to newest on the landscape (Figure 5.1). This comes from the California Department of Parks and Recreation (http://www.parks.ca.gov/?page_id=23545):

Hokan Stock

This stock represents the oldest language groups in California. The broken chain of Hokan language areas around the margins of California presumably represents survival from an ancient continuous distribution.

Penutian Stock (endemic)

This language group occupied nearly half of California upon the arrival of the Spanish. It was a solid block of about thirty groups in the cistmontane interior of California. Their expansion came after the Hokan languages become established in the state. Penutian developed independently in California and thus represents an **endemic** linguistic root language.

Yukian Family (endemic)

This family represents another supposedly **endemic** language group that may have entered the state separately (three thousand years ago). It is the smallest of the language groups, only representing two languages: Wappo (Napa valley and Clear Lake areas) and Yuki (Eel river country).

Uto-Aztecan Stock

Research indicates that this language group entered California earlier than c. 2000 B.C. and began to diversify in California after Hokan and Penutian were present, but before all of the Penutian languages achieved their final distribution.

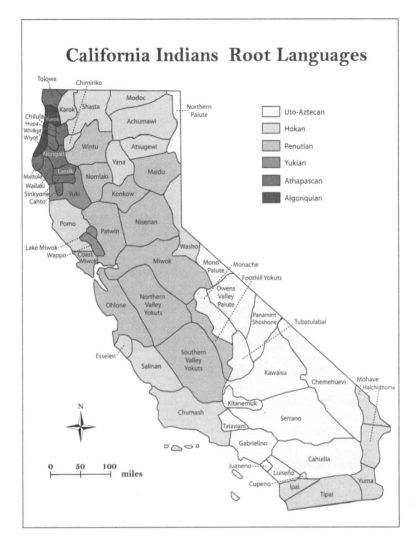

FIGURE 5.1 *California language stock map. (Courtesy Kirstyn Pittman)*

Algonquian Stock

This family is linked to the Algonquian languages of the Great Plains and Northeast. There are links to two tribes, Yuork and Wiyot, which arrived in the northwestern Klamath coastal area of California as distinct languages. On archaeological grounds, the Wiyot arrived circa 900 A.D. and the Yurok arrived circa 1100 A.D.

Athapascan Stock

This language group arrived circa 1250–1350 A.D. It represents a widespread family, more diversified in the Pacific Northwest than the southwest. California Athapascan languages may be related to Navajo and Apachean of the southwest.

In summary, the California Native Americans exhibited a bewildering geographic mosaic of regional differences in language stocks and families. In addition, within a family area there could be many small groups with distinct dialects. For example, the North Valley Yokuts tribal group found in the northern San Joaquin Valley was part of the Yokustan language family. This supposed "tribal" group, however, was represented by forty to fifty smaller groups with their own distinct dialects. The complexity and diversity of the California Native American landscape was truly extraordinary.

5.1.2 LIVING WITH A HIGHLY VARIABLE CLIMATE AND WITHOUT AGRICULTURE

California Native Americans represented a collection of societally complex **hunter-gatherer** and fishing people who, while lacking the knowledge of the wheel, the use of metal and systematic writing, and having no absorption in material goods, were able to live off the landscape which repeatedly challenged them with cyclical droughts and El Nino/La Nina events.

After living in an unpredictable environment for thousands of years many tribal groups were able to develop successful coping strategies. In many ways, archaeologists have come to understand that the idea that California Native Americans lived in an environment with abundant resources is a myth; in reality the Native Americans struggled with the environmental challenges. In context, however, environmental variability is quite variable across California if comparing the coping strategies between interior desert dwelling groups versus coastal groups or even northwestern versus south western Native societies. The challenges would be significant in the interior southeast compared to more resources abundant areas along the coast or the Sacramento-San Joaquin Delta region.

The development of socio-political complexity among California Native Americans was a coping mechanism to extended droughts and cyclical El Nino/La Nina events which posed considerable risks to the more denser populations across the region. California archaeological scholars have shown through the last 30 years of study that many of these coping mechanisms included elaborate economic systems (i.e., use of shell bead currency, exchange of items from other ecological zones using their extensive trail networks), physical storage of food, subsistence diversification, environmental modification (i.e., use of fire), intercommunity marriage ties, and sometimes loosely organized federations overseen by paramount chiefs, who could coordinate food security efforts.

For most of the tribal groups found in California a **sedentary** life in a substantial sized village was the

FIGURE 5.2 *Indian Grinding Rock State Historic Park located near Pine Grove in Amador county. This was home for the Sierra Nevada foothills Miwok. (Courtesy California Department of Parks and Recreation)*

norm, unless the group was located in a more unpredictable environment such as the deserts or the Basin and Range province. Two opposite ends of the food production spectrum exist within California: The majority were hunter-gatherers practicing landscape management (Figure 5.2), and there was a very small minority of classic **proto-agriculturalists** found along the Colorado River (i.e., Mojaves and Yumas) planting corn, bean and squash seeds in the alluvial mud after the annual spring floods. In the context of the former majority, California Native Americans harvested resources in a manner that ensured that desirable plants continually thrived in the same locales. This was accomplished by leaving whole or partial plants behind to ensure adequate harvests in the future. They also understood that by disturbing plants through gathering of seeds, berries and other plant parts this often increased the yield of the plants. In many ways while they were not classic agriculturalists, the California Native Americans can be viewed as cultivators on the basis of practices such as pruning, tilling, weeding, coppicing and burning; they essentially practiced horticultural techniques to encourage desirable plants and give them a competitive edge. In addition, many taboos and social constraints typical in these complex societies served to discourage overexploitation in a resource.

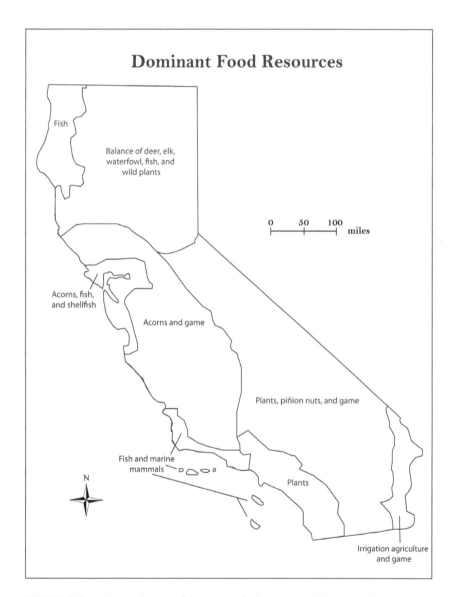

FIGURE 5.3 *General map of dominant food resources. (Courtesy Kirstyn Pittman)*

Living in diverse natural environments can lead to significant adaptive regional differences in resource utilization. California's Native Americans had a vast knowledge and understanding of hundreds of species of plants (i.e., with seeds of native grasses, forbs, bulbs and oak acorns as food staples) and animals particular to each tribe's resource area (i.e., salmon in the Klamath or dolphin in the south coast), and they knew how to exploit these resources on a seasonal basis (Figure 5.3). In order to survive, this usually required exploiting more than one environment—land, water or marine. At the same time, these exploited environments led to the employment of some fairly sophisticated hunting (i.e., salmon weirs), gathering and preparing (i.e., acorn leaching) and storage techniques (i.e., raised granaries with natural pesticides derived from herbs and plant roots).

One major aspect of Native American landscape management was the documented use of **fire** to intentionally manipulate the environment (Figure 5.4). Evidence has been difficult to attain on the use of fire, largely due to the decimation of the majority of California's Native Americans after Spanish conquest and especially through Gold Rush era racist holocaust-like conditions. Nevertheless, researchers have shown that the Native Americans used fire widely in almost all of California's ecosystems. They burned these ecosystems to promote a diversity of habitats (i.e., patch mosaics and resource diversity) for food and fiber production and to provide greater security, stability and predictability in their lives. The use of fire differed from natural fires by the seasonality of burning, the frequency of burning and the intensity of the fire, compared to modern fire conditions in the state. For example, tribes tended to burn during different times of the year, sometimes in the early spring or summer, and at other times in the late fall. Purposely burning during the mid-summer or early fall was hardly ever done due to the extreme drought and wind conditions (i.e., the Santa Anas), making these times most vulnerable to catastrophic fires. Therefore, burning by California's Native Americans was a controlled event to keep the fires at a low level and fuels manageable by burning selected areas yearly, every other year, or at intervals as long as five years. There are at least seven documented reasons for burning by California's Native Americans:

- Hunting—Burning opened areas to allow new grasses to sprout favored by big game (deer, elk) or waterfowl. Burning large areas could also divert big game to smaller unburned areas for easier hunting.

- "Natural" crop management—Burning promoted annual plants (grass seed collection, greens and bulbs) and perennial plants (tarweed, yucca). Clearing ground of grass and brush around oaks to facilitated the gathering of acorns. Generally, their fire manipulations improved growth and yields of important food plants.

- Fireproof areas—Clearing areas around settlements created a safety zone from catastrophic wildfires and opened the areas to deter dangerous wildlife (i.e., grizzly bear and mountain lion).

- Insect collection—Using a "fire surround" they could collect and roast crickets, grasshoppers and moths in pine forests, or gather honey from bees.

- Pest management—Burning could reduce insects (black flies and mosquitos), as well as kill mistletoe that invaded oak trees and thus dampened acorn production.

- Clearing areas for travel—They used fires to clear trails (for visibility, ease of mobility and safety from dangerous animals) for travel through areas overgrown with tall grass or chaparral.

- Clearing riparian areas—They used fire to clear brush from riparian areas and marshes for new grasses (to benefit waterfowl) and tree sprouts (basketry).

Finally, there was a significant network of **trading trails** that had evolved among these isolated tribal groups. The California Native American trails map (Figure 5.5) shows clearly how coastal groups connected with interior groups. In addition to the trail networks that went west to east, there were significant north to south trail networks. The trail map makes it strikingly apparent how California's modern highway and interstate highway systems have used these ancient trail networks. This makes sense as the Native Americans would have developed trail networks that covered landscapes with the least resistance to walking. This would especially include the most likely mountain passes that modern transportation engineers would also want to use. Trade along these routes largely consisted of shells and abalone from the coast going to interior tribes to barter for food unavailable to coastal dwellers or to the majority of California Natives in general. For example, there is archaeological evidence that the proto-agriculturalists along the Colorado River traded their maize (corn), beans and squash resources

FIGURE 5.4 *Picture of the interior of the Central Coast Range near King City showing past evidence of chaparral shrub management using fire by Native Americans. (Photo courtesy of U.S. Geological Survey, http://www.werc.usgs.gov/news/2002-04-24a.jpg; Locator map: Courtesy Kirstyn Pittman)*

CHAPTER 5 California Native American Landscapes

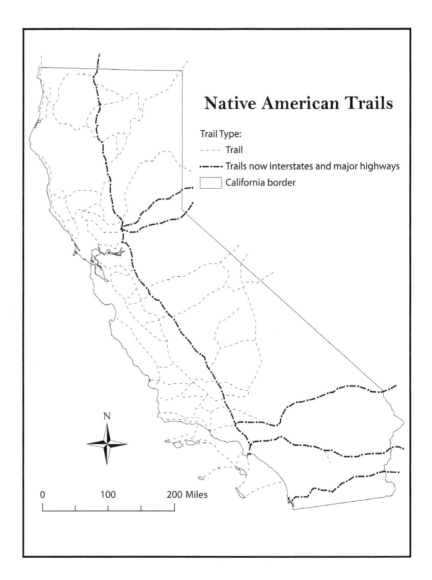

FIGURE 5.5 *Dominant Native American trail network throughout California with trails that are now Interstates and major Highways displayed in bold-dashed line. (Adapted from Hornbeck, 1983 and Davis, 1963; Courtesy Kirstyn Pittman)*

for coastal shells, thus introducing new foods to the major California tribes who could not grow these foods themselves. On the other hand, the interior northern California Native Americans of the Cascade and Modoc areas had extensive trading with the coastal and southern California tribal groups in volcanic obsidian to create arrowheads.

The last part of this discussion leads us to reasons why California's Native Americans did not have the ability to conduct classical proto-agriculture. One basic reason is that they did not need it with the amount and diversity of natural resources already available to them to support their populations and cultural growth. The other is the glaring detail of the Mediterranean climate that they found themselves, dominated by winter rains and a long summer drought period. Unlike the Fertile Crescent of the ancient Middle East, the seed bearing grasses, nuts and fruits found naturally in California's own version of a Mediterranean ecosystem were not as prolific, large or amenable to dry land production agriculture. At the same time, any maize (corn), beans or squash seeds traded into the heartland of California from the Colorado River, Northern Mexico or southwestern tribes would provide a frustrating planting experience for the California tribes because if they planted in the early spring, the rains would stop by early summer and cause automatic crop failure. Those plants require irrigation during the summer months or cultivation in areas of summer rainfall. California was clearly different, and yet the Native Americans here thrived and developed into very unique cultural groups based on their geographical isolation from each other and the rest of North America, from the landform diversity and from immediate resource availability and diversity.

5.2 California Native American "Tribes," Tribelets, and Cultural Areas

In order to piece together any clear meaning regarding California's Native American landscape, we must both understand the physical and the cultural environments synergistically and then realize that no two areas or regions are the same. Patterns of occupancy reflect a diverse mixture of social institutions integrated with the physical environment for long-term survival. To use the word "**tribe**" or tribal area is not straightforward or necessarily correct in the context of California's Native Americans because they did not organize themselves with any "tribal" unity as we would find in other tribes of North America. Revisiting the linguistic origins is really the key to understanding the cultural landscape patterns, where the approximately sixty to eighty "tribes" actually reflect language dialect groups. These then extend to around five hundred separate political divisions known as the "**tribelet**," when finally understood by Alfred Kroeber (Professor of anthropology, U.C. Berkeley), one of the early California Indian anthropologists who tried to understand these people. A tribelet represents a small, self-governing and autonomous socio-political group. Geographically, we would describe this as a group of people living in separate villages with one village regarded as the residence of the acknowledged leader of the group. These village tribelet communities could range in size from less than fifty to greater than one thousand people. Kroeber provides this example of the Pomo "tribe" (see page 9 in *California Atlas*, polygon 24) found in the Sonoma/Mendocino County region (approximately seventy-two tribelets ranging in size from one hundred and twenty-five to fifteen hundred people) to better understand the situation:

> "In any strict usage, the word 'tribe' denotes a group of people that act together, feel themselves to be a unit, and are sovereign in a defined territory. Now, in California, these traits attached to the Masut Pomo, again to the Elem Pomo, to the Yokaia Pomo, and to the 30 other Pomo tribelets. They did not attach to the Pomo as a whole, because the Pomo as a whole did not act or govern themselves, or hold land as a unit. In other words, there was strictly no such tribal entity as 'the Pomo' . . ." Alfred Kroeber, *The Nature of Land-holding Groups in Aboriginal California* (1962).

This discussion must now come back to the use of boundaries, since linguists and anthropologists have derived a set of "tribal" territory boundaries (see page 9 in *California Atlas*). If we look at the California Native landscape on three spatial scales it becomes apparent that while there were the larger ethnic or cultural areas (Figure 5.6), these can be divided further into language/tribal group boundaries (see page 9 in *California Atlas*), and finally down to tribelet areas. Tribelet areas typically represented several settlements often having differing dialects in use representing the organized corporate body whose basic role was landholding. In essence, the basis of land control was occupancy and continued use, where the tribelet areas varied in size based on individual population (Figure 5.7) and the quantity or quality of the resource base. The boundaries of the tribelet were very particular and represent an extreme form of "**localism**" to outline the areal extent in which a group lived, hunted and gathered. Stephen Powers describes this for a tribelet in the Klamath region:

> "The boundaries of all the tribes on Humboldt Bay, Eel River, Van Dusen's Fork, and in fact everywhere, are marked with the greatest precision, being defined by certain creeks, canyons, boulders, conspicuous trees, springs, etc., each one of which has its own individual name. It is perilous for an Indian to be found outside of his tribelet boundaries, wherefore it stands him well in hand to make himself acquainted with the same early in life. Accordingly the [women] teach these things to their children in a kind of sing-song . . . Over and over, time and again, they rehearse all these boulders, creeks, trees, etc., describing each minutely and by name, with its surroundings. Then when the children are old enough, they take them around to beat the boundaries' . . . and so wonderful is the Indian memory naturally, and so faithful has been their instruction, that the children initiates generally recognize the objects from the descriptions of them previously given by their mothers." Stephen Powers, *Tribes of California* (1877)

Therefore, the California Native landscape represents a mosaic of hundreds of small, autonomous groups each differing from neighbors in speech and cultural ele-

FIGURE 5.6 *California Native American cultural areas. (Adapted from Heiser, 1978; Courtesy Kirstyn Pittman)*

ments and based on a defined set of landscape boundaries that were orally and physically set to memory.

In an effort to better summarize the differences across such a vast settlement mosaic, researchers can cluster the linguistic boundaries along with the subsistence patterns, population and settlement patterns into geographical-**cultural areas**. By dividing up California's Native American cultures into geographical-cultural areas, we can come to understand their way of life, **worldviews**, and differences within the state. California terrain, regional climates and biological resources come together to influence the lifestyle of the "tribal" groups through the process of geographical **determinism** (as discussed in Chapter 1). In addition, we can discuss cultural elements, as they depended on the availability and abundances of various resources: water, ample terrestrial or aquatic wildlife, plant food-gathering or wood, which then provided the possibilities for other non-basic livelihood elements to develop. For example, a lack of abundant food and thus constant food searches to sustain life would hinder the development of craft production, whereas an ample food supply could lead to sophisticated manufacturing activities becoming part of local culture, such as boat building (Figure 5.8), basketry and jewelry.

5.2.1 NORTHWESTERN CULTURAL AREA

The Yurok, Hupa, Tolowa and Karuk are the main "tribes" that represent this region (see page 9 in *California Atlas*, polygons 1, 2, 6, and 13).

Defining features:

- Languages—Athapascan and Algonquian root languages. Adjacent groups often spoke completely dif-

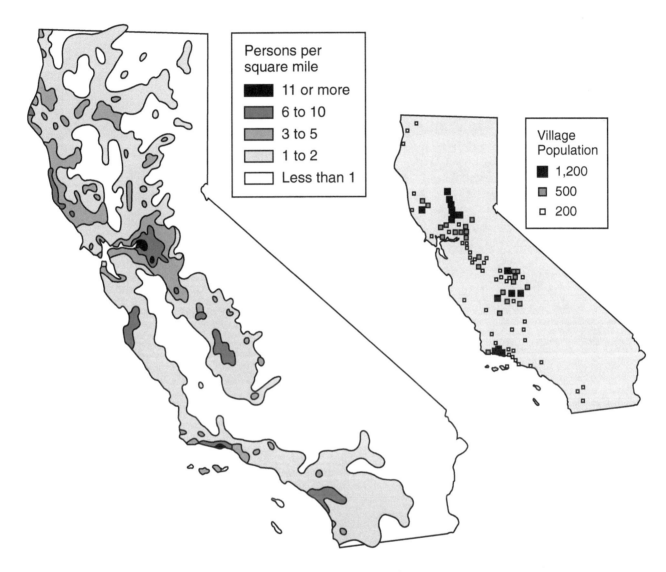

FIGURE 5.7 *California Native American population density at time of European contact.*

ferent languages; extreme isolation along steep sided river valleys of the Klamath mountains.

- Subsistence—Riverine and maritime oriented: salmon, trout, smelt, perch, seals, shellfish, acorns, and berries
- Settlement patterns—Sophisticated redwood and/or cedar plank houses (Figure 5.9), gable roofs and plank walls (sweathouses), permanent villages situated along river banks and the ocean coast at stream mouths, lagoons or bays.
- Politics—"Big Man" politics, precise legal codes governing villages and tribelet areas. Occasionally militant.
- Religion—Animistic and shaman based. Strict taboos and puberty rights of passage.
- Material Culture—seagoing redwood dugout canoes (Figure 5.8), wood carvers, precision tools and ornate baskets, dances and extravagant dress; capitalistic.
- Trade—Coastal tribelets had some limited trade with interior tribelets: Dried smelt and tooth shells (in context of shell money) were traded for soaproot (provided food, soap, glue, brushes, medicine and a potent fish poison) and pine nut beads (for jewelry).

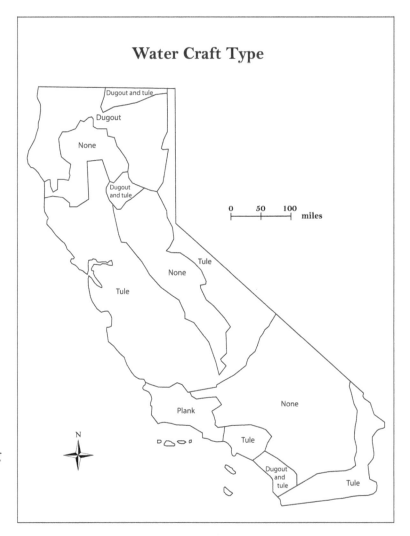

FIGURE 5.8 *Water craft type map showing areas of canoe construction from dugout logs, Tule reeds and sown planks caulked with oil tar. (Adapted from Heiser, 1978; Courtesy Kirstyn Pittman)*

5.2.2 CENTRAL CULTURAL AREA

This vast area covers the Northern and Central Coastal Ranges, the San Francisco Bay-Delta, the Great Central valley and the west slope of the Sierra Nevada. Major "tribes" representing this area include the Pomo, Wintu, Yana, Maidu, Miwok, Costanoan (Ohlone), Konkow, Nisenan, Salinan, and Yokuts (see page 9 in *California Atlas*, polygons 18–20, 23, 24, 30–32, and 34–36).

Defining features:

- Languages—Dominated by Penutian, with limited areas of Yukian, and Hokan root languages.

- Subsistence—Highly variable, but acorns were all important, berries and other vegetation (root crops, like the potato tuber of the Mariposa lily, seeds), fish (salmon and trout), waterfowl (along Pacific migration flyway. See page 20 in *California Atlas*); high command of fire to manage landscape for hunting (deer, bear and small game) and maintain high abundance of food vegetation.

- Settlement Patterns—Permanent villages occupied year round along major Sierran rivers and the Delta area although some **transhumance** was practiced by most groups, especially in the Great Central Valley with summer retreats into the Sierra foothills. Housing types were varied depending on the climate and resources available (Figure 5.9): domed thatched grass structures in the Central Coast; cone-shaped unhewn, timber framed structures covered with grass and soil in the Cascadian foothills; Semi-subterranean, cone-shaped structures with pole frames,

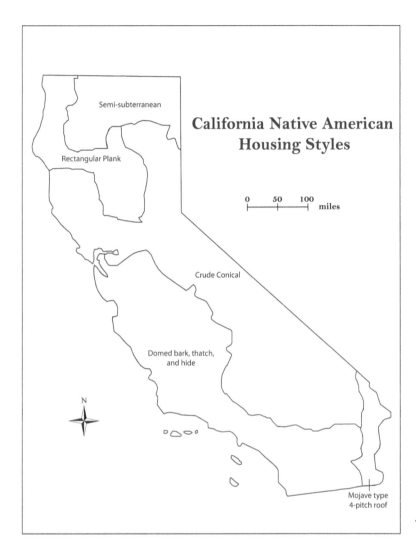

FIGURE 5.9 *Housing style map. (Adapted from Durrenberger, 1976; Courtesy Kirstyn Pittman)*

covered with bark or tule reeds in the Delta and northern San Joaquin Valley; semi-subterranean, a central post with radiating rafters, covered in bark or tule reeds, resting on the surrounding edges of an earth pit in the Sacramento Valley; semi-subterranean, A-framed structures covered with tule reed in rows all covered by a thatched pergola; conical houses of bark slabs with poles for support in the North Coastal Range.

- Politics—"Tribelet" politics, peaceful, communal.
- Religion—Kuksu religion, which featured dance cycles aimed at hunting/gathering, good health and weather, led by a shaman as contact with the spirit world. Strict taboos and puberty rights of passage. Includes annual mourning ceremony and secret male societies.
- Material Culture—Tule balsa canoe (Figure 5.8), simple tools, highest level of technical artistry in basketry, colorful costumes in ceremonies, in San Joaquin Valley tribes men and women pierced ears and noses and women had tattoos for decoration.
- Trade—Intertribal trade; San Joaquin tribes engaged in pine nut trade over the Sierra Nevada with Basin and Range tribes. North and Central Coast range tribes traded coastal products (dried abalone, abalone shell, tooth shells) with interior tribes for salt and obsidian.

5.2.3 SOUTHERN CULTURAL AREA

The Chumash, Tongva (Gabrielino), Kumeyaay (Luiseno, Ipai, Tipal), Serrano, and Cahuilla are the dominant "tribes" representing this large region (see page 9 in *California Atlas*, polygons 39, 42–45, 47, and 48)

Defining features:

- Languages—Hokan and Uto-Aztecan root languages.

- Subsistence—Same as for Central cultural area, but increased emphasis on marine resources: fish, abalone, mussels, seals, dolphins, and sea birds.

- Settlement Patterns—Ranged from fully sedentary villages of 1,000 or more persons on the coastal mesas to temporary winter encampments in the interior mountains and valleys. Housing styles (Figure 5.9) were represented by large, circular, willow framed domed brush houses (wickiups) separating multiple family areas for entire area—from San Luis Obispo to San Diego.

- Politics—"Tribelet" politics (women could be head of villages). "Tribal" ceremonies, elder councils and strict laws assured sustainable well-being. Peaceful and communal.

- Religion—Chungishnich religion (i.e., monotheistic lawgiver or god) and shaman based. Strict taboos and puberty rights of passage. Includes annual mourning ceremony and bird songs that remember history, morals, law, creation, and migrations.

- Material Culture—artistic tools and baskets, highest level of technical artistry in ocean-going watercraft (i.e., tomol, sown-plank canoe (Figure 5.8) built by the Chumash and Gabrielinos—highly probable Polynesian connection; did ancient Hawaiians teach Chumash how to make ocean-going canoes? The name Tomol is a Hawaiian root word. In addition the fish hooks used by both groups are related to those found throughout Polynesia—especially Hawaii) and polychromatic rock art linked to drug induced shamanistic ceremonies.

- Trade—maritime with offshore Santa Barbara Channel Islands as well as interior groups.

5.2.4 COLORADO RIVER CULTURAL AREA

Three tribes represent this region, Mohave, Yuma (Quechan) and Halchidhoma, which extended into Nevada, Arizona and northern Mexico (see page 9 in *California Atlas*).

Defining features:

- Languages—Hokan root language.

- Subsistence—Actively engaged in irrigated agriculture (beans, corn, squash and melons) along the banks of the Colorado River via knowledge and technology brought north from Mexico. Hunting, fishing, trapping and limited gathering.

- Settlement patterns—Large villages strongly tied to the banks of the Colorado River. Four-posted structures built over a circular excavation, thatched with brush and covered with mud or dome-shaped brush thatched over excavation with an extended porch-type structure to provide shade and food storage.

- Politics—"Big man" politics, bellicose militant disposition. Communal.

- Religion—Dream song and shaman based. Strict taboos and puberty rights of passage. Sang bird songs.

- Material culture—basketry and pottery, colorful costumes, tule balsa canoes (Figure 5.9).

5.2.5 GREAT BASIN EASTERN CALIFORNIA CULTURAL AREA

The various Paiute groups, the Shoshone and Washo, are the "tribes" representing this area that extends from the southeastern Mojave Desert and includes the transmontane areas all the way to the northeastern California border with Oregon (see page 9 in *California Atlas*). All of these "tribes," however, find their larger areas of dominance in Nevada and eastern Oregon.

Defining features:

- Languages—Hokan and Uto-Aztecan root languages.

- Subsistence—gathering and hunting: small animals, limited fishing, seeds, pine nuts, berries and roots.

- Settlement patterns—variable: from seasonal round pattern to full sedentism; highly **nomadic** with perhaps the simplest culture of California Native Americans. Housing style ranged from round structures made of long pine poles tied together in the center, covered with cedar bark and pine boughs made by the Washo to small simple brush covered circular lean-tos created by the Paiute groups (Figure 5.9).

- Politics—variable: from male head of nuclear family to "tribelet" chiefs. Generally peaceful and communal.
- Religion—Bear dance and shaman based. Strict taboos and puberty rights of passage.
- Material culture—simple baskets, hardwood digging sticks and tule balsa canoe for lake fishing (Figure 5.8).
- Trade—salt, pine nuts and rabbit-skin blankets to tribelet groups in the Central California cultural area for acorns, soaproot, dried abalone, shell beads, abalone shell and baskets. Also tribes from this cultural area traded with the Colorado River groups for beans, corn, squash, and melons.

5.2.6 NORTHEASTERN CULTURAL AREA

Three tribes, the Modoc, Achumawi and Atsugewi, represent this small region that largely covers the Modoc plateau region but the Modoc tribe extends into central Oregon (see page 9 in *California Atlas*, polygons 16 and 17).

Defining features:

- Languages—Hokan and Penutian root languages.
- Subsistence—Roots, seeds and various terrestrial animals.
- Settlement patterns—Sedentism in winter then seasonal movement. Housing style varied from cone-shaped structures covered with tule reeds in the summer to wood-frame, semi-subterranean houses covered with grass, bark and dirt in winter (Figure 5.9).
- Politics—Male heads of extended families as well as shaman. The Modoc were regarded as extreme defenders of their territory and were warlike.
- Religion—Animistic and shaman based. Strict taboos and puberty rights of passage. Includes annual mourning ceremony.
- Material elements—Simplistic basketry, obsidian tools and log dugout and tule balsa canoes for lake fishing (Figure 5.8). Petroglyph rock art conducted on basalt outcrops.
- Trade—Slaves (especially women), furs, blankets and clothing

5.3 ECOLOGICAL IMPACTS OF SPANISH CONTACT

In order to understand the indirect and direct devastation brought on California's Native Americans it is important to explain the advantages that the **Spanish** or any European (Eurasian) would have had coming to the **New World** of North and South America. As related by geographer Jared Diamond in *Guns, Germs and Steel* (1996), the Eurasian continent had both geographical and historical advantages. These include the continent being the planet's largest land mass with a broad east-west extent that allowed for large regions with roughly similar climates to exist. Within these large regions plant and animals were able to evolve in strong competition with each other, which led to the chance existence of easily domesticated plants and animals with which early humans could develop a tight relationship. Archaeologically there has been a long history of human evolution in the **Old World** represented by the Eurasian continent, starting with the cultural heart of the Fertile Crescent on the eastern shores of the Mediterranean basin. Here agriculture developed over 10,000 years ago, but more importantly the long co-existence with domesticated animals bred partial immunity to animal-borne diseases (i.e., gene resistance over time). These include gonorrhea, smallpox, measles, influenza, typhus, whooping cough, the mumps, and diphtheria, to name just a few of the important ones critical to human health and hygiene.

> "If a species evolves a superior adaptation to a given environment on one land mass, it may outcompete native organisms once it is transported to analogous ecosystems overseas." Mark A. Blumler, *Invasion and Transformation of California's Valley Grassland* (1995).

In contrast, New World ecosystems, especially that of California, are smaller and more isolated—and inherently more fragile. As already stated in Chapter 4 and discussed in this chapter, California is characterized by a high rate of endemism in biodiversity and in human cultural diversity. Credit for these patterns goes to its isolation and insulation from outside disturbance. In addition, the migration of humans into the New World at the end of the last Ice Age occurred before the development of classical agriculture in the Old World, which means that not many Old World pathogens would have been brought over to California, therefore not allowing for the development of gene resistance in the newly migrated New World population. This is further compounded by the lack of many animals to domesticate in the New World (i.e., only two, the Guinea pig and the llama in Peru were ever successfully domesticated), even though agricultural development did occur in Mexico, Central America and Peru. The tight human-farm animal relationship developed in Eurasia did not occur in the majority of New World cultures, and especially not in California.

Even though California's Native Americans were sheltered from major direct contact with Europeans for at least 275 years, there is evidence that some European **pests and pathogens** arrived in geographically isolated California early and started to exact a toll on its sensitive ecosystems and people. William Preston (Professor of geography, California Polytechnic University, San Luis Obispo) long explored this, and his research notes that "long before the foundations of Mission San Diego were established in 1769, a host of Old World pathogens traveled along terrestrial and maritime pathways to penetrate California."

There is evidence to suggest that rather than the Spanish explorers finding a pristine wilderness and pristine natives, they instead came upon an environment already being changed by **exotic species** and pathogens leading to a weakened and reduced human population. Several elements led to this revised view of what California experienced before Spanish settlement:

1. Other tribal contacts from the southeast brought in disease (i.e., via tribal trade routes; they were already in direct contact with the Spanish in Santa Fe, New Mexico and Northern Mexico).

2. There was direct contact from indigenous vectors such as lice, mosquitoes and fleas that would have carried pathogens from northern Mexico.

3. Early Spanish explorers, such as Cabrillo in 1542–1543 and Vizcaino in 1602–1603 or the English explorer Sir Francis Drake in 1579 would have had brief contact with the natives and thus provided direct contamination and left behind highly invasive alien annual grasses that studies show were in the state before the establishment of Mission San Diego.

4. It is also possible that migrating birds along the Pacific flyway (see page 20 in *California Atlas*) would have allowed for the dispersal of alien weed seeds gathered from stops in Mexico. (For example, in examining the modern age of the H5N1 bird flu virus dispersing via bird migration in Asia or the spread of West Nile virus from the U.S. East Coast to the West Coast in less than six years, this dispersal thesis seems reasonable.)

Since California's Native Americans were superior big game hunters, and thus created significant pressure on game populations and their habitat ranges, increasing Native American deaths from new diseases would have triggered an imbalance in game. The reduced hunting pressure on the game populations would have resulted in a game population eruption noted by early Spanish explorers to the state and during the Spanish settlement period.

5.3.1 IMPACTS OF DIRECT SPANISH SETTLEMENT

When the Spanish finally decided to partially colonize California via the **mission-presidio-pueblo system**, it ushered in the beginning of the end for the fragile California Native American populations and their co-evolved sensitive ecosystems. For the Spanish, California represented the first land area in the New World that resembled the climate and landscapes of their Mediterranean European homeland. This allowed for the rapid intentional and inadvertent alteration of the California environment through the introduction of Old World livestock and crops already adapted to a Mediterranean type climate. This major step also expanded the entry of alien "weed" plants and parasites, and radically changed approaches to landscape management (i.e., the reduced use of fire). While the plant, animal and Native American communities along the coastal areas bore the brunt of the changes from Spanish settlement, interior cultural regions and environments beyond the margins of direct settlement also began to show significant alterations via indirect affects (see page 10 in *California Atlas*).

Specifically, increasing outside contact during the Spanish era ushered in the following profound changes in Native population and environment:

1. A much smaller Spanish population severely reduced Native American populations, especially along the coast from San Diego to Sonoma, via viruses and disease.

2. European annual grasses and weeds replaced native perennial bunch grasses.

3. European cattle rapidly spread along the coast, and many escaped to the Central Valley, becoming the dominant large land mammal.

4. Chaparral shrubs and coastal sage scrub spread after Native fire management practices stopped.

5. European fur trapping severely impacted native small mammal populations.

5.4 Summary

Cultural complexity and co-evolved unique/sensitive ecosystems derived through diverse landscapes, isolation and time. This was the situation in California before direct European settlement. This combination of culture and environment also led to the unraveling of its fragile balance. The diversity of California's Native languages led to a lack of unity among tribelets, "tribes" or cultural areas; thus there was no united effort to thwart Spanish colonization. The Natives' learned landscape management that provided all that was required to sustain them. Therefore, the lack of plants and animals for domestication in a Mediterranean-type climate made classical agriculture a non-issue. In fact, sophisticated hunter-gatherer operations under rather sophisticated land management methods led to one of the largest populations of Native Americans north of the Valley of Mexico-Aztec cultural area.

> "They [Spanish] had neither the foresight nor the desire to utilize the more useful values and institutions [and landscape management practices] which California Native Americans had established over centuries of settlement. The white man in his wisdom brought in a new concept of California life, and the ensuing race for material goods, political power and social status indicated that the California Dream would soon become a flourishing myth."
> Howard DeWitt, *The California Dream* (1999).

Pre-Spanish contact and later direct contact led to viruses and diseases with which an isolated human population with no evolved resistance could contend, and thus rampant death ensued. The increasing settlement of the California landscape by successive colonial groups and their various land settlement systems resulted in the California Native American population going from a population high of approximately 350,000 to an estimate of less than 10,000 persons by 1910. The northern counties and isolated mountain areas (i.e., areas of escape) were largely dominated by the remaining population distribution. Decimation and cultural loss came rapidly.

Bibliography

Anderson, K. 1993. Native Californians as ancient and contemporary cultivators. In *Before the Wilderness: Environmental Management by Native Californians*, Eds. T.C. Blackburn and K. Anderson, pp. 151–174. Ballena Press, Menlo Park.

Baumhoff, M.A. (1963). Ecological Determinants of Aboriginal California Population. *University of California Publications in American Archaeology and Ethnology*, 49(2): 155–236.

Beals, R.L., and J.A. Hester, Jr. (1971). A New Ecological Typology of the California Indians. In *The California Indians*, edited by R.F. Heizer and M.A. Whipple. University of California Press, Berkeley.

Bean, L.J. (1972). *Mukat's People: The Cahuilla Indians of Southern California*. University of California Press, Berkeley.

Bean, L.J. and T.F. King (1974) *Antap: California Indian Political and Economic Organization*. Anthropological Papers 2. Ballena Press, Ramona, CA.

Bean, L.J. and H. Lawton. (1976). Some explanations for the rise of cultural complexity in native California with comments on proto-agriculture and agriculture. In *Native Californians: A Theoretical Retrospective*, edited by L.J. Bean and T. Blackburn, Ballena Press, Menlo Park, CA, pp. 279–308.

Blackburn, T.C. (1974). Ceremonial Integration and Social Interaction in Aboriginal California. In *Antap: California Indian Political and Economic Organization*, edited by

L.J. Bean and T.F. King, pp. 93–110. Anthropological Papers 2. Ballena Press, Ramona, CA.

Booth, S. (2008). *California Geography*. Course taught at Sierra College. [online] http://geography.sierra.cc.ca.us/booth/California/cal_index.htm.

Brand, D.D. (1938). Aboriginal Trade Routes for Sea Shells in the Southwest. *Yearbook of the Association of Pacific Coast Geographers* 4:3–10.

Broadbent, S.M. (1972). The Tumsen of Monterey: An Ethnography from Historical Sources." In *Miscellaneous Papers on Archaeology*, pp. 45–93. University of California Archaeological Research Facility Contributions, 14. Berkeley, CA.

Brown, A.K. (1967). The Aboriginal Population of the Santa Barbara Channel. *University of California Archaeological Survey Reports* 69: 1–99.

California Department of Parks and Recreation (2008). California Indians Root Languages and Tribal Groups. Sacramento, CA [online] http://www.parks.ca.gov/?page_id=23545

California State University San Francisco (2008). The California Native American Page. California Studies Program. [online] http://bss.sfsu.edu/calstudies/NativeWebPages/ca%20web%201.html

Calisphere (2008a). California Cultures. University of California [online] http://www.calisphere.universityofcalifornia.edu/calcultures/

Calisphere (2008b). A world of California Primary Sources. University of California [online] http://www.calisphere.universityofcalifornia.edu/

Campbell, P.D. (2005). *Survival Skills of Native Californians*. Gibbs-Smith, New York, NY.

Chagnon, N.A. (1970). Ecological and Adaptive Aspects of California Shell Money. *Annual Reports of the University of California Archaeological Survey* 12:1–25.

Cook, S.F. (1957). The Aboriginal Population of Alameda and Contra Costa Counties, California. *University of California Anthropological Records* 16(4):131–56.

Cook, S.F. (1955). The Aboriginal Population of the San Joaquin Valley, California. *University of California Anthropological Records* 16(2):31–80.

Cook, S.F. (1956). The Aboriginal Population of the North Coast of California. *University of California Anthropological Records* 16(3):81–130.

Cook, S.F. (1964). The Aboriginal Population of Upper California. In *Proceedings of the 35th International Congress of Americanists*, pp. 397–403. Mexico.

Cook, S.F. (1976). *The Population of the California Indians, 1769–1970*. University of California Press, Berkeley.

Cook, S.F. and R.F. Heizer (1965). The Quantitative Approach to the Relation between Population and Settlement Size. *University of California Archaeological Survey Reports* 64:1–97.

Cook, S.F. and R.F. Heizer (1968). Relationships among houses, settlement areas, and population in aboriginal California. In *Settlement Archaeological*, edited by K.C. Chang, pp. 79–116. National Press Books, Palo Alto, CA.

Cox, C.J. (2008). *California Geography*. Course taught at Sierra College. [online] http://faculty.sierracollege.edu/ccox/california_geography/index.html.

Davis, J.T. (1963). Trade Routes and Economic Exchange among the Indians of California. In *Aboriginal California: Three Studies in Culture History*, edited by R.F. Heizer. University of California Press, Berkeley, CA.

DeWitt, H.A. (1999). The *Fragmented Dream: Multicultural California*. 2nd edition, Kendall/Hunt Publishing Co., Dubuque, IA.

DeWitt, H.A. (1999). *The California Dream*. 2nd Edition, Kendall/Hunt Publishing Co., Dubuque, IA.

Diamond, J. (1996). *Guns, Germs, and Steel: The Fates of Human Societies*. Penguin Press, New York, NY.

Donley, M.W., Allan, S., Caro, P., and C.P. Patton (1979). *Atlas of California*. Pacific Book Center, Culver City, CA.

Durrenberger, R.W. and R.B. Johnson (1976). *California Patterns on the Land*. 5th edition, Mayfield Publishing Company, Mountain View, CA.

Dutschke, D. (1988). A history of American Indians in California. In: *Five Views: An Ethnic Historic Site Survey for California*, pp. 3–55. California Department of Parks and Recreation, Office of Historic Preservation, Sacramento, CA.

Edgar, B. (2005). The Polynesian Connection. *Archaeology*, March/April: 42–45.

Fagan, B. (2003). *Before California: an Archaeologist Looks at Our Earliest Inhabitants*. Rowman & Littlefield Publishers, New York, NY.

Forbes, J.D. (1969). *Native Americans of California and Nevada: A Handbook*. Naturegraph Publishers, Healdsburg, CA.

Gould, R.A. (1975). Ecology and Adaptive Response among the Tolowa Indians of Northwestern California. *Journal of California Anthropology* 2(2):148–70.

Heizer, R.F. (1941). Aboriginal Trade between the Southwest and California. *Masterkey* 15(5): 185–88.

Heizer, R.F. (1960). California Population Densities, 1770 and 1950. *University of California Archaeological Survey Reports* 41: 1–9.

Heizer, R.F., ed. (1978). *Handbook of North American Indians*. Vol. 8. Smithsonian Institute, Washington, D.C.

Heizer, R.F. and Albert B. Elsasseiz (1980). *The Natural World of the California Indians*. University of California Press, Berkeley, CA.

Heizer, R.F., Nissen, K.M., and E.D. Castillo (1975). *California Indian History: A Classified and Annotated Guide to Source Materials*. Publications in Archaeology, Ethnology, and History, 4. Ballena Press, Ramona, CA.

Hinton, L. (1994) *Flutes of Fire: Essays on California Indian Languages.* Heyday Books, Berkeley, CA.

Hornbeck, D. (1983). *California Patterns: A Geographical and Historical Atlas.* Mayfield Publishing Company, Mountain View, CA.

Kroeber, A.L. (1925). *Handbook of the Indians of California.* Bureau of American Ethnology Bulletin no. 78. Washington, D.C.

Johnson, J.R. 2000. Social responses to climate change among Chumash Indians of south-central California. In *The Way the Wind Blows: Climate, History, and Human Action,* Eds. R.J. McIntosh, J.A. Tainter, and S.K. McIntosh, pp. 301–327. Columbia University Press, New York.

Jones, T.L., and K.A. Klar. 2009. On Linguistics and Cascading Inventions: A Comment on Arnold's Dismissal of a Polynesian Contact Event in Southern California. *American Antiquity* 74:173–182.

Jones, T.L. and K.A. Klar. 2007. *California Prehistory: Colonization, Culture, and Complexity.* Alta Mira Press, New York.

Kroeber, A.L. (1962). The Nature of Land-holding Groups in Aboriginal California. In *Two Papers on the Aboriginal Ethnography of California,* edited by D.H. Hymes and R. F. Heizer, pp.19–58. *University of California Archaeological Survey Reports* 56. Berkeley, CA.

Kunkel, P.H. (1974). The Pomo Kin Group and the Political Unit in Aboriginal California. J*ournal of California Anthropology* 1(1): 7–18.

Landberg, L.C.W. (1965). *The Chumash Indians of Southern California.* Southwest Museum Papers, 19. Los Angeles, CA.

Lawton, H.W. and L.J. Bean. (1968). A Preliminary Reconstruction of Aboriginal Agricultural Technology among the Cahuilla. *The Indian Historian* 1(5):18–24.

Lewis, H.T. (1973). Patterns of Indian Burning in California: Ecology and Ethnohistory. In Thomas C. Blackburn and Kat Anderson (eds.) *Before the Wilderness: Environmental Management by Native Californians.* Ballena Press, Menlo Park, CA.

Margolin, M. (1993) *The Way We Lived: California Indian Stories, Songs and Reminiscences.* Heyday Books, Berkeley, CA.

Michaelson, J. (2008). *Geography of California.* Course taught at UC Santa Barbara, Dept. of Geography. [online] http://www.geog.ucsb.edu/~joel/g148_f08/.

Miller, C.S. and Hyslop, R.S. (1983). *California: The Geography of Diversity.* Mayfield Publishing Company, Mountain View, CA.

Moratto, M. (1984). *California Archaeology,* Academic Press, San Diego, CA.

Powers, S. (1975). The Northern California Indians. Edited by R.F. Heizer. *University of California Archaeological Research Facility Contributions,* 25. Berkeley, CA.

Preston, W. (1998) Serpent in the Garden: Environmental Change in Colonial California. In R.A. Gutierrez and R.J. Orsi (eds.) *Contested Eden: California Before the Gold Rush.* University of California Press, Berkeley, CA, pp. 260–298.

Preston, W. (2002) Portents of Plague from California Protohistoric. *Ethnohistory* 49(1): 69–121.

Raab, L.M. and T.L. Jones. 2004. *Prehistoric California: Archaeology and the Myth of Paradise.* University of Utah Press, Salt Lake City.

Rawls, J.J. and W. Bean (2008). *California: An Interpretative History.* 9th Edition, McGraw-Hill Publishing Co., New York, NY.

Rice, R., Bullough, W., and R. Orsi (2001). *The Elusive Eden: A New History of California.* 3rd edition, McGraw-Hill.

Sample, L.L. (1950). Trade and Trails in Aboriginal California. *University of California Archaeological Survey Reports,* 8. Berkeley, CA.

Starr, K. (2005). *California: A History.* The Modern Library, New York, NY.

Ward, G.C. (1996). *The West.* Little, Brown and Company, Boston, CA.

Sutton, I. (1975). *Indian Land Tenure: Bibliographical Essays and a Guide to The Literature.* Clearwater Publishing Company, New York.

Wuertele, E. (1975). Bibliographical History of California Anthropological Research. *University of California Archaeological Research Facility Contributions,* 26. Berkeley, CA.

CHAPTER 6

Spanish Exploration and Settlement Pattern in Alta California

Key terms

Acculturation	Kino	Red clay tiles
Adobe bricks	Laws of the Indies	Russian
Alcalde	Los Angeles	San Diego
Alta California	Manila Galleon	San Francisco
American fur trappers	Mercantilism	San Francisco Bay
Branciforte	Mestizo	San Jose
British Northwest territories	Mexican War for Independence	Santa Barbara
Cabrillo, Juan	Mexico City	Santa Fe
California	Mission	Secularization
Californio	Mission-presidio-pueblo (MPP)	Serra, Junipero
Cermenho	Monterey	Spanish land grant
Cortes, Hernando	Neophytes	Strait of Anián
Criollo	Nuova Albion	Stucco
Drake, Sir Francis	Peninsulare	Ulloa, Francisco de
Fermin Lausén	Philippines	Viscaino
Franciscans	Presidio	Zambo
Garci Ordonez de Montalvo	Pueblo	Zanjero
God, Gold and Glory	Queen Califia	
Irrigation agriculture	Racism	

Introduction

DISCOVERY AND RELUCTANCE ON THE SPANISH FRONTIER

God, Gold and Glory were the motivating themes behind the Spanish exploration and colonization of the New World for the 275 years before the founding of Mission San Diego in 1769. The three G's mostly applied to Mexico and South American regions, and even to the remote northerly outpost of Santa Fe, New Mexico, laden with silver mines, but it turned out that

95

only the first G, as well as the terms fantasy and fear, represent the exploration and the later reluctant colonization of California.

California represented the last frontier for the Spanish colonies and the last Spanish settlement in the New World. The discovery of "the Californias" marked the final stage of the great European myths developed through popular romantic literature, the native tales of cities of gold (El Dorado or Cibola) and the idle chat regarding gold lust among explorers in New Spain (i.e., Mexico). Both Baja and **Alta California** represented a mistaken and disappointing identity for the Spanish, but the position of these lands on the western flanks of the Spanish empire and on the newly developed "**Manila Galleon**" spice import/export trade route from Acapulco, Mexico to Manila in the Philippines brought their value back into focus. Protection of these galleon's as they made their way southward back to Acapulco after reaching the Mendocino coastline from their long voyage across the Pacific became a concern as their British enemies, as well as Russian fur trappers, became increasingly prevalent in the far northern Pacific. The focus on Alta California turned into the development of one of the most systematic processes for controlling the landscape and its natives in the New World. The **mission-presidio-pueblo** (MPP) system represents the original planned sketch of the future for modern California. The MPP presented a systematic control of the land and, as discussed in Chapter 5, became the undoing of the fragile coastal world of Native California and accelerated changes to the environment. Within a hundred years—from the founding of Mission San Diego in 1769 to the completion of the Transcontinental railroad in 1869—California's isolation disintegrated and its uniqueness was repackaged by boosters as a dream—the California dream.

6.1 INDIANA JONES AND THE KINGDOM OF QUEEN CALIFIA!... OR MAYBE NOT...

What's in a name? If the name we are questioning is California then there is plenty to unpack to understand its origin and meaning. While the title of this section is tongue-in-cheek, the point made of adventure and flights of fancy in connection with the name California is all too real. The name originates from a popular piece of Spanish romantic myth literature from the early sixteenth century called *The Exploits of Esplandian* by **Garci Ordonez de Montalvo**. This pulp fiction tale of Christian conquest expresses many misconceptions of the New World that the Spanish had just started to explore at the time of its publication. This book would have been very popular with any literate ship captains as they made their way across the Atlantic Ocean for the distant shores of Mexico. The lonely days at sea with limited literature and excitable frantic minds made for all manner of strange beliefs about these new lands.

Early Spanish exploration was part of a larger settlement scheme that was to run from the Gulf of Mexico to the Pacific Ocean and into the interior as far north as Colorado, the area considered the Spanish northern territories. To the mind of the Spanish explorer at the time, fed on pulp fiction and exotic talk among themselves, a mysterious and fabulous island lay off the Pacific coast just as Garci Ordonez de Montalvo described in his novel—an "Island called **California** near to the region of the Terrestrial Paradise." A mountainous land of gold and jewel-covered slopes, pearl-lined beaches and agriculturally productive valleys, but ruled by the beautiful, strong, brave and black Amazon **Queen Califia** and her fellow female warriors who ride griffins for transportation, California, meaning the lands of Queen Califia, has its representation (toponym) in a fantasy character representing the highest order of misconceptions of the New World.

It was this search for the fabled Isles of the Indies and the Amazons that eventually led to the initial discovery of the California coast. **Hernando Cortes**, the gold-lusting conqueror of the Aztecs, made the first attempt to reach these lands. He pushed for voyages of discovery from the Mexican West coast and discovered Baja California in 1533. The Baja Peninsula and the Gulf of California easily fooled Cortes and other explorers into thinking it was an island.

6.1.1 NORTHWARD EXPLORATION AND SETTLEMENT

Fellow explorer **Francisco de Ulloa** cast doubt on or disproved this island belief (Figure 6.1) in his 1540

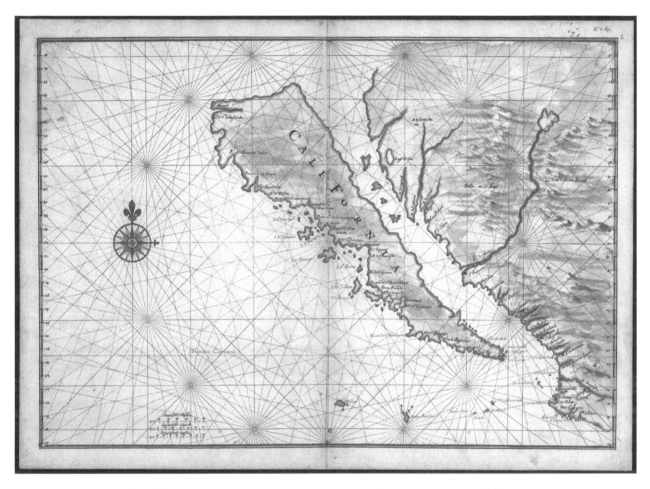

FIGURE 6.1 *Map of California as an island, circa 1650. (Courtesy of The U.S. Library of Congress)*

exploration of the entire Gulf of California and part of the west coast of Baja California. But, although there was no large island, myth was greater than reality, and for the next sixty-three years the flame of mystery and intrigue drove exploration.

Further explorations included **Juan Cabrillo**'s (Figure 6.2) expedition in 1542. He attempted to find the mythical **Strait of Anián** or the Northwest Passage, which many believed connected the Pacific and Atlantic Oceans and thereby being an ideal route to the Orient for trade. Cabrillo therefore became the first to survey the coast up to Alta California. Even though he died during the return voyage his crew was able to return to Mexico with valuable information, which unfortunately was lost to future expeditions.

Practical considerations, however, became part of the drive towards the end of the sixteenth century as the Spanish controlled city of Manila in the **Philippines** (colonized in 1565) became involved in the galleon trade developed to bring spices and Oriental luxury goods to the New World for further transport to Spain. It did not amuse the Spanish to have the British privateer **Sir Francis Drake** escape from the Spanish ships he robbed on the southern West coast of Mexico, by going north past Baja California to the Alta California. He stopped for repairs somewhere between San Francisco and Point Reyes (though no one really knows the exact location) and then declared Alta California as **Nuova Albion** (New England) for the British crown (Figure 6.3). His initial glimpse of the enormous San Francisco Bay revived the island myth and continued a tradition into the seventeenth century whereby mapmakers showed the lands of California as an island (Figure 6.1). In any case, the Spanish crown decided to conduct two more surveys of this strange new coast by using the voyage of **Cermenho** in 1595. He used his Manila galleon to explore and map the coast on its return voyage from Manila to Acapulco, Mexico. Since the

FIGURE 6.2 *Juan Rodriguez Cabrillo. (Courtesy California State Library, Howard DeWitt image)*

galleons left from Acapulco and caught the eastern Trade winds and currents to sail west across the Pacific to their port at Manila, the return journey used the mid-latitude Westerly winds and ocean currents of the clockwise rotating North Pacific gyre to bring them back across the Pacific (see gyre figure in Chapter 3). They made landfall off the region of Mendocino/Sonoma County, thereby catching the southward flowing California current and coastal winds back to Acapulco. Cermenho charted down the coastline looking for good harbors for Spain to use. Later in 1602, **Vizcaino** took Ceremenho's charted information and went up the coast still looking for the Strait of Anián, as well as for good harbors, which the Spanish government were interested in for their Manila Galleons on their return trips to Acapulco. He thereby re-discovered (as the earlier Cabrillo expedition had already noted these places albeit with different names) the Monterey and San Diego harbors, Carmel valley and Catalina island, but not San Francisco Bay.

FIGURE 6.3 *Sir Francis Drake meeting the California Miwok Indians, a 1599 engraving. (Courtesy the Bancroft Library, Howard DeWitt image)*

The development of empire faded for California in the new century. From 1603 to 1769 (167 years) California became a forgotten land as the development of the **Santa Fe** colony in New Mexico in 1610 to extract riches from the silver mines became paramount in Spain's New World plans. By 1710 the first mission became established in southern Arizona. **Father Kino**, who established several missions, explored the land from Arizona to Baja California, finally putting an end to the myth of the island of California around 1700. It was not until Spain learned of **Russian** plans to extend their settlement for fur trapping into the northern section of Alta California that it felt the need to reaffirm its claim to the Pacific coast south of the border of the **British Northwest territories**, encompassing Oregon, Washington and British Columbia. Since Spain was also not on good terms with the British crown, these two "foreign" usurpers had to be stopped via a landscape development model that used the mission-presidio-pueblo (MPP) system.

In 1769, Spanish settlement of California began in earnest with both sea and land expeditions (led by Gaspar de Portola, Figure 6.4) to establish a settlement at **San Diego** harbor and then one in the **Monterey** Bay, described earlier by Vizcaino as a wonderful protected bay. Ironically, explorers had not investigated **San Francisco Bay** to its full extent and thus not considered it in the early occupation years until 1776. Thus began the MPP system of settlement in Alta California, which occurred not only in response to the Russian and British presence, but especially was a shield for the rich silver mines of the Sonora region (Arizona and New Mexico). It was also a system of protective coastal forts and fresh food supplying missions to support the rich cargo-laden Manila galleons as they made their way

FIGURE 6.4   *Gaspar de Portola discovering San Francisco Bay. (Courtesy of TheBancroft Library, Howard DeWitt image)*

down the coast for Acapulco after turning south from the Mendocino/Sonoma region (right where the Russians had established their fur trading post, Ft. Ross). As the next section will address, despite the Spanish system of settlement put in place in Alta California the region remained isolated from the rest of the empire. The isolation continues into the Mexican rancho era, and up until the U.S. take over and the Gold Rush changed everything about this mysterious Western shore.

6.2   Spanish Settlement Pattern

The settlement system set up by the Spanish was under tight control by **Mexico City** and the prevailing **mercantilistic** system that discriminated against the colony in favor of dependence on Mexico City and Spain. An entire body of laws set down for the New World by the Spanish crown also controlled the system, called the **Laws of the Indies** (*Leyes de Indias*). These laws regulated social, political and economic life across the Spanish empire and were useful for regulating the interactions between Spanish settlers and Native Americans. In terms of settlement geography, they represented some of the first attempts at a general plan by codifying the city planning process with wide-ranging guidelines towards the design and development of **pueblos** (civilian communities). In addition, they provided guidance and regulations for the establishment of **presidios** (military forts) and **missions** (centers of conversion for Natives to Christian and Spanish cultural values).

Specifically, when the laws were finally completed in 1680, they provided a comprehensive guide that comprised over a hundred ordinances to aid the military, the church and the colonists in locating, building, populating, and administrating settlements. The following are some examples of the diverse geographical and cultural rules outlined in the Laws of the Indies:

- Having made the selection of the site where the town (pueblo) is to be built, it must, as already stated, be in an elevated and healthy location (i.e., reduce malaria problems); with means of fortification; have fertile soil and with plenty of land for farming and pasturage; have fuel, timber, and resources; fresh water, a Native population, ease of transport, access and exit; and be open to the north wind; and, if on the coast, due consideration should be paid to the quality of the harbor and that the sea does not lie to the south or west; and if possible not near lagoons or marshes in which poisonous animals and polluted air and water breed. In addition, a pueblo must adhere to Spanish religious law by not being established within a two mile radius of a mission.

- Colonists shall try as far as possible to have the buildings all of one type for the sake of the beauty of the town. For example: white washed plaster/stucco and red tiled roofs.

- Within the town, a commons shall be delimited, large enough that although the population may experience a rapid expansion, there will always be sufficient space where the people may go for recreation and take their cattle to pasture without them making any damage.

- Must have a central plaza surrounded by important buildings with portales or arcades, and from which the principal streets, laid out in a grid pattern, shall begin. Smaller secondary plazas were also called for as well as narrow streets, in hot climates, in order to provide shade (Note: this type of town layout is based on the North African Arabic medina model adopted in Spain via its long Arabic settlement and cultural heritage).

- The site and building lots for slaughter houses, fisheries, tanneries, and other business which produce filth shall be so placed that the filth can easily be disposed of down stream of civilian users of water.

- The **alcalde** system was used to both provide administration and judicial operations. In this system one person is elected each year to administer the pueblo and to deal out justice when required. By requiring two major functions, the alcalde wielded a lot of power.

In Alta California, these laws continued after Spain lost possession to Mexican independence and into the American colonization of the state. The Spanish considered many of the rules practical and good organization for the young region.

6.2.1 MISSIONS

Missions represented the preeminent frontier institution in the Spanish strategy to secure its northern territories. In contrast to the views of British colonists in the New World region, the Spanish regarded the Native Americans of paramount importance in their settlement plans and used the mission system to transform them into Spanish citizens—in religion, language, farming practice and gradually through bloodlines. Since there were few Spanish colonists to settle these northern areas, transformation of Native Americans (since they were considered official subjects of the Spanish crown) was an ingenuous way of saving souls with required contributions of labor. This type of mission was called the *reducción* or *congregación*. It attracted Natives living in their smaller tribelet villages to congregate at the mission and reduce their pagan ways in order to become Spanish-cultured farmers.

In AltaCalifornia, the mission represented a method of securing and consolidating an isolated frontier. In order to do this, the goal of the Spanish was to remove from the Native Americans (known as **neophytes**) any pre-Christian practices considered blocks to integration within the greater Spanish society and economy. Thus, the mission was not only for religious conversion but also for agricultural production and some manufacture of goods (Figure 6.5a). While the process of conversion was based on a voluntary model rather than upon coercion, once the neophytes were in the care and education of the mission fathers they were not allowed to leave without severe punishment. The Spanish called the entire process **acculturation** (where a cultural group adopts the beliefs and behaviors of another culture). They required the missions to complete this process within ten years after establishment at which point the missions were to become **secularized** to the civil sector.

The mission system in California confined itself to the coast or within thirty miles of the coast and spread

CHAPTER 6 *Spanish Exploration and Settlement Pattern in Alta California* **101**

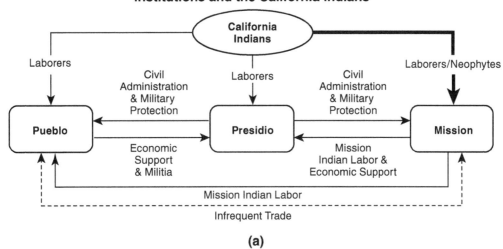

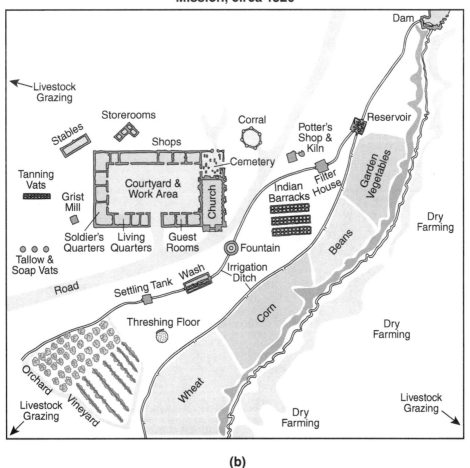

FIGURE 6.5 *(a) Interaction amongst the mission, presidio and pueblo entities and the California natives; and (b) a typical plan view of a California mission with its extensive and intensive irrigated agriculture integrated with a sophisticated water transfer system (adapted from Hornbeck, 1982).*

from San Diego to as far north as Sonoma (see page 10 in *California Atlas*). Using the siting guidelines established by the Laws of the Indies (typically on elevated land just inland but visible to a galleon from the coast), they were spread out a few days' walk apart or one horse day's journey apart. The Spanish control of the coast ensured the re-supplying of the Manila galleons on their way to Acapulco with fresh provisions and water, as well as safety from any British or Russian interlopers.

The original plan was to build only five missions. However, this expanded to a total of twenty-one missions, with the last one established in 1823 in what is now Sonoma County. The original leader of the mission systems was Father **Junipero Serra** (Figure 6.6), who oversaw the first nine missions. Father **Fermin Lasuén** took over after Serra's death and oversaw the construction of the final eleven missions. The expansion occurred as the Spanish discovered new Native populations, especially when the original contact populations succumbed to early deaths from further introduction to exotic diseases, as chapter 5 explained. Under the settlement strategy, missions had a more flexible approach to establishing a site than the other two institutions, the presidio and the pueblo. For a site to work there had to be a substantial population of neophytes to convert, the built structure had to be visible by passing ship if directly on the coast and have a good water supply, plenty of wood for fires and appropriate building materials. They were allowed to take up as much land as necessary to function for agricultural production, i.e., to feed the neophyte population and to provide provisions to passing galleons and provisions to soldiers if near a presidio.

Since missions had to be self supporting, there was a strong interplay between site selection and the design of the mission estate. This included the building and site design, which generally followed a standard pattern. The main structure tended to be rectangular and built around a central courtyard, with the mission church the major element of the main structure. The more prosperous missions (i.e., Carmel and Santa Barbara) produced relatively elaborate churches (Figure 6.7). While the buildings were initially very crude (Figure 6.8), the final typical construction consisted of **adobe bricks**, which were sun-dried, unfired bricks of clay and straw. Available forests (even at quite a distance) provided logs for creating the roofs, which they then covered in oven-fired, **red clay roof tiles** (Figure 6.9). Finally, the builders used available limestone deposits (especially found along the coastal hills and mountains) or clam and abalone shells from Native American shell mounds to make lime. They then created plaster or **stucco** for the walls, which they whitewashed. In stark contrast, the neophyte quarters were generally rough thatched, wattle and dab barracks outside the main structure.

No discussion of the mission system would be complete without reviewing its substantial water storage and delivery systems and the extensive irrigated garden plots and orchards that surrounded the main structure (Figure 6.5b). These working components of the missions (Figure 6.10a,b,c) provide a foreshadowing of future irrigation agriculture in California. In fact, with the latitudinal breadth that the final mission system encompassed, as well as the location of missions on the coast and in inland valleys, the missions had a variety of climatic situations in which to try various crops. Many of these crops eventually became part of the rise in commercial **irrigation agriculture** in the state, i.e., oranges, almonds, grapes, figs and olives, as well as dryland agriculture, i.e., wheat, pinto beans, etc. Essentially we can view the missions as California's first set of agricultural experiment stations, ones from which the later American population in the late 1800s benefited. In a Mediterranean type of climate, irrigation is the key to success. Therefore we can view the missions as reasonably successful at agriculture and in raising

FIGURE 6.6 *Father Junipero Serra. (Courtesy of The Bancroft Library, Howard DeWitt image)*

FIGURE 6.7 *Santa Barbara Mission. (Image © aspen rock, 2012. Used under license from Shutterstock, Inc.)*

FIGURE 6.8 *The Mission of St. Carlos near Monterrey.*

FIGURE 6.9 *An example of mission construction at Mission La Purisma, Lompoc. (Courtesy Portia Ceruti)*

FIGURE 6.10(A) *Using gravity flow, drinking and cooking water was delivered in closed, underground clay pipes from a established spring house. (Mission La Purisma, Lompoc; Courtesy Portia Ceruti)*

FIGURE 6.10(B) *Excess water from the first water fountain (A) went to the lavendaria for washing clothes. (Mission La Purisma, Lompoc; Courtesy Portia Ceruti)*

FIGURE 6.10(C) *The soapy water from the lavendaria drained into a cistern, cleared, then traveled by aqueduct to be used for irrigation agriculture. (Mission La Purisma, Lompoc; Courtesy Portia Ceruti)*

livestock, especially while neophyte labor was readily available. However, after about 1820 agricultural production in the missions started to decline, with part of this due to a long term drought that affected California during this time.

6.2.2 PRESIDIOS

The presidio was the frontier fort for the Spanish who built four of them along the California coast to establish and maintain military control. The sites include **San Diego**, **Santa Barbara**, **Monterey** and **San Francisco**. The presidio acted as the secular authority until the Spanish could establish a civil government. In the meantime, it acted as defense from the ever feared British, the nearby Russians (who never caused any problems) or hostile Native Americans. The military's direct duty was to make sure that the missions were well protected and to enforce mission rules, thereby ensuring a supply of neophytes to make the mission-presidio system as successful as it was in other parts of the Spanish Empire (Figure 6.5a).

Siting a presidio involved selecting the best placement for coastal defense, locating the availability of a good harbor and making sure it was a reasonably short distance to aid a mission in case of hostility. Thus the Spanish formed the following presidial districts to make the mission-presidio arrangement a formal landscape development strategy:

- San Francisco
  - San Francisco Solano
  - San Rafael
  - San Francisco de Asis
  - Santa Clara
  - San Jose
  - Santa Cruz
- Monterey
  - San Juan Bautista
  - San Carlos de Monterey
  - Soledad
  - San Antonio
  - San Miguel
  - San Luis Obispo
- Santa Barbara
  - La Purisima de la Concepcion (Lompoc)
  - Santa Ynez
  - Santa Barbara
  - San Buenaventura
  - San Fernando Rey
- San Diego
  - San Gabriel
  - San Juan Capistrano
  - San Luis
  - San Diego

As far as building design and quality were concerned, the presidio was definitely a poor match against the cannons of a British man-o-war ship. The Spanish tended to construct the buildings around a rectangular central courtyard similar to those of the missions, but they tended to be quite primitive compared to many of the larger missions. In addition, they never sent many soldiers to California—fifty in the early years and never more than two hundred in the final years. Soldiers saw being sent to Alta California as one of the least popular tours of duty within the Spanish Empire.

While missions relied on the presidios for protection, the presidios were dependent on missions for food. The goals of the military and those of the religious **Franciscan** fathers were opposites, which tended to lead to frictions between the two groups (Figure 6.11). However, in the isolated lands of California they were forced to make it work under conditions different than those experienced in other parts of the Spanish Empire, as we will explain later.

FIGURE 6.11 *The cross and the sword: Franciscans and Spaniards in California. (Howard DeWitt image)*

6.2.3 PUEBLOS

The pueblo was the Spanish Empire's third settlement strategy for Alta California, representing the civil sector. The success of the pueblo was considered limited, if not rather disappointing. The Spanish founded only three pueblos in California: **San Jose** (1777), **Los Angeles** (1781) and **Branciforte** (1797, which was actually established as villa. See case study box for details).

The Spanish established the pueblos to entice settlers from Mexico or they retained retiring soldiers from the presidios (Figure 6.5a). The reality is that most of the settlers were poor and some were undesirables (debtors and petty criminals) no longer wanted in Mexico City. The enticement package from New Spain's government included a house, land, livestock, tools, and a clothing all along with an allowance for five years. In return, they required the settlers to build the pueblo by helping till the public agricultural lands and build the public buildings and water supply system. While the pueblos originally added a further boost to Spain's colonization efforts, their success turned out to be less than expected because they never seemed able to produce the needed agricultural surplus. The plan was for the pueblo citizens to set an example for ex-mission Native Americans to further the cause of Spanish culture and civilized farming. In addition, the pueblo settlers would help supply the military with agricultural products as mission supplies were not always reliable and supplies brought in from Mexico were infrequent and expensive. In the early years these lofty goals did not always work out, but by 1820 the pueblos had become an important settlement institution.

The siting of a pueblo was similar to the process used for the mission and the presidio, except there was more emphasis on easy tilled flat land, on being near a port, on being under a presidio's administration and upon having access to a mission's neophyte laborers. The rule was to lay out four square leagues of land (or 36 square miles), which is similar to the township sizes apportioned under the later policy of the American Homestead Act (1862). The builders then laid out a central plaza and constructed all public buildings, homes and the civil church around it. They situated a

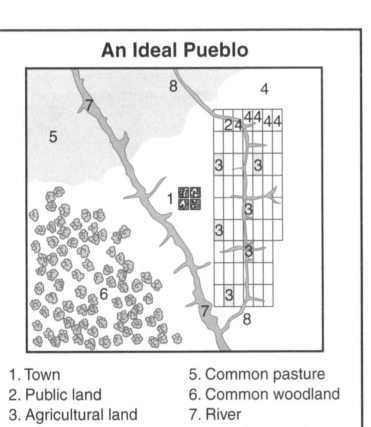

FIGURE 6.12 *An idealized plan for the pueblos based on the Spanish Laws of the Indies. (Adapted from Hornbeck, 1982)*

California In Focus

VILLA DE BRANCIFORTE, UNDESIRABLE AMONG THE PUEBLOS?

Branciforte? Where was that pueblo placed? Alongside San Jose and Los Angeles—the other well remembered and still very much part of California settlements—the name Branciforte is not familiar. It never gets much respect, often being called a failure from the beginning—non-flourishing, unsavory, obscure and even extinct on the landscape. Despite the negative sentiments, Branciforte was very much a part of the landscape that is now Santa Cruz, across Monterey Bay north from Monterey. The site and region were ideally suited, encompassing abundant pasturelands and year round sources of water. It contained a variety of construction materials, but most important it had a harbor with defensible perimeters. Unfortunately, it was not a good choice of location, as it was not situated along the main thoroughfare that connected all of the stops along the California mission chain (i.e., El Camino Real), making accessibility difficult. The Monterey governor broke Spanish religious law by placing a civilian settlement within a two-mile radius of a mission, which would result in a negative atmosphere between settlers and religious clergy at mission Santa Cruz.

One of the biggest differences between Branciforte and San Jose/Los Angeles was its designation of "pueblo" or civilian town. The latter two were pueblos; however, Branciforte was in fact a "villa." Villas were innovative soldier-settler communities established on the frontier with dual purposes: colonization and defense. Villa de Branciforte on the north shore of Monterey Bay was part of New Spain's plan to firm up defenses in case the English should attack, along the coast. It is the often-disregarded settlement because of the way the Fathers at mission Santa Cruz maligned the inhabitants as unskilled criminals (some were, but they were petty criminals, and non-violent). While the villa did also receive retired soldiers for the community as planned, the settlement never took off, as constant problems with promised resources from the government and fights with Mission Santa Cruz made for haphazard development and little local government control. Villa de Branciforte became part of the City of Santa Cruz and was forgotten among the original Alta California settlements. When California became a state in 1850, Santa Cruz County was first named Branciforte County, until the American inhabitants of Santa Cruz rebelled and had it changed 90 days later.

public agricultural commons outside the central area and surveyed individual agricultural plots (Figure 6.12). They also developed a public water works of ditches (*zanjas*) and clay pipes from a perennial water source to the pueblo and agricultural zone, all managed by the **zanjero** (water master). In essence though, the structures tended to be mostly one-story adobe buildings that were primitive, dusty and generally unsanitary when compared to the presidios or especially some of the richer missions.

Later as the pueblos evolved, cattle became a natural outgrowth of abundant grazing lands, available Native American labor and a rapid reproduction of the initial small herd started at Mission San Diego. One of the first **Spanish land grant** ranchos enacted in California occurred in 1784. This was Rancho San Pedro in the Los Angeles pueblo district (it covered San Pedro, the Long Beach harbor and the Palos Verdes peninsula). Under the Spanish rancho system the land grants usually went to retiring soldiers in return for their service protecting Alta California. Therefore, only twenty-five to thirty Spanish ranchos were ever granted in California, but they were a harbinger of things to come in the 1830s during the Mexican government's era of settlement building strategy.

6.3 Postcards from a Decaying Empire

California's distance from the seat of the Spanish Empire in Mexico City, and thus its isolation, made the MPP system difficult to employ and manage. While landscape development was a constant struggle, there is one lasting legacy that the Spanish established in California and that is the seeds of **racism**. At the time of Alta California's settlement there were two types of Spaniards: the **Peninsulares**, who were pure blooded Europeans, and the **Criollos**, who were mixed blood Spaniards-Africans who held a lower social status. The Peninsulares controlled the government, church and mercantile system in the New World. The Criollo class tended to be the soldiers who interacted with the local mission neophyte population and had mulatto children called **Zambos**—"half civilized" in the eyes of the shaming Franciscan priests. This social class structure and the barbaric cruelty of the Franciscan order towards the Zambos and neophytes confined to a small number of missions, as well as the overall genocide of the California Native American from diseases, changes in food consumption, family structure and enslavement in the missions led to Native American cultural disintegration.

Other legacies are the lasting presidios and missions that are reminders of the Spanish conquest. The cities of San Jose and Los Angeles have both retained a Spanish character up until modern times. Taken even further, the pueblos represent the development of a mixed blood cultural landscape that California came to represent well into modern times. As Spanish California developed, Black-Spaniards dominated the pueblos as opportunity for land and economic freedoms attracted discontented people from Mexico with mixed bloods: Spanish-African and Spanish-Aztec also known as **Mestizo**. Later under the Mexican government era a local "nationalism" formed that prompted many in California to call themselves "**Californios**." This is a term referring to native-born, Spanish speaking Californians and is the first significant expression of a locally evolved voice and culture.

In the end, Spanish California became a quagmire of political problems, economic conflict and class warfare during the decaying time frame of 1790 to 1820. California was isolated, which always represented its major problem to the Spanish crown's settlement strategy, so when the **Mexican War for Independence** started between Spain and New World-born Spaniards, Mestizos, Criollos and Native Americans between 1810 and 1821, foreign traders intruded upon California. The Russians established Fort Ross on the Sonoma coast in 1812 to trap local beavers and sea otters. **American fur trappers** made their first forays into Alta California and insensitively interacted with the pueblo populations; and finally, the British actively sought trade with the Californios. All of these outside intrusions came at a difficult time for New Spain, but the mixed blooded Spaniards of California demanded change from the rigid social structure and rich mission system. This came in the form of rising local nationalism. As a result, the local-born Californio demanded home rule and forced the church into a position of defending its dying mission system. By the end of the Mexican War for Independence, Californios had become independent and didn't necessarily share newly independent Mexico's attitudes. Loyalty, the development of foreign trade and the dominance of church wealth created the major issues which were carried forward into the Mexican Era of Alta California.

Bibliography

Archibald, R. (1978). *The Economic Aspects of the California Missions*. Academy of American Franciscan History, Washington, D.C.

Baer, K. (1958). *Architecture of the California Missions*. University of California Press, Berkeley, CA.

Bannon, J.F. (1979). The Mission as a Frontier Institution: Sixty Years of Interest and Research. *The Western Historical Quarterly*, 10(3).

Bannon, J.F. (1970). *The Spanish Borderlands Frontier, 1513–1821*. Holt, Rinehart and Winston, New York, NY.

Beebe, R.M. and R.M. Senkewicz (2001). The Invention of 'California', from 'The Labors of the Very Brave Knight Espladian' by Garci Rodriguez de Montalvo. In *Lands of Promise and Despair, Chronicles of Early California, 1535–1846*, edited by R.M. Beebe and R.M. Senkewicz., pp. 9–11. Heyday Books, Berkeley, CA.

Beebe, R.M. and R.M. Senkewicz (2001). A Secularization-Oriented Proposal for Alta California, from 'A Statistical Description of Alta California' by Juan Bandini. In *Lands of Promise and Despair, Chronicles of Early California, 1535–1846*, edited by R.M. Beebe and R.M. Senkewicz, pp. 375–385. Heyday Books, Berkeley, CA.

Beebe, R.M. and R.M. Senkewicz. (2006). *Testimonios: Early California through the Eyes of Women, 1815–1848*. Heyday Books, Berkeley, CA.

Beck, W. (1974). *Historical Atlas of California*. University of Oklahoma Press, Norman, OK.

Bolton, H.E. (1917). The Mission as a Frontier Institution in the Spanish-American Colonies. *American Historical Review*, 23(3): 42–61.

Bolton, H.E. (1939). *An Outpost of Empire: The Story of the Founding of San Francisco*. Russell & Russell, New York, NY.

Booth, S. (2008). *California Geography*. Course taught at Sierra College. [online] http://geography.sierra.cc.ca.us/booth/California/cal_index.htm.

Bowman, J.H. (1964). Weights and Measures of Provincial California. *The California Historical Quarterly* 30.

Calisphere (2008a). California Cultures. University of California [online] http://www.calisphere.universityofcalifornia.edu/calcultures/

Calisphere (2008b). A world of California Primary Sources. University of California [online] http://www.calisphere.universityofcalifornia.edu/

Campbell, L.G. (1972). The First Californios: Presidial Society in Spanish California, 1769–1822. *Journal of the West* 11(4).

Campbell, L.G. (1977). The Spanish Presidio in Alta California During the Mission Period, 1769–1784. *Journal of the West* 16(4).

Chapman, C. (1916). *The Founding of Spanish California: The Northward Expansion of New Spain, 1687–1783*. Macmillan, New York, NY.

Cleland, R.G. (1944). *From Wilderness to Empire: A History of California, 1542–1900*. Alfred A. Knopf, New York, NY.

Cook, S.F. (1976). *The Conflict Between the California Indian and White Civilization*. University of California Press, Berkeley, CA.

Cook, S.F. and W. Borah. (1979). Mission Registers as Sources of Vital Statistics: Eight Missions of Northern California. In *Essays in Population History: Mexico and California*. Vol. 3. University of California Press, Berkeley.

Costello, J.G. (1977). Lime processing in Spanish California. *Pacific Coast Archaeological Society Quarterly*, 13(2):22.

Costello, J.G. (1989). Variability among the California Missions: the economics of Agricultural Production. In *Columbian Consequences, Vol. 1: Archaeological and Historical Perspectives on the Spanish Borderlands West*, D.H. Thomas, Ed., pp. 345–350. Smithsonian Institution, Washington, D.C.

Costello, J.G. (1992). Not peas in a pod: documenting diversity among the California missions. In *Text-Aided Archaeology*, B.J. Little, Editor, pp. 67–83. CRC Press, Boca Raton, FL.

Cox, C.J. (2008). California Geography. Course taught at Sierra College. [online] http://faculty.sierracollege.edu/ccox/california_geography/index.html.

Cuttler, D. (1947). Spanish Explorations of California's Central Valley. Ph.D. dissertation, University of California, Berkeley.

Davis, C. and D. Igler (2002). *The Human Tradition in California*. Scholarly Resources, Wilmington, DEL.

DeWitt, H.A. (1999). *The Fragmented Dream: Multicultural California*. Kendall/Hunt Publishers.

DeWitt, H.A. (1999). *The California Dream*. 2nd edition. Kendall/Hunt Publishers.

Donley, M.W., Allan, S., Caro, P., and C.P. Patton (1979). *Atlas of California*. Pacific Book Center, Culver City, CA.

Durrenberger, R.W. and R.B. Johnson (1976). *California Patterns on the Land*. 5th edition, Mayfield Publishing Company, Mountain View, CA.

Engelhardt, Z. (1929). *Missions and Missionaries of California*. 2 vols. Santa Barbara, Calif.: Mission Santa Barbara.

Engstrand, I.H.W. (1998) Seekers of the 'Northern Mystery': European Exploration of California and the Pacific. In *Contested Eden: California Before the Gold Rush*, edited by R.A. Guttiérrez and R.J. Orsi, pp. 78–92. University of

California Press, Berkeley, CA.

Fages, P. (1937). *A Historical, Political, and Natural Description of California*. University of California Press, Berkeley, CA.

Francis, J.D. (1976). *An Economic and Social History of Mexican California*. Arno Press, New York, NY.

Garr, D.J. (1972). Planning, Politics and Plunder: The Missions and Indian Pueblos of Hispanic California. *Historical Society of Southern California* 54(4).

Guest, F.F. (1979). An Examination of the Thesis of S. F. Cook on the Forced Conversion of Indians in the California Missions. *Southern California Quarterly* 61(1).

Guest, F.F. (1967). Municipal Government in Spanish California. *California Historical Society Quarterly* 46(4).

Haas, L. (1996). *Conquests and Historical Identities in California, 1769-1936*. University of California Press, Berkeley, CA.

Hackel, S. (1998) Land, labor, and production: the colonial economy of Spanish and Mexican California. In *Contested Eden: California Before the Gold Rush*, edited by R.A. Guttiérrez and R.J. Orsi, pp. 111–146. University of California Press, Berkeley, CA.

Hackel, S. (2005). *Children of Coyote, Missionaries of Saint Francis: Indian-Spanish Relations in Colonial California, 1769-1850*. University of North Carolina Press.

Hayes, D. (2007). *Historical Atlas of California*. University of California Press, Berkeley, CA.

Heizer, R.F. and A.J. Almquist (1971). *The Other Californians: Prejudice and Discrimination under Spain, Mexico, and the United States to 1920*. University of California Press, Berkeley, CA.

Hornbeck, D. (1983). *California Patterns: A Geographical and Historical Atlas*. Mayfield Publishing Company, Mountain View, CA.

Hornbeck, D. (1989). Economic growth and change at the missions of Alta California, 1769–1846. In *Columbian Consequences, Vol. 1: Archaeological and Historical Perspectives on the Spanish Borderlands West*, D.H. Thomas, Ed., pp. 423–433, Smithsonian Institution, Washington, D.C.

Jackson, R.H. (1988). Patterns of demographic change in the missions of central Alta California. *Journal of California and Great Basin Anthropology* 9(2): 251–272.

Knowland, J.R. (1941). *California, a Landmark History*. Tribune Company, Oakland, CA.

Larson, D.O., Johnson, J.R., and J.C. Michaelsen (1994). Missionization among the Coastal Chumash of Central California: A Study of Risk Minimization Strategies. *American Anthropologist* 96(2): 263–299.

McClure, J.D. (1948). *California Landmarks: A Photographic Guide to the State's Historic Spots*. Stanford University Press, Stanford, CA.

McGovern, C.G. (1978). Hispanic Population in Alta California: 1790 and the 1830s. Master's thesis, California State University, Northridge.

Michaelson, J. (2008). *Geography of California*. Course at UC Santa Barbara, Dept. of Geography. [online] http://www.geog.ucsb.edu/~joel/g148_f08/.

Monroy, D. (1993). *Thrown Among Strangers: The Making of Mexican Culture in Frontier California*. University of California Press, Berkeley, CA.

Ogden, A. (1941). *The California Sea Otter Trade, 1784–1848*. University of California Press, Berkeley, CA.

Oliver, V.L. (1978). Foreign Images of Alta California. Master's thesis, California State University, Northridge.

Perry, F.A. (2007). In search of Spanish lime kilns. In: *Lime Kiln Legacies: The History of the Lime Industry in Santa Cruz County*, F.A. Perry, R.W. Piwarzyk, M.D. Luther, A. Orlando, A. Molho, and S.L. Perry (Eds.), pp. 15–21, The Museum of Art & History, Santa Cruz, CA.

Phillips, G.H. (1974). Indians and the Breakdown of the Spanish Mission System in California. *Ethnohistory* 21.

Polk, D. (1995). *The Island of California: A History of the Myth*. University of Nebraska Press, Lincoln, NE.

Preston, W.L. (1998) Serpent in the Garden: Environmental Change in Colonial California. In *Contested Eden: California Before the Gold Rush*, edited by R.A. Guttiérrez and R.J. Orsi. Berkeley: University of California Press, pp. 260–298

Rawls, J.J. and W. Bean (2008). *California: An Interpretative History*. 9th Edition, McGraw-Hill Publishing Co., New York, NY.

Reese, R.W. (1969). *A Brief History of Old Monterey*. City of Monterey Planning Commiesion, Monterey, CA.

Rice, R., Bullough, W., and R. Orsi (2001). *The Elusive Eden: A New History of California*. 3rd edition, McGraw-Hill.

Rowntree, L.B. (1985a). A crop-based rainfall chronology for pre-instrubmental record Southern California. *Climate Change* 7:327–341.

Rowntree, L.B. (1985b). Drought during California's mission period, 1769-1834. *Journal of California and Great Basin Anthropology* 7(1): 7–20.

Servin, M.P. (1965). The Secularization of the California Missions: A Reappraisal. *Southern California Quarterly* 47.

Starr, K. (2005). *California: A History*. The Modern Library, New York, NY.

Watkins, T.H. (1983). *California: An Illustrated History*. American West Publishing Company, New York, NY.

Webb, E.B. (1952). *Indian Life at the Old Missions*. Warren F. Lewis publishers, Los Angeles, CA.

CHAPTER 7

Mexican Rancho Era
Settlement and Development by "Los Californios"

Key terms

- Acculturation
- Casta system
- Castizo
- Cattle
- Cholo
- Criollo
- Diseño
- Dons
- Gente de razon
- Gente sin razon
- Hides and Tallow
- Indios
- International trade
- Kanakas (Hawaiians)
- Los Californios
- Mercantilism
- Mestizo
- Mexican Republic
- Mulatto
- Quasi-feudal society
- Racisim
- Rancho
- Russians
- Secularization
- Spaniard (peninsulares)
- Vacquero
- Zambo

Introduction

MEXICAN INDEPENDENCE

When the ancient lands of Mexico won their independence from the Spanish crown in 1821, the **Mexican Republic** began. During the conflict between 1810 and 1820, Spanish Alta California was nearly forgotten. The new Mexican Republic, however, immediately attempted to solidify its hold on Alta California by trying to expand settlements at the two barely flourishing pueblos of San Jose and Los Angeles.

The inhabitants of Alta California were caught in a vacuum of neglect and isolation and eventually came to define their life on their own terms. In less than a generation, Alta California's population began to identify completely with California—not with Spain or Mexico—despite the fact that the central Mexican government continuously sent governors to the colonial captial of Monterey and issued orders. For example, the economic organization of California under Spanish rule was the system of **mercantilism** whereby colonial economies were kept in a state of dependency that supplemented Spain's economic growth at home. By the mid 1810s the Alta California missions were able to produce surpluses in maize, wheat, cattle and sheep, but private trading, either foreign or Spanish, was not allowed. Only officially sanctioned trade between Mexico and Alta California was allowed. This stunting of the regions potential growth makes sense if we remember that the design of the original mission-presidio-pueblo colonization system was to stem potential foreign aggression and to Christianize the Native Americans while becoming self-sufficient. In 1810, Mexico's struggle for independence from Spain changed everything for Alta California's internally focused sanctioned trade, as regular ship visits from Mexico with supplies ended. Therefore, both the

113

FIGURE 7.1 *Mission era cattle still being raised at Mission La Purisima, Lompoc. (Courtesy Portia Ceruti)*

presidio and the pueblo residents came to rely on the successful missions for farming, manufacturing, trading, banking and so forth. This reliance on the missions made the missions and the church very powerful and sowed the seeds to future problems. In addition, smuggling with foreigners began to become a lucrative process despite the Alta California governor outlawing it. The Santa Barbara Channel Islands far to the south of Monterey, out of the reach of the governors authority, were used as active smuggling grounds with British and American ships.

Despite the important role that missions were forced into, their decline accelerated due to a diminishing supply of neophyte Native Americans, a series of neophyte revolts over their poor treatment within several of the missions, development of a questionable (to their original reason for being) commercially focused enterprise scrutinized by the growing civilian population, and no support from Mexico City. Their biggest problem was that they were still the main institutions in Alta California. They controlled most of the valuable **cattle** (Figure 7.1) grazing land, still had a substantial Native American labor force and had been allowed after Mexican independence to contract with U.S. and British trading companies as their main suppliers for all of their needed cattle **hides and tallow** (beef fat) in exchange for hard currency and manufactured goods. This capitalist power evolving in a non-secular institution did not bode well among the growing Alta California native citizens—**Los Californios**.

7.1 Trade and Economic Growth

Mexican independence did not grant Alta California any new or extra help from Mexico City, but the new government did decide to rid the colony of the Spanish mercantilism system and to open the area to **international trade** in 1823 with visiting U.S., British, Russian and other ships that were increasingly visiting the eastern Pacific waters. As mentioned earlier, at first only the missions were able to take advantage of this open-

FIGURE 7.2 *Cattle hides drying in the old manner at Mission La Purisima, Lompoc. (Courtesy Portia Ceruti)*

ing of the Alta California market. The missions represented California's economic engine with their supply of acculturated Native Americans on the twenty-one missions, their development of superior irrigation systems and their vast tracts of lands for grazing cattle. Goods manufacturers in Boston and England began to readily engage the missions for vast supplies of cattle hides (Figure 7.2 and 7.3) and tallow (e.g., for the manufacture of leather goods (shoes), soap, candles, etc.) in trade for consumer goods in a barter economy. This led by 1830 to the missions being more profit oriented than "spiritually" oriented. Even though as noted in Chapter 6, missions were to complete their neophyte acculturation process within ten years of establishment and become secularized to the civil sector, Spains ignoring behavior and the civil war allowed them to do what they wanted without complaint from the governor. Missions became California's first large scale commercial farms that were also becoming short on neophyte labor. The Californios were not amused by the missions' land holdings which essentially monopolized and prevented civilians from participating in the lucrative hide and tallow trade because the scale of the mission ranching system was so much larger to what tracts of land a pueblo settler was allocated. Growing unrest and complaints to Mexico City by some of the more powerful and wealthy Californio families located in the pueblo of Los Angeles together with a liberal Mexican congress forced through the **secularization** of the missions between 1834 and 1836. This instantly opened up eight million acres of high quality grazing land and irrigated agricultural holdings along the California coast, close to useful harbors.

7.2 Mexican Ranchos and Mission Secularization

The mission era was over in a single act by the Mexican government in 1834. While political pressure from the Californios helped, it really coincided with a parallel rise of liberal, anti-church ideology in Mexico, which forced the Catholic Church and its vast holdings into a losing defensive position. Secularizing the missions in Alta California would be easier and less controversial than doing something similar in Mexico because Alta

FIGURE 7.3 *Piles of cattle hides ready to be picked up by American hide droughing ships from Mission La Purisima, Lompoc. (Courtesy Portia Ceruti)*

California was isolated and had a small church clergy population. Opening up eight million acres of land led Mexico City to produce a new settlement strategy for Alta California, one that the Spanish had experimented with conservatively—the **rancho**.

While the original rancho concept was a product of Spain and was used as a parting gift (concessions) for soldiers of higher ranks that had served in the Alta California presidios, it was the Mexican government that, besides promoting the pueblos, used ranchos to colonize California. Under the Spanish, those retiring soldiers that received rancho concession lands did so unofficially and received no formal titles to the land. We could consider the deals gentlemanly handshakes with the presidio generals, the California governor and the retiring soldiers. Therefore under the Spanish era of occupation, the government only issued twenty-seven ranchos. The secular authorities did not, however, consider these lands as prime as those owned by the missions.

Under Mexican rule, however, the rancho became an institutional strategy to encourage settlement with firm officially recorded titles to the land. This strategy allowed Californios, as well as an increasing number of foreigners from the U.S. and England, to take advantage of Mexico's colonization laws. By 1840, the rancho had become the dominant way of settlement and cultural life in the Alta California colony.

The push for ex-mission lands by Los Californios created a network of cattle ranchos along the coast. The rancho owners came largely from the original pueblo dwellers who petitioned for lands, retiring soldiers, families migrating from Mexico, and finally American Yankees that became citizens of Mexico in order to partake of rancho ownership. The rancheros prized ex-mission lands because they were already developed, represented prime grazing lands, and usually were stocked with numerous mission cattle (Figure 7.4). Furthermore, they were able to typically convince ex-mission neophyte laborers (though rancheros did use

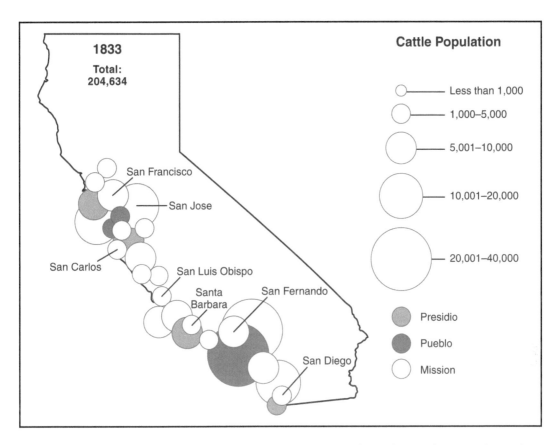

FIGURE 7.4 *The distribution and estimated abundance values of the Alta California cattle population in 1833, one year before secularization by the Mexican government in 1834. (Original from Hornbeck, 1983)*

indentured Native American slavery as well) to work on these ranchos as vacqueros, i.e., cowboys. The rapid transferal of land to private hands along the coast and inland lands covering the well-watered Sacramento River Valley, the Sacramento-San Joaquin Delta and the upper San Joaquin River Valley led to a total award of eight hundred land grants between 1832 and 1846 (Figure 7.5). In comparison, the Mexican governors only granted twenty ranchos between 1822–1832. The latter period really reflects the influence of the newly arriving Americans who willingly became citizens of Mexico through **acculturation** to receive a land grant. The two most common cultural elements adopted by foreigners was the ability to speak and write in Spanish and become Catholic.

The coastal cattle rancho network led to two major development issues. First, the missions were in decline and decay as the loss of land and the neophyte labor force (from disease and running away) accelerated their demise, as most were unable to continue as churches alone. The buildings rapidly deteriorated and generally became unsafe but glorified barns and ranging areas for the huge herds of cattle. Santa Barbara and San Diego are examples of the few mission churches that remained active through the Franciscans, providing Catholic services to what was left of their local Native American populations and the civilian sector. Second, while San Jose and especially Los Angeles grew larger as pueblos, the old presidio nodes of San Diego, Santa Barbara and Monterey became pueblo developments in their own right. During this period, however, the major development nodes became Monterey and Los Angeles—Monterey because it was the seat of administrative and international trade power (i.e., the Mexican governor's home and customs house) and Los Angeles because it had become the largest hide and tallow trading port. This was due to the fact that the Los Angeles basin had the largest tracks of grazing lands with powerful families on very large rancho holdings.

By 1846, as a precursor of things to come, the government granted 20 percent of all ranchos to foreign

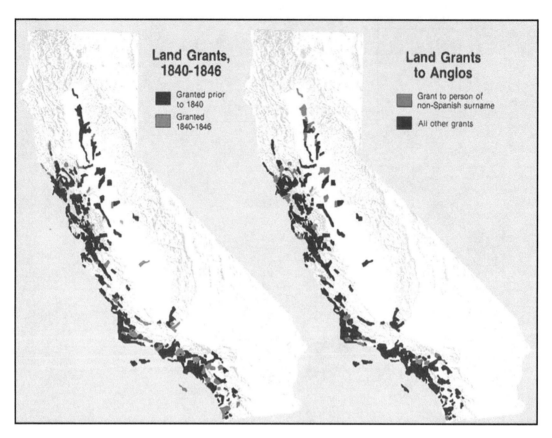

FIGURE 7.5 *Distribution of Mexican land grants to Californios and to foreigners. (Original from Hornbeck, 1983)*

immigrants (mostly American "Yankees") (Figure 7.5). While it considered these people legally acculturated into Mexican society, it turned out not to be the case from the immigrants' point of view. Nevertheless, if one takes the entire secular rancho process of land development, it really represented Mexico's systematic attempt to impress on the land a set of cultural and economic values that would come to symbolize land and society in Alta California.

7.2.1 RANCHO DISEÑOS

One of the more fascinating sets of images that this era provides us is the rancho diseño sketch map. Starting in 1828 land applicants had to submit rough sketch maps of the areas that they wished to acquire, which then accompanied rancho petitions to the governor in Monterey who used these oblique landscape images of the area contained in petitions to ascertain that the requested land was not part of existing ranchos. During this time there were two types of land being petitioned: one for new lands and the other to validate lands that had already been settled on in the hope of making it legal under the new rancho system. The **diseño** (meaning design in English) became the key simple surveying tool to allow ranchos to become the central institutions in California. We can consider them geographic sketches of the natural landscape via horseback field analysis, which led to official demarcations of ranchos that sometimes were upwards of 100,000 acres in size when combined by a family, but typically could not be larger than 48,000 acres.

The rancho diseños allowed the settlers to graphically describe their landscapes and its environmental components (Figure 7.6). There were no rules about how to make them, no set scale, no cartographic convention, no size or medium used to draw them. In many respects, the diseños represented what the potential owner found important to record. At the same time the perspective often was from several points of view.

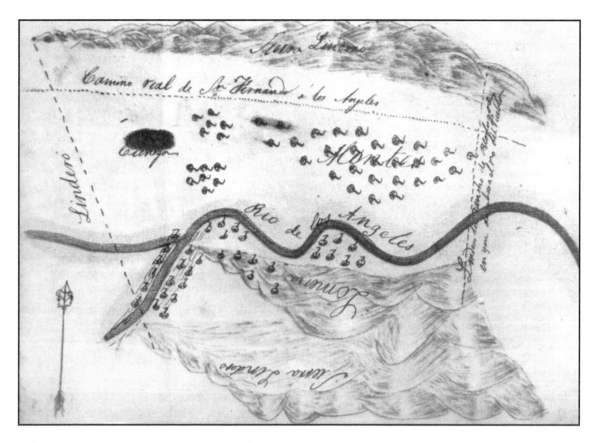

FIGURE 7.6   *Sketch map or diseño of Rancho Providencia, now part of Burbank, California from the 1840s. Note that the rancho contains a section of the Los Angeles River. (Courtesy of The Library of Congress)*

The most dominant environmental feature placed on diseños was the topography of the ranchos. The sketchers of these images made sure that they clearly described valleys, ridges, hills and high mountains, sometimes even showing grayscale shading to bring about some 3D relief. All items to do with water were the next features placed on a diseño. The old Californios and the new settlers learned that California experienced a six to eight month drought period during the year. Since the focus on economic activity during this time included everything to do with cattle, a permanent water source somewhere on the property was a must have natural resource. Therefore, they duly noted and identified by name all natural lakes, wetlands, springs and perennial streams. The next natural feature was the vegetation of the area. In this case two vegetation types were important: woodlands/forests and grasslands. The settlers required the trees for building structures and providing wood fuel for cooking and the grasslands for the much needed natural feed for the huge herds of ex-mission cattle. Lastly, any cultural features found on the landscape were important map components, especially if the new rancho included ex-mission lands or the features represented previously settled but unofficially registered land. Many of these common features included roads, houses, mills, agricultural plots, missions, irrigation works or neighboring ranchos (though boundaries were imaginary as there was no such thing as barbed wire fences at this time!).

Rancho diseños provide us with an important set of images and geographical information on land tenure during the Mexican era of development. Despite the many different ways the settlers created these sketches and the fact that no standardization for symbols, perspective or naming applied, we can at least gather information from these maps on how they read the landscape in order to create a ranch that planned itself based on the environmental features and constraints presented to the ranchero. In many respects, the eight hundred ranchos created along coastal California and in the northern Great Central Valley show a natural demarcation of boundaries, i.e., their layout is in tune with the spatial patterns found in the environment.

Finally, many of these ranchos or at least their names are still with us today: San Pedro, Los Alamitos, Santa Anita, La Brea, San Ramon, Chico, etc. Some of their original boundaries form our current county boundaries for the coastal counties (e.g., Marin to San Diego). Even some very famous street-names are genuine derivations from actual ranchos. Very hip, Rodeo Drive in Beverly Hills, came from the title of the land grant encompassing the area: Rancho Rodeo de las Aguas "gathering of the waters", which represented the streams coming from Benedict and Coldwater canyons. Landscapes hold memories, and the California of today still reflects these rancho landscapes of the past—in place names and in boundaries.

7.3 A Rigid Cultural and Economic Landscape in a Mixed Blood Land

A local aristocracy dominated Mexican Alta California who ruled through wealth, class and by race. This was a seigniorial, **quasi-feudal society** with power and control resting with rancho owners known as Californio **dons**. Californio dons oversaw vast land holdings, which became largely autonomous and self-sufficient. Again, the immensity of the operation led to 800 land grants averaging 40,000 acres in size, which totaled 14 million acres of land all held by only 200 Spanish speaking families. For example, Californio dons such as Don Pablo de la Guerra received land and bought other ranchos, which brought his total to 14 ranchos totaling 488,000 acres surrounding Santa Barbara.

A don's goal was to control and provide for all family members as well as for the rancho's laborers' families. Many of these laborers were former mission neophytes who had been adrift since secularization, and mestizos who became part of the cowboy or **vacquero** class. Despite the lucrative hide and tallow trade that the dons controlled in their trade with American and British companies, there was always a lack of full economic development in Alta California. The hide droughering (i.e., receivers and preparers of cattle hides) American traders found, as described by Henry Dana in *Two Years Before the Mast*, the Californio population lazy about manufacturing work and the landscape a cultural void. The rather simple and leisurely ranchero economy led to only the production of the bare essentials, everything else that was needed or desired in the form of manufactured goods bartered from the merchant ships.

What the aristocratic Californios found important was social class distinctions, which were not too different from the race caste system made infamous in Apartheid-era South Africa during the latter half of the twentieth century. In Alta California, there were strong distinctions drawn between **gente de razon** (people of reason) and **gente sin razon** (people without reason). Californios maintained self-serving negative images of the lowest classes in their *casta* **system** as part of their control of the landscape. Ethnic identity contributed to a person's social status during the Mexican era and was really a refinement of the already rampant **racism** began during the Spanish era. These social classes include the following in a descending scale of socio-economic/political power:

- **Spaniard** (*peninsulares* born in Spain) or American (under acculturation)
- **Criollo**—Spanish (pure) blood, but born in the New World
- **Castizo**—Spanish and Mestizo
- **Mestizo**—Spanish and Ameri-Indian mix
- **Mulatto**—Spanish and African mix
- **Cholo**—Mestizo and Ameri-Indian mix
- **Zambo**—African and Ameri-Indian mix
- **Indios**—Native American (Ameri-Indian)

The power the dons in the Spaniard or Criollo classes had on the landscape was overwhelming. Despite the fact that they represented less than 20 percent of the Alta California population, the Americans moving in and the foreigner traders they represented viewed them as the typical Californio because of the power they held at the racial, economic, and political levels. The level of relaxed life that the dons' families had in contrast to

FIGURE 7.7 *Mexican California: a pastoral scene. (Howard DeWitt image)*

that of the majority of the population that worked on their ranchos or in the pueblos has led to a view of this era as a pastoral one of halcyon days (Figure 7.7).

The relaxed atmosphere was defined simply as calm, relaxed and prosperous, with no need to engage in competitive capitalism. This was a bit misleading though as the Californios were strongly engaged in business with foreign traders and with local pueblo commerce, in addition to illegal smuggling activities within the safety of the Santa Barbara Channel Islands. The commerce system, illustrated at its height during the early 1840s, provided for a pleasant, relaxed lifestyle for large rancho owners at the top of the social hierarchy in contrast to hard work and ridicule for the lower classes, especially ex-mission neophytes. The trade with the Yankee ships was lucrative as the ships brought manufactured goods from Boston to trade for hides at the main ports from San Francisco to San Diego. The ships would work the harbors along the coast (i.e., San Francisco, Monterey, Avila, Santa Barbara, Ventura, San Pedro, Dana Point, and San Diego) for up to two years or until a full load of over 100,000 hides was in the ship's hold before making the return journey around South America's Cape Horn. The hides collected along the coast were stored and processed for the long voyage home at company warehouses lined up along the beach in San Diego harbor. The majority of the American companies used Hawaiian labor, unknown as **Kanakas**, to collect and process the hides. Many of the Hawaiians were later hired as laborers on ranchos held by acculturated foreigners as the available Native American labor pool became scarce.

The ranchos in the Los Angeles basin made the most of this trade because of the rich grassland environment and relatively level topography—hide droughers considered the port of San Pedro the most productive for hides. Therefore the pueblo of Los Angeles quickly became the largest pueblo in Alta California because of the bias in trade with the foreigners. The money that came in allowed many of the dons and their families to become absentee rancho landlords, by instead building adobe townhouses in the pueblos. They left the smaller dwellings found in the pueblos to the less wealthy or lower social classes. The high rancho absenteeism was a harbinger of things to come, for it described the modern industrial agribusiness landscape constructed in California during the twentieth century.

Other cultural elements that describe these people and their era include the flamboyant clothing the men wore, such as embroidered vests with silver buttons, red satin scarves and velvet riding pants. The upper class women would dress in flowing dresses and skirts and were fond of jewelry. The Californios were outstanding horsemen (caballeros) and thus became known as gentlemen cowboys (Figure 7.8). They considered dressing themselves and their horses up for a fiesta or a local fandango (dance party) an important pastime. As far as other Californio entertainment particular to Alta California at this time, there were such strange events as capturing a grizzly bear and tying it to a longhorn bull to watch them tear each other apart—the loser being dinner for the fiesta that evening. Lastly, the even more bizarre carrero del gallo involved burying a rooster up to its neck in the pueblo street's soil. Skilled horsemen would then try to reach down and try plucking him out while galloping by. What fun to be in Alta California at this time?!

One final note must be given with regards to the uniqueness of Californio food and its preparation during this era. Compared to colonial America, food preparation was unique. In Alta California, the kitchens were not inside a house but were built outside (imagine them as cruder versions of the fancy in-built patio BBQs one finds in some modern California homes). The reason for outside kitchens was the mild Mediterranean climate, compared to houses in the Eastern American colonies where the kitchen was an area to warm the house. As one can imagine, with all the hides produced along the coast, beef for breakfast,

FIGURE 7.8 *(a) Californios lassoing a steer (Courtesy of the Bancroft Library); and (b) A rancho don (Courtesy of the Bancroft Library).*

lunch and dinner was normal. Historical notes from this era state that the main meals were beef and pinto beans, which we can view as a dietary regression from the Spanish mission era where the immaculate irrigated mission gardens produced many fruits, vegetables and grains that we now consider part of California cuisine. However, during the Mexican rancho period the missions became non-functional and the farmers of the pueblos grew only the most essential of foods, including irrigated corn, pumpkins and other squashes, while they left dryland farming to wheat and pinto beans. One of the more astonishing facts regards the consumption of wine. Wine used to come from the Franciscan-run missions, but in the era of secularization with its focus on cattle, the Californios imported their table wine from all places . . . Boston!

7.4 Pre-gold Rush Scenario

Foreigners had been visiting Alta California's shores before the Mexican rancho era started, but these were limited to research expeditions, beaver trapping, sea otter and seal hunting. Most notably the **Russians** conducted the hunting, and they had built an outpost on the Sonoma coast at Fort Ross in 1812. They had a tenuous agreement with the Spanish and Mexican governments to conduct their otter hunting in the northern region of California at which they were successful until they left in 1841. One of their lasting cultural impressions upon the landscape was the adoption of Russian names (i.e., the Russian River, Sebastopol) from some environmental features and settlements, and the adoption of some Russian terms by the local Pomo "tribelets" with whom they conducted business.

In addition, by the beginning of the Mexican rancho era American fur trappers or "mountain men" like Jedediah Smith had made it to California (1826) via the southern desert route and made the first recorded crossing of the Sierra Nevada. These American fur trappers became a constant visiting group during the Mexican rancho era and later become part of the great struggle for Alta California during the Mexican-American war.

The foreigners that did become significant figures on the landscape came in search of land. Such people as Peter Lassen and John Sutter became citizens of the Mexican Republic through the process of acculturation that then allowed them to petition for ranchos. By the 1840s, the government had granted 20 percent of all ranchos to foreign immigrants. These foreigners, along with increasingly dissatisfied Californios, became part of the home rule movement, which took a strange turn to American occupation via the Bear Flag revolt and the Mexican-American War.

In summary, the Mexican rancho represented a systematic attempt to impress on the land a set of economic and cultural values. The rancho arrangements were largely agreeable with the environmental opportunities and constraints found in California. This era still typifies land and society in California today along its central and southern coasts. To witness it, take a trip to Santa Barbara for Fiesta Days in August and gaze on the eloquence and pageantry of the caballero don during the parade.

Bibliography

Beck, W. (1974). *Historical Atlas of California*. University of Oklahoma Press, Norman, OK.

Becker, R.H. (1969). *Designs on the Land: Disenos of California Ranchos*. Book Club of California, San Francisco, CA.

Becker, R.H. (1964). *Disenos of California Ranchos: Maps of Thirty-Seven Land Grants, 1822–1846*. Book Club of California, San Francisco, CA.

Beebe, R.M. and R.M. Senkewicz. (2006). *Testimonios: Early California through the Eyes of Women, 1815–1848*. Heyday Books, Berkeley, CA.

Booth, S. (2008). *California Geography*. Course taught at Sierra College. [online] http://geography.sierra.cc.ca.us/booth/California/cal—index.htm.

Calisphere (2008a). *California Cultures*. University of California [online] http://www.calisphere.universityofcalifornia.edu/calcultures/

Calisphere (2008b). A world of California Primary Sources. University of California [online] http://www.calisphere.universityofcalifornia.edu/

Cleland, R.G. (1944). *From Wilderness to Empire: A History of California, 1542–1900*. Alfred A. Knopf, New York, NY.

Cleland, R.G. (1951). *The Cattle on a Thousand Hills*. Huntington Library: San Marino, CA.

Cook, S.F. (1962). Expeditions to the Interior of California: Central Valley, 1820–1840. *University of California Anthropological Records* 2, no. 5. Berkeley.

Cowan, R.G. (1956). Ranchos of California: a list of Spanish concessions, 1775–1822, and Mexican grants, 1822–1846. Academy Library Guild, Fresno, CA.

Cox, C.J. (2008). *California Geography*. Course taught at Sierra College. [online] http://faculty.sierracollege.edu/ccox/california—geography/index.html.

Dana, R.H. Jr. (1840). *Two Years Before the Mast*. P.F. Collier and Son Corp., New York, NY.

Davis, C. and D. Igler (2002). *The Human Tradition in California*. Scholarly Resources, Wilmington, DEL.

DeWitt, H.A. (1999). *The Fragmented Dream: Multicultural California*. Kendall/Hunt Publishers.

DeWitt, H.A. (1999). *The California Dream*. 2nd edition. Kendall/Hunt Publishers.

Donley, M.W., Allan, S., Caro, P., and C.P. Patton (1979). *Atlas of California*. Pacific Book Center, Culver City, CA.

Durrenberger, R.W. and R.B. Johnson (1976). *California Patterns on the Land*. 5th edition, Mayfield Publishing Company, Mountain View, CA.

Dutschke, D. (1988). A history of American Indians in California. In: F*ive Views: An Ethnic Historic Site Survey for California*, pp. 3–55. California Department of Parks and Recreation, Office of Historic Preservation, Sacramento, CA.

Haas, L. (1996). *Conquests and Historical Identities in California, 1769–1936*. University of California Press, Berkeley, CA.

Hackel, S. (1998) Land, labor, and production: the colonial economy of Spanish and Mexican California. In *Contested Eden: California Before the Gold Rush*, edited by R.A. Guttiérrez and R.J. Orsi, pp. 111–146. University of California Press, Berkeley, CA.

Hayes, D. (2007). *Historical Atlas of California*. University of California Press, Berkeley, CA.

Heizer, R.F. and A.J. Almquist (1971). *The Other Californians: Prejudice and Discrimination under Spain, Mexico, and the United States to 1920*. University of California Press, Berkeley, CA.

Hornbeck, D. and M. Tucey (1975). Agriculture in Hispanic California. *California Geographer* 15: 52–59.

Hornbeck, D. (1978). Land Tenure and Rancho Expansion in Alta California. *Journal of Historical Geography* 4(4): 371–390.

Hornbeck, D. (1983). *California Patterns: A Geographical and Historical Atlas*. Mayfield Publishing Company, Mountain View, CA.

Hutchinson, C.A. (1969). *Frontier Settlement in Mexican California: The Hijar-Padre's Colony and Its Origins, 1769–1835*. Yale University Press, New Haven, CN.

Marinacci, B. and R. Marinacci. 1980. *California's Spanish Place-Names: What They Mean and How They Got There*. Presidio Press, San Rafael.

Michaelson, J. (2008). *Geography of California*. Course at UC Santa Barbara, Dept. of Geography. [online] http://www.geog.ucsb.edu/~joel/g148—f08/.

Miller, C.S. and Hyslop, R.S. (1983). *California: The Geography of Diversity*. Mayfield Publishing Company, Mountain View, CA.

Monroy, D. (1998). The creation and re-creation of Californio society. In *Contested Eden: California Before the Gold Rush*, edited by R.A. Guttiérrez and R.J. Orsi, pp. 173–195. University of California Press, Berkeley, CA.

Monroy, D. (1993). *Thrown Among Strangers: The Making of Mexican Culture in Frontier California*. University of California Press, Berkeley, CA.

Pitt, L. (1970). *The Decline of the Californios: A Social History of the Spanish-Speaking Californians, 1846–1890*. University of California Press, Berkeley, CA.

Pitti, J., Castaneda, A., and C. Cortes (1988). A history of Mexican Americans in California. In: *Five Views: An Ethnic Historic Site Survey for California*, pp. 207–264. California Department of Parks and Recreation, Office of Historic Preservation, Sacramento, CA.

Rawls, J.J. and W. Bean (2008). *California: An Interpretative History*. 9th Edition, McGraw-Hill Publishing Co., New York, NY.

Reese, R.W. (1969). *A Brief History of Old Monterey*. City of Monterey Planning Commision, Monterey, CA.

Rice, R., Bullough, W., and R. Orsi (2001). *The Elusive Eden: A New History of California*. 3rd edition, McGraw-Hill.

Robinson, W.W. (1979).Land in California: the story of mission lands, ranchos, squatters, mining claims, railroad grants, land scrip [and] homesteads. University of California Press, Berkeley, CA.

Salvatore, R.D. (1991). Modes of Labor Control in Cattle-Ranching Economies: California, Southern Brazil, and Argentina, 1820–1860. *The Journal of Economic History*, 51(2): 441–451.

Starr, K. (2005). *California: A History*. The Modern Library, New York, NY.

State Land Commission. 1983. *Grants of Land in California Made by Spanish or Mexican Authorities*. Sacramento.

Ward, G.C. (1996). *The West*. Little, Brown and Company, Boston, MA.

Watkins, T.H. (1983). *California: An Illustrated History*. American West Publishing Company, New York, NY.

CHAPTER 8

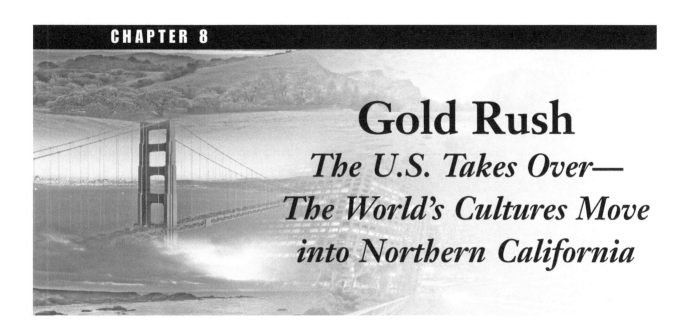

Gold Rush
The U.S. Takes Over—The World's Cultures Move into Northern California

Key terms

1849	Guandong province	Preemption Act of 1841
African Americans	Gum Shan	President Polk
Alcalde	Homestead Act of 1862	Prior appropriation
American period	Hydraulic mining	Property rights
Andres Pico	Isthmus of Panama	Public Land Survey System
Argonauts	John C. Frémont	Quarter section (160 acres)
Bear Flag Republic	John Sutter	Rio Buenaventura
Bear River	José Castro	Sacramento
California Land Act of 1851	Kanakas	San Francisco
California trail	Latinos	Section
Cape Horn	Lode mining	Slickens
Chinese	Manifest Destiny	Stanislaus River
Common law	Mariano Vallejo	Stockton
Dryland farm	Mexican-American War	Survey meridians
Feather River	Mining laws	Texas Game
Foreign Miners License Tax of 1850	Mokelumne River	Township
	Native Americans	Treaty of Cahuenga
Foreign Miners License Tax of 1852	Nativist	Treaty of Guadalupe Hidalgo
	Natural resource law	Trinity River
Free mining	Northwest Passage	Tuolumne River
French	Opium wars	WASP
Germans	Oregon trail	Wheat farming
Gold	Pio Pico	Yuba River
Gold rush	Placer	

125

Introduction

AMERICAN OCCUPATION

The **American period** that came into play during this chapter's timeframe brought to the fore the concept of landscape co-evolution. The interplay among the sectors of values, knowledge, organization, environment and technology to form the path in how the geography of land will develop (where, when, and why) over time. The various sectors become clearer in the American period and present a stronger role in the formation of California's modern layout. Early in the Mexican rancho era the values and organization were encased in the ideas of easy living and rule within a quasi-feudal society. As such, the settlers did not push technology or broach further knowledge of the territory's potential, while environmentally the value of cattle changed the landscape into one of exotic annual grasses used for hide and tallow production purposes. What we witness during the American period is how quickly a change in any one of the sectors can change the geographic layout of the entire landscape and give birth to a new development path.

8.1 BEAR FLAGGERS, THE TEXAS GAME, AND A GAME CHANGING CHARACTER—JOHN FRÉMONT

The Mexican government had always feared the influx of Americans expanding their way westward. In 1821, what was know as the continental U.S. became split in a three-way ownership. The U.S. consisted of the East Coast and parts of the Midwest, Britain owned the Northwest Territories (Oregon, Washington and Idaho), and Mexico owned the West/Southwest. The Mexican government's desire to further the permanent settlement of Alta California allowed them to let in immigrants from the U.S. as long as they converted to Catholicism, learned to speak and write in Spanish, and gave their allegiance to Mexico. This then allowed these people to petition for ranchos. The fact is that thousands of American squatters came and didn't follow the rules, and they literally became a fringe society of traders, ranchers and farmers within Alta California. This process was also occurring in the Mexican territory of Tejas (Texas).

Other pressures associated with American people moving into the West included the development of the **Oregon Trail**, which allowed Protestant missionaries into the Pacific Northwest without British permission. Then there was the exodus of the American religious group, the Mormons, who left Illinois and detoured off the Oregon Trail to found their colony in the Mexican territory of Utah around the Great Salt Lake. In all, we can sum up the desire to move West by noting the availability of cheap land, the motivation to convert Native Americans to Protestant Christianity, the hope of avoiding religious persecution, and the persistent fur trade.

8.1.1 TEXAS GAME, A PRECURSOR OF THINGS TO COME

By 1836 the Mexican settlement scheme in Texas was not going well. Americans, who had willingly become Mexican citizens to receive cheap land reneged on their allegiance to Mexico after feeling neglected by Mexico City. Texas tried for independence and wrote a constitution for a republic. After several skirmishes with the Mexican army (remember the Alamo?), Texas was allowed to become a republic (which included parts of Oklahoma and Colorado as well). Mexico realized from this incident that New Mexico, Arizona and Alta California were vulnerable. Interestingly, the leaders of the new Texas Republic wanted the U.S. government to take them over and to declare them a U.S. territory to make way for statehood, but the U.S. government thought this would cause problems with its neighbor Mexico and that it would lead to an imbalance in the non-slave vs. slave state negotiations in Congress. Thus Texas remained an independent republic until 1845.

8.1.2 JOHN CHARLES FRÉMONT, MANIFEST DESTINY, AND THE BEAR FLAG REVOLT

John C. Frémont (Figure 8.1), "The Pathfinder," was an ambitious captain in the U.S. Army Corp of Topographical Engineers. He and his men had been out to Alta California three times during the early to

FIGURE 8.1 *John C. Fremont. (Courtesy of The Bancroft Library)*

mid 1840s (see page 11 in *California Atlas*). During this time American explorers still believed in the mysterious **Rio Buenaventura**—the "Great River of the West" or the **Northwest Passage** (Strait of Ainan to the Spaniards)—a supposed large waterway flowing from the Rocky Mountains to the Pacific Ocean and thus an easy route to the Orient for trade. This represented a further rehashing of a centuries old myth sought by the Spanish and British. In fact, the American West had been home to many landscape myths of marvelous places:

- Cibola—a land rich in gold, silver, etc.
- El Dorado—seven great cities, prosperous because of gold.
- Quivira—a city of gold with great hoards of game.
- Land of the Madocians—a fair-skinned race said to be descended from a Welsh prince whose "supposed" discovery of America in 1170 gave the British evidence that they had more claim to the continent than the Spanish!

All of these myths of place represented speculation of the highest magnitude. The Rio Buenaventura, however, remained on popular American made maps (usually shown flowing into San Francisco Bay) up until the 1840s. Frémont had been caught up in this myth, and thus in 1843 after completing an expedition to the British Oregon territories he disobeyed orders and marched his men into Mexican Alta California rather than return along the Oregon Trail. Frémont's rationale was to once and for all put an end to the speculation about this great river. His team's survey of the Sierra Nevada and the Sacramento River put the idea to rest.

More important, geographically speaking Frémont was tied to the idea of **Manifest Destiny**, which was the nineteenth-century doctrine that the U.S. had the right and the duty to expand throughout the North American continent, "from sea to shining sea." Several earlier presidents had strongly promoted this idea, but none were as passionate about it as **President Polk** (1845–1849). Interestingly, Frémont's wife was the daughter of a U.S. senator who was one of Polk's biggest backers for supporting "Manifest Destiny."

Therefore, in 1846 Frémont led another group back into Mexican Alta California (over the Sierra Nevada, also becoming the first White person to discover Lake Tahoe) on possible secret orders from the U.S. government to cause disruption in the region and to rile the acculturated American rancho owners (remember the problems noted earlier that occurred in Texas? Reneging on alliances is known as the "**Texas game**"). Alta California commandante **José Castro**, based in Monterey, allowed Frémont's troop to winter in the San Joaquin Valley, but Frémont moved into the Salinas area and started causing trouble. They were ordered to leave after a standoff with Castro and went northward to Klamath Lake in Oregon. Frémont was aware of the start of the Mexican-American War down along the Texas border near the New Mexico territory by President Polk via letters from his wife and father-in-law that he received through the U.S. consul, which had been established earlier, in Monterey.

As the last chapter discussed, the Californios were having their own internal troubles over power and whether Mexico City really cared about them and their territory. During Frémont's arrival into Alta California, Commandante Castro was trying to overthrow Governor **Pio Pico** (Figure 8.2). However, they quickly made amends once they figured out what Frémont was up to. Frémont didn't make it to Klamath Lake but instead encouraged more foreigner-run ranchos in the Sacramento Valley to rebel against Mexico City and the

128 CHAPTER 8 *Gold Rush*

FIGURE 8.2 *Pio Pico: the last Mexican Alta California governor. (Courtesy of the California Department of Parks & Recreation)*

FIGURE 8.3 *General Mariano Vallejo and his daughters. (Courtesy of The Bancroft Library)*

Alta California government. He headed over to Sonoma and took over the home of General **Mariano Vallejo** (Figure 8.3), the commander of the presidio de Sonoma (a defense against the Russians). Vallejo surrendered and actually throws them a party as he was unhappy with Mexico City's governance of Alta California anyway. In fact, he secretly favored the annexation of Alta California to the U.S. Frémont and other Californio foreigners declared California a republic, the **Bear Flag Republic**, in June 1846. Oddly enough, the Bear Flag (which is now California's state flag, with the red star acknowledging the Texas Republic) was designed and painted on cloth by one of Frémont's men, a first cousin to Mrs. Abraham Lincoln. Several days later Castro sent a military force north to stop the rebellion, which ended up with the defeat of the Californios in the minor battle of Olompali (northern Marin County), the only force of arms witnessed during the less than 30 day existence of the Bear Flag Republic.

8.1.3 THE MEXICAN-AMERICAN WAR REACHES CALIFORNIA

The **Mexican-American War** of 1845–48 had two very different rhetorical approaches. In one corner we had a country that believed in its manifest destiny of being composed of lands from "sea to shining sea." In the other corner, however, we had an even younger country that viewed the aggression of its neighbor to the north in defensive terms. The following list provides the variety of titles Mexican historians have given, to what U.S. historians call the Mexican-American War, as they appear in contemporary Mexican high school textbooks:

- *La Intervención Norteamericana* ("The North American Intervention")
- *La Invasión Estadounidense* ("The United States Invasion")
- *La Guerra de Defensa* ("The Defensive War")

Declaration of war finally reached California officially through the arrival of U.S. Navy Commodore Sloat in Monterey and then onto Presidio San Francisco and the settlement of Yerba Buena (San Francisco). The Californios loyal to Mexico retreated south to Los Angeles without resisting. After being a independent republic for only 25 days, Mexican Alta California was declared a U.S. territory under military occupation in July 1846. Frémont's Bear Flaggers entered the army as the First California Battalion, and Frémont took over the San Francisco presidio (Figure 8.4). Newly arrived Commodore Stockton replaced Commodore Sloat, who then took Frémont's troop south to occupy the San Diego presidio and then the pueblo of Los Angeles.

Meanwhile Castro and Pico raised a troop of Californios (mostly rancho dons) in Los Angeles to confront the U.S. aggression. Before Frémont's troops arrived in Los Angeles, the Californios headed for the mountains (the Hollywood hills area) to avoid contact with the Yankees and to wait for Castro and Pico who had gone to Mexico to try to get support from the Mexican government. After securing Los Angeles and leaving some troops behind, Frémont returned north to Monterey in August 1846. The Californios came down from the mountains and chased out the California Battalion troops, who immediately sought support from Commodore Stockton anchored down at San Pedro harbor, but Stockton's naval troops were defeated. Frémont returned to Los Angeles in January 1847 and retook the pueblo, as the Californio troop had moved out again and was heading south towards San Diego under the direction of **Andres Pico**. The Californios met U.S. General Kearny coming west to secure Mexican Alta California after recently leaving victories over the Mexican army in New Mexico. Pico's troop of Californios badly defeated the U.S. Army at the battle of San Pascual (northeast of San Diego), but the Californios were unable to press the advantage and instead retreated to Los Angeles only to surrender to Frémont. The **Treaty of Cahuenga** declared the end of the Alta California hostilities in January 1847.

FIGURE 8.4 *San Francisco, 1846–47, after American occupation but before the Gold Rush. Note: Montgomery St. is in the foreground on the shore. It is currently five blocks from the in-filled shoreline, which represents the Embarcadero. (Courtesy of The Bancroft Library)*

In summary, the U.S. government did not send American immigrants out to take foreign lands; immigrants brought the U.S. with them and in doing so forced the government's hand in wresting away the Mexican and in addition the British Northwest Territories—each time the "Texas game" played out well to a U.S. imperialist victory. Under the **Treaty of Guadalupe Hidalgo** the U.S. agreed to pay Mexico for Alta California, Arizona and New Mexico; guaranteed the rights of Californios as U.S. citizens, upheld pueblo rights (administrative and ownership rights granted to pueblo lands under the Law of the Indies) and administration; and agreed to preliminarily honor the rancho ownership system until tenure could be fully researched within U.S. land ownership laws. The continental area of the United States—"sea to shining sea"—was now essentially complete.

8.2 Before the Gold Rush—California Before the Dawn

In summary, the geographical character of California under the rule of the United States can be described by the following: The population was around 10,000 (not including the remaining Native Americans). Of this number the language was still mainly Spanish, the U.S. territorial capital was at Monterey, most of the settlement and activity was still in the old mission-presidio-pueblo belt along the coast and the main commerce was still hide and tallow via the vast tracts of cattle ranchos. Then in early 1848, at a sawmill being built on the American River (Coloma, CA) by Central Valley rancho owner **John Sutter** (Figure 8.5), his construction manager James Marshall discovered **gold** in the river's gravel deposits. While they initially tried to keep the find secret it soon leaked out to San Francisco and then to the rest of the world (see page 12 in *California Atlas*).

To understand **Gold Rush** fanaticism under U.S. occupation, we need to link it to the value system of the American culture at this time, which was very different from the Spanish-Mexican culture-value system. The value system of the Spanish-Mexican culture had been linked to vast quantities of easily obtainable pre-processed gold and silver, won over from the Ameriindian miners they took it from earlier in Mexico, South America and New Mexico. Therefore it is not surprising that the Spanish-Mexican Californios had ignored earlier discoveries of gold in the Colorado River, in an area near San Diego, and in the San Fernando Valley near Los Angeles as insignificant or not worth the effort, in relation to what they had become used to in their larger society since the early sixteenth century in the New World. Basically they were unfazed by these earlier discoveries of gold in California; it did not seem worth their time and energy to mine or look for more deposits. The Americans on the other hand were of a completely different character and value system. These were a people based on independent thinking White Anglo-Saxon Protestant (**WASP**) Puritan hard-working stock with no knowledge or often times having a limited connection to riches, but who were easily lulled into the American dream of making it big if possible under the banner of "life, liberty and the pursuit of happiness." Having that happiness generated through monetary means fit in with

FIGURE 8.5 *John Sutter, early foreign schemer and founder of Sacramento, which was originally his rancho, Nuevo Helvetica. (Courtesy of The Bancroft Library)*

the American psyche, no matter how empty, crass and cultureless it can be as an end in itself.

Therefore the discovery of gold touched off a firestorm of public relations and boosterism among the Americans in California that spread across the young nation (Figure 8.6). These were riches that a lone miner in the early part of the Gold Rush could literally pick out of the riverbed gravels with bare hands (Figure 8.7). The initial rush was limited to those in Northern California, based in San Francisco and Monterey, and then to those south in Santa Barbara, Los Angeles and San Diego. Thus the initial rush was limited to people

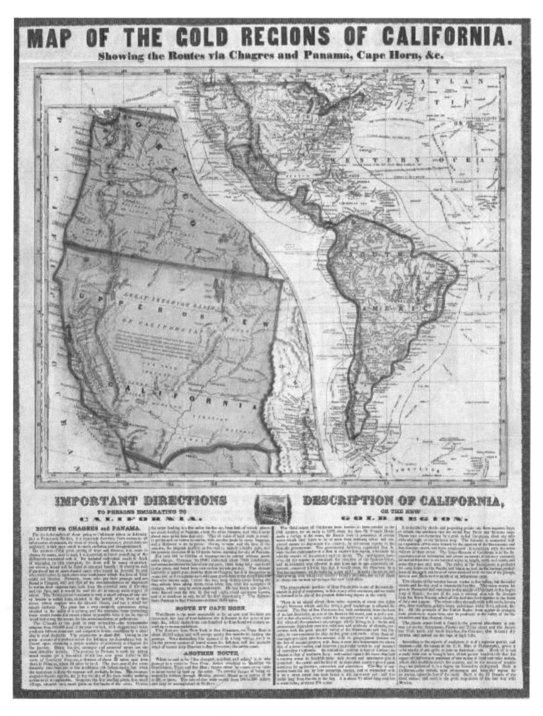

FIGURE 8.6 *Map of the gold regions of California, showing the routes via ship that was published as an insert in East Coast newspapers in 1849. (Courtesy of The Bancroft Library)*

FIGURE 8.7 *Panning for gold on the Mokelumne River. (Courtesy of The Bancroft Library)*

already in California or able to get there quickly. Those cultural groups included Californians, Oregonians, Mexicans, Russians, South Americans (Peruvians and Chileans) and Hawaiians. The pickings were relatively rich and easy to get in the form of placer deposits. **Placer** is a Spanish word meaning "alluvial sands," and it represents areas along river bends where deposition occurs (i.e., point bars) and the alluvium is rich in accumulated valuable minerals, i.e., in this case gold. The large amount of gold found in these Sierran rivers (see Chapter 2 for geological details) made people abandon much of the rest of California to the point of nearly deserting established pueblos and ranchos (see page 13 in *California Atlas*). However, by late 1848 word reached the East Coast of the "fabulous gold diggings" in California. Every newspaper announced it (Figure 8.6), as did U.S. President Polk with fanfare in a joint session to Congress … the Rush Is On! Immediately word made it to Europe and to the Orient—the magical lands of Queen Califa were truly lined with gold riches after all.

8.3 The Rush Is On: California the Distant Outpost

Everyone was promoting the Gold Rush in California on a global scale, but there is a simple fact about the region: It is a difficult place to reach. Despite the fact that California was now a U.S. territory, it was not well connected with the rest of the U.S. or the major population centers of the world. While explorers' accounts of the West existed via earlier reports by Frémont, Protestant missionaries going to Oregon, mountain men fur trappers, and the Mormon colony in Utah, there were still copious amounts of misinformation and ignorance. The bottom line was that California was still an isolated place with formidable natural barriers. There were, however, three fully developed ways to get to San Francisco: around **Cape Horn** (South America), across the **Isthmus of Panama** and overland via the Oregon and **California Trails** (Figure 8.8).

The voyage from Boston or New York around Cape Horn in South America was relatively safe but long (20,000 miles) and unpleasant (five to eight months at sea). If one could get on one of the newly designed clipper ships (Figure 8.9), the voyage could take place in three or four months, but usually they only carried expensive freight and passage was prohibitively expensive. Those Americans that took this route or the next ocean route to be mentioned largely came from the Eastern seaboard or east of the Appalachian Mountains.

A second option was to take a ship from Boston or New York to the port of Chagres on the Caribbean Sea side of the Isthmus of Panama. In Chagres the traveler boarded a small boat and went up the Chagres River until it became impassable, at which point he had to travel the rest of the way to the Pacific coast port of Panama City on donkeys and by foot. Once in Panama City the traveler gained passage on a ship to San Francisco. This route to California was hot and malaria-ridden with the danger of confronting unfriendly natives. And while the whole trip from Boston to San Francisco could take two to three months, it generally took longer. The problem was usually a shortage of small boats to go up the Chagres River and a sometimes an even longer stay at the port of Panama City waiting for a ship to go the rest of the way to San Francisco.

Finally, the overland routes via ox-pulled wagons were the most popular way to get to the gold fields, but they were dangerous. If a traveler could make it to the edge of the U.S. frontier at St. Joseph, Missouri on the Missouri River, then he could join or create a group that left by May to avoid the fall snowfalls in the Sierra Nevada. The Americans living west of the Appalachians and in the Deep South used the overland routes. Wagon

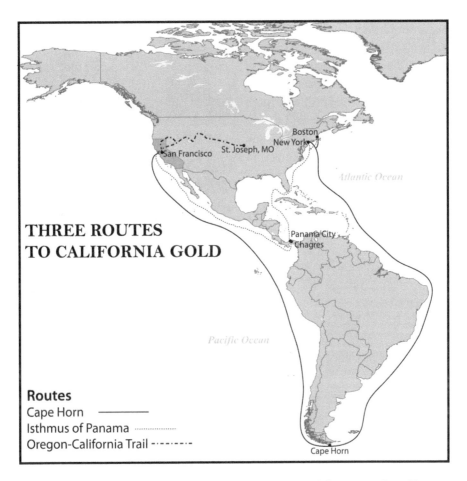

FIGURE 8.8 *Map displaying the three major routes to California: via Cape Horn, across the Isthmus of Panama and overland from St. Joseph, MO. (Courtesy Curtis Page)*

FIGURE 8.9 *New York handbill advertisement for clipper ship transportation to San Francisco.*

train parties followed the Oregon Trail first to Fort Laramie, Wyoming, which, at the peak, had some six hundred wagons passing through it per day. At South Pass, Wyoming or City of Rocks, Idaho parties would get onto the California Trail either by passing through the Mormon colony at the Great Salt Lake in Utah or by immediately dropping down into Nevada. In any case, once in Nevada there were two possible routes into California. One was the Lassen Trail, which dropped down through the far northern corner of California and came around Mt. Lassen (this was the trail developed by rancho owner Peter Lassen in 1840). The other choice cut through the Sierra Nevada along what is today's Interestate-80 route through Truckee and the Donner Pass down to John Sutter's fort on the Sacramento River (the trail first crossed by the John Bidwell party in 1841 and it was an original California Native American trading trail). Even though the overland routes became the most popular, they were difficult and dangerous, with precarious river crossings, large cholera outbreaks, hostile Native Americans, freezing weather and scurvy, to name a few major problems. The travel parties consisted of young inexperienced males (Figure 8.10; that could lead to big problems, i.e., the Donner party disaster) with overloaded wagons trying to make the roughly two thousand mile journey from the Missouri River to the Central Valley in four or five months, leaving no later than May to avoid getting stuck in the snows. Despite the risks, however, around six thousand wagons (approximately 25,000 people) came through these overland routes during 1849.

FIGURE 8.10   *Caricature of an independent gold hunter on his way to California. (Courtesy of The Bancroft Library)*

8.4   1849ERS—THE ARGONAUTS AND RAPID GROWTH IN NORTHERN CALIFORNIA

The Gold Rush represents a group of people of exceptional character—largely single, young men (usually under the age of twenty-five). They came to be known as "**Argonauts**." Suggestive of the mythic Greek tale of *Jason and the Argonauts* who venture forth to claim the golden fleece, the name suitably conveys the spirit that surrounded this mass migration of young men. The total that reached California in **1849** was approximately 40,000. Broken down: 15,000 came by ocean and around 25,000 traveled overland. The importance of this mass migration is the cultural diversity it represented. Some of the largest foreign-born groups to arrive during the Rush included Mexicans, English, Germans, French, Irish, and Chinese. By 1860 the Chinese were the largest foreign-born group (around 40,000). In what seemed like an overnight event, the territory of California reached a population of 100,000 by 1850, the magic number for requesting statehood, and 200,000 by 1852.

It was champagne days in **San Francisco**, where near instant wealth and irrational exuberance dusted social life and customs with gold. The forgettable village of Yerba Buena had been renamed San Francisco (linked to the name of the local mission and the presidio), but when the Gold Rush came it exploded as the main port of entry to the goldfields (Figure 8.11). The captains and crews of many ships that came into San Francisco Bay abandoned their ships, wanting to try their luck in the rich pickings. Left beached on the San Francisco waterfront known as the Embarcadero, these ships eventually became used as buildings (i.e., warehouses, hotels, saloons) or were dismantled to provide instant building supplies to the growing town (Figure 8.12). Of course with such rapid expansion there was little spatial urban planning (unlike the rules in place during the establishment of Spanish pueblos), and the people who populated the city, being "punch drunk" literally and figuratively on the riches of the gold rush, tended towards civic negligence. So much so in fact, that the growing city burned down several times during 1849 and 1850, only to be rapidly rebuilt each time. Who cared when every day felt like summer and the living was easy? More than seven hundred ships had arrived by late 1849 to provide the resources to rebuild.

Once the intrepid miners left the ships, smaller boats took them up through the Carquinez Straits into the Sacramento-San Joaquin Delta to either **Sacramento** or **Stockton**—the inland ports. In fact, early on entrepreneurs brought river steamboats around Cape Horn to start a very lucrative passenger service in late 1849. During the early months of the Gold Rush the focus was on Sacramento and its access to the northern mining areas on the **Trinity**, **Feather**, **Bear**, and **Yuba** rivers as well as in the Grass Valley and Nevada City area. Soon though, the development of the inland port of Stockton situated on the San Joaquin River allowed miners to supply themselves to work the southern mines located on the **Mokelumne**, **Stanislaus** and **Tuolumne**

FIGURE 8.11 *San Francisco Bay with clipper ships in 1851. View looking down Market Street out toward Yerba Buena Island (artficially created Treasure Island would be later be added to the left in 1936/37) and the Berkeley Hills. (top: Library of Congress; bottom: Bancroft Library)*

136 **CHAPTER 8** *Gold Rush*

FIGURE 8.12 *San Francisco at the waterfront in the 1850s. Note that several of the buildings are beached converted ships. (Courtesy of The Bancroft Library)*

rivers as well as around Mariposa, Jamestown, and Columbia—the region known as the Mother Lode (see page 13 in *California Atlas*).

What is clear geographically is that the population center of California made a rapid and significant shift northward from San Diego and Los Angeles. Until the end of the nineteenth century the domination of California's affairs rested in the northern part of the state in the hands of San Francisco, Sacramento and Stockton, as well as in the widely populated Sierra Foothills from Oroville to Sonora.

8.4.1 EARLY MINING PATTERNS

In the early months miners told extravagant stories about the amount of gold they could obtain and the ease with which they could obtain it. These stories and tall tales (i.e., 273 pounds of gold taken from the Feather River in seven days) maintained many miners' hopes that the next day of working the claims would be the big payout and a return home. In fact, many of these Argonauts had only planned on staying a year or maybe two in the California goldfields, just long enough to make their fortunes and then to return home to settle down and buy a farm. The classical description of an Argonaut is the stereotypical lone miner (Figure 8.13) with simple placer tools (i.e., pick, shovel and washing pan). This type of person was only viable for a short period in 1848; however, even in the Mother Lode country (the area with large veins of gold) by 1849 working a claim soon required teamwork, and eventually very large group efforts.

As mentioned earlier, the gold had been laid down in the modern alluvial deposits called placers. The min-

FIGURE 8.13 *The lone miner panning for gold soon became obsolete.*

FIGURE 8.14 *Group effort mining painted in 1851. See images on page 12 of the California Atlas as well. (Courtesy of The Bancroft Library)*

ers would then shovel this alluvium into a large shallow pan with water. Panning or gently swirling the pan washed out the lighter clays, silts and sand, which then left behind the heavier gold nuggets. As soon as the miners had panned out these easy surface gold deposits, the real work of digging into these large placer deposits began—thus the need for group efforts (Figure 8.14). Miners began using cradles or rocker boxes to separate larger quantities of sediment. Two and three miners would pick and shovel sediment into the rocker worked by another miner. Eventually placer mining technology moved on to long sluice boxes known as long toms, which required a lot more work and required six to eight men to operate. Unfortunately, most prospectors toiled long hours to merely maintain their basic expenses in the goldfields.

8.5 CALIFORNIA LIFE IN THE 1850s

In this period California, as a U.S. territory, was rapidly transitioning to statehood as it reached the magic number of 100,000 people. Another way of putting it is: "Instant statehood: Just add gold crazed people!" During this time the historical Spanish and Mexican government capital of Monterey was running the territorial government. In addition the **alcalde** system of running a settlement (see Chapter 6 for review) was still the only system known, and thus carried on until statehood. However, with so many young men immigrating into California from all over the world, the flimsy territorial bureaucratic system was virtually ineffective to control the situation. Political affairs were confused and chaotic at the territory level, while all civil and judicial life was being handled by elected alcaldes at the local level. Therefore a constitutional convention was conducted in Monterey in 1849 to skip being a territory in favor of statehood. Four important items came from this convention. First, it was a unanimous vote that California be a slave free state. Second, Spanish common law was adopted that upheld separate ownership of property by a married woman. Third, they made provisions to build and fund public schools and a university. Fourth, the legislators established the state's boundary, with the eastern boundary as the Sierra Nevada. Finally, after being sent to Congress, California became a U.S. state in 1850 with 27 defined counties (Figure 8.15), but its political affairs remained unsettled as they could not settle on a permanent state capitol. It roved around Northern California residing in Monterey for some months, then San Jose,

138 CHAPTER 8 Gold Rush

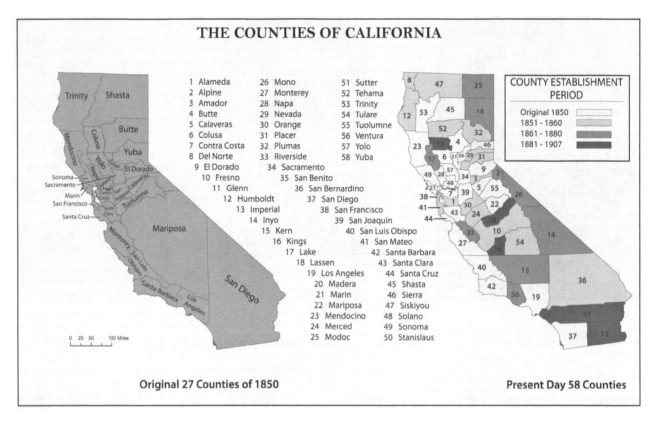

FIGURE 8.15 *Original twenty-seven county geographical configuration after obtaining statehood in 1850. The state now has fifty-eight counties. The smaller counties represent the early population distribution which was concentrated in the gold regions and San Francisco Bay area. (Courtesy Curtis Page)*

Benicia, Sacramento and back to Monterey. Finally Sacramento agreed to build the statehouse and essentially bought the capitol for itself in 1854. To the highest bidder goes the spoils . . . or the problems.

There were some precursors for things to come: Governmental control remained weak to non-existent on most important issues, such as land title, natural resources, planning, etc. Essentially, the miners ignored the government and established their own codes governing claims, natural resource use, courts of law, business, etc. (a form of the alcalde system persisted). The problem was that across these makeshift mining camps in the Sierra Nevada foothills (Figure 8.17) anarchy and vigilantism were widespread. Mining disappointment coupled with a large, young testosterone-infused male population represented as a rapidly changing cultural population, led to ugly examples of racism and nativism.

8.5.1 CULTURAL DIVERSITY AND ETHNIC CONFLICT

In terms of emigration potential and timing, the California Gold Rush occurred at the right time for many countries and regions under strife. Eager gold seekers from outside the U.S. coincided with (push factors) conflict (i.e., the European revolutions of 1848), economic hardship (i.e., the Chinese-British Opium Wars), famine (i.e., the Irish potato famine) and repression in Europe (i.e., France and Germany), Chile, and China. California's potential riches (pull factor) gave the young men leaving these areas hope for a better life for them and their families back home. The result of this was that California became the most ethnically diverse region in the United States (see Figure 8.16 for a regional example). Unfortunately, unsuccessful gold

FIGURE 8.16 *Mining life in the Sierra Nevada. (Courtesy of The Bancroft Library)*

seekers vented their frustration on ethnic 'scapegoats', making the region famous for its brutal ethnic conflicts. The following is just a sampling of how various ethnic groups were treated during this dynamic period:

- **Native Americans**—They were forced to abandon their traditional lands in the Sierra Nevada foothills and the Klamath. They fell victim to genocidal campaigns and some forms of forced work slavery in the mining areas. In order to open up the mining lands, legalized and subsidized murder on a mass scale occurred as sanctioned by local, state and the federal governments. In some situations the Native Americans fought back against the miners (Figure 8.18) only to be doubly attacked by the authorities later. Therefore, their overall population fell from an estimated ~150,000 in 1846 to ~30,000 by 1870.

- **Latinos**—This group of people covers the Californios, Mexicans and other Hispanics from the Americas. These miners endured violent opposition as well as discriminatory taxes. The Californios suffered the destruction of their ranchos as squatters came in and took them over; even John Sutter suffered as he lost Fort Sutter and the lands of Sacramento. First, there was the passage of the **Foreign Miners License Tax of 1850** (an impossible $20/month) directly aimed at all Latinos (derogatorily called "greasers"), but especially at Californios, Mexicans and Chileans who were very successful in the gold country since they were the first to arrive. In the end, many Latinos refused to pay the miner's tax and frustrated by the hostile racial oppression they moved into the San Joaquin Valley to work on small cattle ranches or went back down to Southern California, which at this point had become a dead-end backwater compared to the rapidly developing north. Their gold country population was estimated at 15,000 in 1850, but number decreased to 5,000 after the passage of the tax. Next, Congress passed the **California Land Act of 1851**,

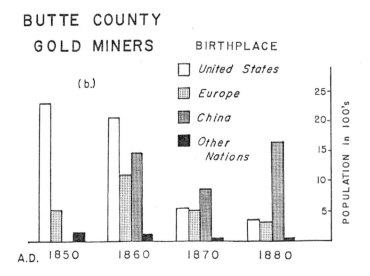

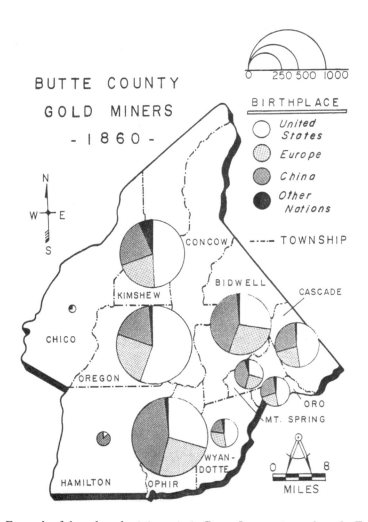

FIGURE 8.17 A and B *Example of the cultural mining mix in Butte County situated on the Feather River mining district: (a) Ethnic mix 1850–1880; (b) Ethnic mix within the Butte County administrative areas, 1860.*

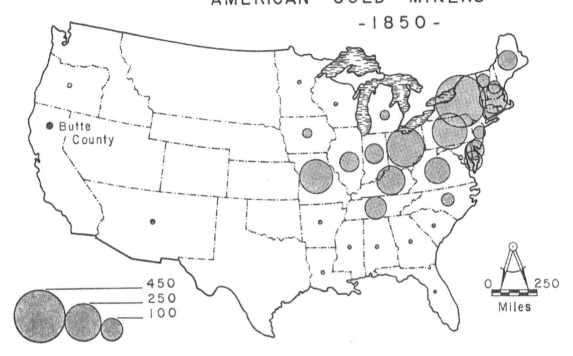
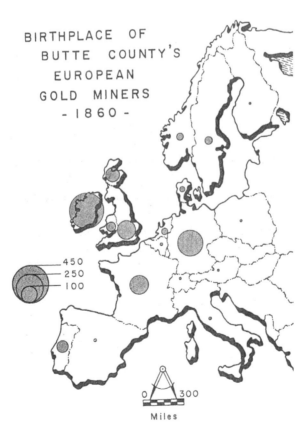

FIGURE 8.17 C and D *(c) U.S. origination of miners coming to Butte County during the Gold Rush; and (d) European origination of miners coming to Butte County during the Gold Rush. (Courtesy James Monaco)*

FIGURE 8.18 *Native Americans attacking placer miners.*

which attempted to sort out the Californio rancho land claims versus immigrants from other parts of the U.S. who contested the rancho owners claims. The U.S. technocratic approach to land tenure via title deed was legally superior to the crudely drawn rancho diseño maps deemed legal under the era of the Mexican Republic government. The net result was a loss of rancho lands by the Californio dons of Latino descent. Some of these dons fought the U.S. government's land claims commission for fifteen years and became penniless paupers.

- **French and Germans** (Figure 8.19)—They were also compelled to pay extra fees as foreign miners. **Nativist** Americans did not appreciate them as they tried to support the Latinos in their fight against oppression and taxation. Despite the hostility, the French, for example, numbered 32,000 miners by 1853. In addition, many place names found today in California reflect their mining camps: French Camp, French Corral, French Gulch, etc.

- **Hawaiians**—This group, commonly called **Kanakas**, had been in California before the Gold Rush as part of ship crews during the Mexican era hide and tallow trade. Many had also decided to stay in California and work in for the ranchos directly; John Sutter had used them exclusively on his rancho. They actively worked their own mining claims, many up in the Klamath mountains region along the Trinity River. Some California place names still reflect their early presence on the landscape: Kanaka Creek, Kanaka Bar, etc.

- **Chinese**—They came to seek their fortune in the fabled land they called **Gum Shan** ("gold mountain"). Many of the Chinese miners originated in the **Guandong province** on China's south coast. Young men left after the humiliation and devastation brought on by the British-instigated **Opium Wars**. They composed the largest foreign group to reside in California; for example, 25 percent of the miners were Chinese by 1870. Racial hostility towards this hardworking group of young men boiled over into a second **Foreign Miners License Tax of 1852**, which they obediently paid to keep working their claims (Figure 8.20). In 1854, state testimony laws were amended to prohibit Chinese persons from testifying for or against a White person. From this point on, these young Chinese men tried to avoid open confrontation or direct competition with White miners. The often reworked older, supposedly, exhausted claims rather than openly compete with other miners for new claims.

- **African Americans**—As a group they were a small minority, but they were bounded by a set of unfair laws and practices that made them unique. On one hand, California statehood was admitted as a slave free state, which afforded African Americans freedom. On the other hand, harsh federal fugitive slave laws bound California's government. As an example, if a slave escaped the South and made it to California, the state government had to send the slave back to his rightful owner. However, if a slave owner came to California, bringing his slaves, to

FIGURE 8.19 *An advertisement for German miners. (Courtesy of The Bancroft Library)*

FIGURE 8.20 *Chinese miners in the Sierra Nevada. (Courtesy of The Bancroft Library)*

work a mining claim, but then he wanted to return to his state of origin the slaves were freed and did not have to leave California because California was a slave free state. Once here, however, there were additional laws against African Americans. Only "free White males" could hold membership in the state militia. In addition, state testimony laws worked against Native Americans and African-Americans, in that neither could defend themselves or be witnesses in a court of law.

8.5.2 URBAN GROWTH VS. GOLD RUSH COMMUNITIES

Immigrants from other parts of the U.S. and from Europe settled vast amounts of California land between 1850 and 1890. Because of the unique social, economic and physical geographic problems (i.e., terrain barriers, climate), settlement in California occurred in different areas, at different times and therefore for different reasons. This started with urban growth, which occurred in northern California with the rapid, uncontrolled growth of San Francisco, Sacramento, and Stockton. It led to major fires being a common occurrence as the construction was predominantly wooden. In addition, California's Mediterranean-type climate denoted by long dry summers made for tinderbox conditions among these wooden buildings. The central city governments were rarely in control, which led to poorly equipped volunteer fire departments that relied on inadequate and unreliable water supplies (i.e., a foreshadowing of things to come). In summary, San Francisco had seven major fires between 1849 and 1851, and Sacramento largely burnt down in 1852. If it was not for the large influx of gold and capital to these cities, they would have been difficult to rebuild. This was not the case though as each time they were resurrected only to expand because of the constant flow of new incoming miners and miners returning from the gold fields.

The other major settlement region was the gold mining areas of the Sierra Nevada foothills and the Klamath Mountains. The focus area for many miners was the area of the Sierra Nevada bordering the Sacramento and San Joaquin valleys and the towns founded to service the mining regions: Shasta, Oroville, Marysville, Yuba City, Auburn, Folsom, Sonora, Mariposa, and Columbia, just to name a few. Like the rapid, unplanned development in the larger cities, these mining towns were even more unplanned and chaotic. The construction was generally ramshackle, board and baton wooden construction. This heavy timber resource use for mining and buildings severely impacted the surrounding forest areas (i.e., clearcuts). Any type of basic service (i.e., hotel, saloon, barber, hardware, laundry, etc.) was limited or very expensive. All this led to an economic boom for merchants and suppliers, not the miners.

8.6 U.S. WESTERN SETTLEMENT PROCEDURES—HOMESTEADIN'

California may have become a U.S. state, but it had no systematic plan for settlement and was not well connected to the rest of the country, let alone with the hub of economic and political power on the East Coast. While many of the immigrants came for gold prospecting, others came to provide retail services to the miners: prostitution, mercantile, hardware, hotels, saloons, etc. By the mid-1850s, many miners discovered that instant wealth was not around the next river bend so they left the gold fields and moved into the Sacramento Valley and the Delta region to become farmers in the favorable soils and climate. Families then started to come out to California to become farmers, who would then help supply food to the masses in the urban areas and the Sierra foothill mining camps. As early as 1841 under the **Preemption Act**, the U.S. government had allowed squatters on public lands to purchase up to 160 acres (65 hectares) to farm. There was no set systematic way to draw up the tenure boundaries under this act so in the end it was largely disregarded and land squatting became the norm. In 1862, however, the U.S. government developed a system especially designed for the West (at this time considered from Ohio to the Pacific coast) to give out land under freehold title: the **Homestead Act**.

8.6.1 HOMESTEAD ACT OF 1862— THE GREAT GRID OF THE WEST

In many respects "lines conquer," and the California landscape is a testament to their power. Compasses and

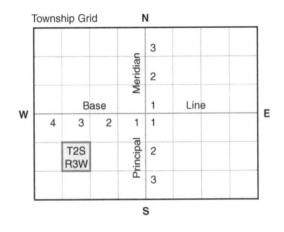
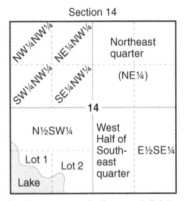

FIGURE 8.21 *Illustration of the Public Land Survey System from the U.S. National Atlas. (Courtesy Curtis Page)*

plumb lines, more than a force of arms, subdue landscape and henceforth demarcate control and change. Early attempts to "civil engineer" the Western landscape for private and public tenure began in the early 1850s, but there was no formal law and procedure until 1862.

To appreciate the monumental changes occurring on California's landscapes under American occupation we have to understand the **Public Land Survey System** (PLSS) and how different it was from the diseno demarcated Mexican rancho system. The PLSS was devised in the late 1780s as a way to survey and identify land parcels, particularly in rural or undeveloped land. Before its inception the original colonies used the British system of metes and bounds, which described property lines based on local markers and boundaries, often based on topography (coincidentally this is similar to how Calfornia Native Americans demarcated their triblet boundaries, although by oral tradition. See Chapter 5). The problem with this system is that it often made for irregular shapes with complex property descriptions and was not useful for the large, newly surveyed lands being opened in the West. The PLSS system's basic units of area are the **township** and the **section**, and it lays out land on a large mathematical cadastral grid (Figure 8.21). Townships are laid out 6 miles on a side (36 square miles) and then subdivided into 36 one square mile (640 acres) sections, which are further subdivided in **quarter sections** of 0.25 square miles (**160 acres**) each. Once California attained statehood in 1850, it enacted the PLSS system, using three new main **survey meridian** points derived for the state (Figure 8.22):

FIGURE 8.22 *Map of the principal meridians established in California to enact the Public Land Survey System which provided structure to the Homestead Act. (Courtesy Curtis Page)*

- the Humboldt Meridian (1853) at the summit of Mt. Pierce (40° 35′ 02″ N, 124° 07′ 10″ W), Humboldt County;
- the Mount Diablo Meridian (1851) at the summit of Mt. Diablo (37° 52′ 54″ N, 121° 54′ 47″ W), Contra Costa County, and;
- the San Bernardino Meridian (1852) at the summit of Mt. San Jacinto (34° 07′ 13″ N, 116° 55′ 48″ W), San Bernardino County.

Up until this time, and therefore during the first part of the Gold Rush, lands in California were under the Mexican rancho system (using the descriptions found on the diseño maps) and prospector claims were similar to the British metes and bounds description system. From 1850 onwards they were supposed to use a grid structure to develop public lands for freehold title to settlers and prospectors. This did not always work though as the lack of government control in California during the 1850s allowed the miners to claim much of the land thereby limiting agricultural immigrants.

The Homestead Act of 1862 made possible and enforceable by federal law the ability to help survey, divide and settle public lands by granting freehold title to 160 acres outside the thirteen original U.S. colonies. A settler could get 160 acres upon payment of a nominal fee after five years of residence as long as the land had improvements. He could also acquire land after six months of residence at $1.25 an acre. This gridding of California's landscape was far different from the development of Mexican rancho lands, as the townships and sections did not follow any natural features but instead cut across them in draconian contrast. Despite the square shapes that controlled the landscape, the HomesteadAct was the important driver towards settlement in California with one hang-up: Settlers could only **dryland farm** the 160 acres; irrigation was not allowed. We shall see that this caused limitations and creative solutions in the young California agricultural industry after the cattle ranching days of yore. This American style of landscape development in California also got another boost as six months after the Act's passage, the Railroad Act was signed, and by May 1869, a Transcontinental railroad stretched to California with homestead development rights attached.

8.7 INDUSTRIALIZATION OF MINING: EVOLUTION OF CALIFORNIA MINING

In the evolution of **mining law**, mining in the U.S. was not widespread before the California Gold Rush. Under the original British colonial charters, the British crown granted the American colonies mineral rights. However, it required them to give a share of what they found to the crown. U.S. law made the dichotomy between private and public lands, where the law reserved one-third of the minerals taken from public land for the federal government. In an early effort to mine lead (for bullets) the federal government attempted a lease system whereby the War Department was supposed to get a royalty. Miners refused to get permits and the cost to administer the whole system was greater than the revenues obtained. Finally, by the 1840s the sales of mining claims on public lands became adopted into law.

8.7.1 THE GOLD RUSH MINING LAW

The California Gold Rush was the instigator for changing the character and operation of **natural resource laws** nationally, but especially in the West. During the Gold Rush, mining in the public domain was open to all, and consequently a large number of claims were filed, almost all on public land. There were no charges or regulations on gold rushers during 1848 to the 1850s, as the government did not have the power to follow through in asserting them. The miners were essentially trespassers on government land, but they decided who would be granted access to gold and under what conditions. Before statehood the territory governor doubted the ability of the army to enforce sales of claims, so the miners adopted their own policies based on **common law**. Thereby, anyone could stake a claim for free. The rights to the claims were dependent on priority in time (i.e., first come, first served), and the miners had to work the claims to retain the right. They also decided the use of other natural resources, especially water. The focus on water and who would be allowed to use it later formed the foundation for the legal doctrine of **"prior appropriation."** The bottom line is that all the mining camp codes that developed during this time were concerned with the allocation of natural resources to individuals, not the state.

Even after statehood, with no hope of enforcing U.S. federal or state natural resource laws, the California legislature was forced to officially sanction **free mining** in 1851. The U.S. Congress also realized that the California Gold Rush had opened Pandora's Box in the West, and it adopted free mining as well in 1866. Both of these state and federal laws covered placer and shaft (lode) mining. Finally, in 1872 a third mining law was adopted that gave the control of precious metals to the miners directly, then to counties, and finally to the states, thereby leaving the federal government out of the mining claim fiasco. California's miners drove all these actions, and essentially the laws remain the same today.

8.7.2 RISE OF INDUSTRIAL HYDRAULIC MINING

As mentioned earlier the stereotype of the lone miner was overblown, as the easily accessed placer deposits that one or two persons could work quickly became exhausted and mining rapidly moved on to be a group enterprise. Miners began forming groups to pool labor and claims, and they built partnerships and joint stock companies. In the beginning, the operations were primarily labor intensive with low capital needs.

Early in the Gold Rush, however, miners discovered much larger placer deposits in a core area north of Sacramento, mainly in the Yuba, Bear and Feather River drainages (see page 13 in *California Atlas*). These tertiary river deposits were 40–50 million years old and were now elevated above the modern incised rivers. The deposits in some places are 600 feet thick and are flush with gold, but the only way to get to the gold was to use water, tremendous amounts of water to create artificial erosion.

In the beginning, the construction of dams and flumes controlled the rivers, exposing the riverbed deposits. The invention of flumes lifted the rivers out of their ancient courses, and miners used the water power rushing past within these flumes in inventive ways to wash away the riverbed sediments and to locate gold (Figure 8.23). They constructed many dredging

FIGURE 8.23 *Sophisticated placer mining by controlling water on the American River. (Courtesy of The Bancroft Library)*

and lifting devices to aid in riverbed excavation. Eventually the use of all this gravity-fed water power allowed the development of **hydraulic mining**, which uses large hoses with pressure intensifying nozzles to allow the miners to essentially erode-wash the hillsides of the ancient placer deposits next to the modern riverbeds. They then guided the slurry of water and sediment, known as **slickens**, towards sluices lined with mercury (mined from the Coast Ranges) which was used for trapping the gold. At first miners tried hydraulic mining on a modest scale in the 1850s. However, the increasing water usage facilitated by construction of large impoundment and delivery systems led to technological advances for higher water pressure to work the hill slopes. The miners were then able to get the laws changed to allow for the consolidation of mining claims, which then in turn consolidated water rights. Mining was allowed to become a corporate activity where large companies consolidated water and mining activities. The latter needed a lot of the former to make hydraulic mining profitable.

8.7.3 RISE OF HARD ROCK LODE MINING

In another mining development, miners followed gold encased in quartz rock veins embedded in Sierran granites as the development of **lode mining** took off in the southern Mother Lode territory (see page 13 in *California Atlas*). In contrast to hydraulic mining, quartz lode mining requires a large input of capital investment to make it work successfully for a profit. It also requires specialized knowledge in mine shaft engineering as the construction of extensive tunnel systems is a complicated enterprise—and don't forget earthquakes. Once miners bring the rock from the tunnels to the surface it goes through a complex extraction process to extract the gold. This includes using stamp mills for pulverizing the rock into smaller pieces (Figure 8.24). Then the miners use a chemical process with cyanide and sulfuric acid to separate the gold from the other rock. They dumped the chemical waste from these processes into the rivers, which eventually with mercury from hydraulic mining came to pollute the sediments of San Francisco Bay.

FIGURE 8.24 *Quartz stamp mill (i.e., rock crusher) used in lode mining, Grass Valley. (Courtesy of The Bancroft Library)*

Despite some very successful hard rock mining operations (i.e., see the Empire Mine State Park in Grass Valley, using Internet search), many early mining corporations generally failed. However, the number of corporations grew steadily through the 1850s, with the height of the development occurring with the discovery of the Comstock Lode in 1859 near Carson City, Nevada. In a clever move, individual miners were convinced into selling their claims of the Comstock Lode via the San Francisco Stock Exchange, which allowed wealthy Californians based in San Francisco to completely own the Comstock Lode and to create the Ophir Silver Mining Corporation. By 1860 more than a thousand mining companies became incorporated in California, which put their ownership and operations in the hands of people outside the mining areas. The miners who sold their claims ended up as day wage laborers for these large corporations.

What started as the lone miner or as partnerships quickly became a corporate activity that led to the establishment of the San Francisco Stock Exchange in 1862. The Gold Rush shifted to mining stocks, and stock ownership became widespread for the first time in the U.S. (even though the New York and Chicago exchanges had been established earlier). The problem was that, in the irrational exuberance of the time, the value of mining stocks traded overshadowed the value of gold produced (e.g., sound familiar? The Dot com market crash of 2001 or housing value crash 2007–2012). The mining stocks became a highly speculative market subject to serious insider manipulation—most people, including many insiders, lost their money (similar to modern stock market crashes).

8.8 GOLD VS. GOLDEN GRAIN: THE DOWNFALL OF CALIFORNIA GOLD MINING

Some researchers would say that the destruction of the gold mining industry, which hydraulic mining methods dominated heavily (Figure 8.25), represented the first of the California water conflicts. Since the claim, use, and abuse of water is a culprit in the downfall of gold mining its use should be noteed for future reference when reading Chapter 11.

Setting the stage: Northern California is still the epicenter of population and economic development in California. While urban populations are high in San Francisco and Sacramento, it is still very high as well in the gold bearing counties of the Sierra Nevada foothills and the Klamath Mountains. In the 1870s hydraulic mining was the main employer in such areas as Yuba city, Marysville, Grass Valley, etc. Both foreign and San Francisco-based capital also had heavily invested in the mining industry. The large land ownership and the big money made the mining corporations a powerful political force in the state. The hydraulic mining corporations were the most powerful, and they dominated the industry. Many of these hydraulic mining operations had consolidated large amounts of water—waterworks that were able to deliver up to 35 to 40 million gallons/day. Using such large amounts of water to artificially erode sedimentary hillsides started to cause a lot of environmental degradation, both at the sites with erosion and in downstream sedimentation pollution (and eventually mercury pollution). One of the more famous places is the Malakoff diggings near Nevada City. Estimates were that it contained 350 million cubic yards of gold-bearing sediment at a mining price as high as $5.60/cu. yd., but it required 69,000 gallons of water to generate $1 worth of gold. Wasted water and environmental degradation, however, are not an issue for the mining companies, as they were considered too big to fail.

As mentioned earlier, however, not everyone that came to California got involved in the mining business. Other settlers started to come to California to claim their 160 acres under the Homestead Act. The intense development of small family farms took place in the Sacramento Valley. Unfortunately, many of these newly arrived American immigrants were not used to or did not understand California's Mediterranean type of climate. Many of these farmers came from the East Coast or from Midwestern farming regions where rainfall occurs during the summer (the time when there is enough sunlight for a plant to grow) and thus were they can be successful in dryland farming. The Homestead Act made it clear that settlers could only use the 160 acres for dryland farming, not irrigation farming. Since

FIGURE 8.25 *Hydraulic mining at the North Bloomfield mining operations Malakoff diggings (near Nevada City). Piping four streams from monitors (giants), with aggregate discharge of 2,500 miner's inches at a hydraulic mine. Material is washed through bedrock cuts to the sluices which are not visible. (Courtesy of The Bancroft Library)*

California does not get summer rain, they had to use crops that worked for California's very different climate regime. That main crop that kept farming families alive was spring **wheat**, a Mediterranean grass native to the Middle East and perfect for an area of winter rainfall and summer drought. The Sacramento Valley region shifted quickly from cattle ranching to wheat farming. To note the enormity of the enterprise: California became the second largest wheat producer in the world by 1890 (Figure 8.26). During the 1870s and 1880s farmers produced wheat mainly for the export market. In essence, the economic emphasis in the Sacramento Valley shifted away from growing food for miners to driving profits for the valley's wheat farmers. This wheat bonanza is a preview of twentieth century agribusiness in California: large land holdings (some are former Mexican land grants), the beginnings of absentee ownership and large seasonal variations in farm labor demand. In fact, the near perfect flatness of the Central Valley lends itself to large fields that encourage entrepreneurs to develop mechanization for plowing, planting and harvesting—some using steam powered farm equipment invented in California for the task.

The problem was that the negative results of hydraulic mining activities were becoming more troublesome. The large quantities of coarse sediment (slickens) being washed into the Feather, Yuba and Bear river drainages was causing the beds of the rivers to rise rapidly. Each winter and spring, floods moved more material into the Sacramento Valley over the natural banks of the rivers, which then covered the extensive tracks of wheat fields in several feet of slickens. Besides the ruined farmland, the increased slickens caused major floods at valley cities, such as Yuba City and Marysville, which were forced to build high levees. Lastly, river navigation became difficult on the Sacramento River, which had been navigable by steamboat to as far north as Red Bluff. These boats were used extensively to ship the wheat harvest each summer to San Francisco for export.

In the 1870s a series of catastrophic floods led to lawsuits by farmers seeking damages from the mining companies for lost crops and silted fields. The mining companies in their hubris claimed that mining was more valuable than farmland. They also claimed that their **property rights** included the rights to dump sediment. They based this on the right of prescription (i.e.,

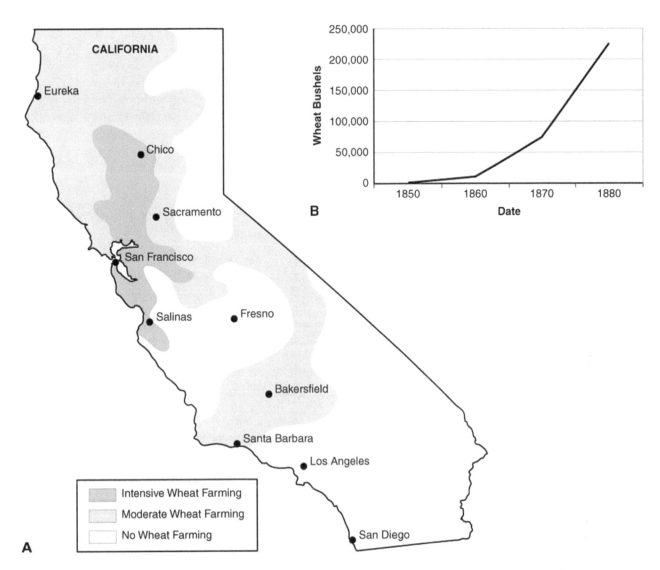

FIGURE 8.26 *(a) Map of estimated wheat growing areas in California based on the U.S. Department of Agriculture census of 1870. The darker areas denote intensive wheat farming. (b) Chart displaying the increase in wheat production in Butte County, an example representing an area in the Sacramento Valley that also was a major gold producing county. (Source data: U.S. Department of Agriculture census 1870 and James Monaco)*

grandfather clause) and said that mines had been dumping tailings long before farmers arrived and complained. The State Supreme Court overturned an initial judgment in favor of the farmers, saying that the farmers couldn't trace the source of debris to specific mines. In 1880, a new government in California appointed a state engineer to conduct a study of the problem. This provided the first quantitative estimates of the volume of slickens and the acreage they covered (which turned out to be an underestimate by a factor of two). The engineer recommended that the state construct a coordinated levee system (the same one California has today) and recommended constructing large debris dams at points along the rivers where they exit from the mountains onto the valley floor. The state government implemented the plans, but quickly discredited the strategy of the debris dams when a flood occurred in 1881, which caused them to be washed away and thus the slickens caused further damages to valley towns and farmland. Of course, the state legislature was too divided to act (sound familiar?), and the powerful mining interests were putting pressure on them to do nothing.

The farmers returned to court to demand that something be done to stop the damage from the reckless mining companies. A district court judge agreed and issued an injunction against the mining companies to stop—which they ignored. Other state court decisions in favor of the farmers in 1882 were ineffective as well, so the scene moved to the federal courts. The following final decision came in January 1884: The federal court detailed the damage done to farmland and the loss of river navigation, rejected the argument that debris dams can stem the flow of sediment and argued that miners' property rights don't allow for such extensive damage to others' property. The federal government issued a permanent injunction, effectively ending hydraulic mining in California.

Geographically this was a tremendous jolt to the landscape, as the effects of the injunction were:

1. The federal government effectively ended the gold mining industry in California, allowing only the expensive shaft (lode) mining corporations to remain.
2. The injunction in context is the first major environmental decision in the U.S., although based solely on property rights.
3. There was an accelerated economic shift away from mining, and essentially away from Northern California dominating the state's affairs.
4. There was a major population shift away from the Sierra Nevada foothills as day wage miners and the people who provided services to them in the mining towns left for the Central Valley to homestead or to the San Francisco Bay area for work, or headed south for the new opportunities being ceated in Southern California. Many of these mining towns/cities have never recovered (i.e., Oroville, Marysville), or have since become part of the "Gold Rush" tourism industry or retirement communities (i.e., Grass Valley, Nevada City, Columbia, Sonora, etc.).

Bibliography

Almaguer, T. (1994). *Racial Fault Lines: The Historical Origins of White Supremacy in California*. University of California Press, Berkeley, CA.

Ambrose, S. (2000). *Nothing Like it in the World: The Men who Built the Transcontinental Railroad, 1863–1869*. Simon and Schuster, New York, NY.

Beck, W. (1974). *Historical Atlas of California*. University of Oklahoma Press, Norman, OK.

Booth, S. (2008). *California Geography*. Course taught at Sierra College. [online] http://geography.sierra.cc.ca.us/booth/California/cal_index.htm.

Brand, H.W. (2002). *The Age of Gold: the California Gold Rush and the New American Dream*. Anchor Books, New York, NY.

Buckley, J. (2000). *Building the Redwood Region: The Redwood Lumber Industry and the Landscape of Northern California, 1850–1929*. UC Berkeley, Dept. of Architecture, PhD Dissertation.

Calisphere (2008a). California Cultures. University of California [online] http://www.calisphere.universityofcalifornia.edu/calcultures/

Calisphere (2008b). A world of California Primary Sources. University of California [online] http://www.calisphere.universityofcalifornia.edu/

Carranco, L. (1982). *Redwood Lumber Industry*. Golden West Books, San Marino, CA.

Chan, S. (2000). A people of exceptional character: ethnic diversity, nativism, and racism in the California gold rush. In *Rooted in Barbarous Soil: People, Culture, and Community in Gold Rush California*, edited by K. Starr and R.J. Orsi, pp. 44–85. University of California Press, Berkeley, CA.

Chinn, T.W., ed. (1969). *A History of Chinese in California*. Chinese Historical Society of America, San Francisco, CA.

Cox, C.J. (2008). *California Geography*. Course taught at Sierra College. [online] http://faculty.sierracollege.edu/ccox/california_geography/index.html.

Davis, C. and D. Igler (2002). *The Human Tradition in California*. Scholarly Resources, Wilmington, DEL.

DeGraaf, L.B., Mulroy, K., and Q. Taylor (2001). *Seeking El Dorado: African Americans in California*. Autry Museum of Western Heritage and University of Washington Press, Los Angeles, CA.

Deverell, W. (1996). Railroad Crossing: Californians and the Railroad, 1850–1910. University of California Press, Berkeley, CA.

Deverell, W. and D. Igler (2008). *A Companion to California History*. Wiley-Blackwell, New York, NY.

DeWitt, H.A. (1999). *The Fragmented Dream: Multicultural California*. Kendall/Hunt Publishers.

DeWitt, H.A. (1999). *The California Dream*. 2nd edition. Kendall/Hunt Publishers.

Donley, M.W., Allan, S., Caro, P., and C.P. Patton (1979). *Atlas of California*. Pacific Book Center, Culver City, CA.

Durrenberger, R.W. and R.B. Johnson (1976). *California Patterns on the Land*. 5th edition, Mayfield Publishing Company, Mountain View, CA.

Dutschke, D. (1988). A history of American Indians in California. In: *Five Views: An Ethnic Historic Site Survey for California*, pp. 3–55. California Department of Parks and Recreation, Office of Historic Preservation, Sacramento, CA.

Haas, L. (1996). *Conquests and Historical Identities in California, 1769–1936*. University of California Press, Berkeley, CA.

Hayes, D. (2007). *Historical Atlas of California.* University of California Press, Berkeley, CA.

Heizer, R.F. (1974). *The Destruction of California Indians.* Peregrine Smith, Inc., Salt Lake City, UT.

Heizer, R.F., ed. (1974). *They Were Only Diggers: A Collection of Articles from California Newspapers, 1851–1866, on Indian and White Relations.* Publications in Archaeology, Ethnology, and History, 1, Ballena Press, Ramona, CA.

Hornbeck, D. (1979). The Patenting of California's Private Land Claims, 1851–1885. *Geographical Review,* 69(4): 434–448.

Hornbeck, D. (1983). *California Patterns: A Geographical and Historical Atlas.* Mayfield Publishing Company, Mountain View, CA.

Kelly, R. (1959). *Battling the inland sea: floods, public policy, and the Sacramento Valley.* University of California Press, Berkeley, CA.

Kimble, J.H. (1943). *The Panama Route, 1848–1869.* University of California Press, Berkeley, CA.

Knowland, J.R. (1941). *California, a Landmark History.* Tribune Company, Oakland, CA.

Jung, M.A. (1999). Capitalism comes to the diggings: from gold-rush adventure to corporate enterprise. In *A Golden State: Mining and Economic Development in Gold Rush California,* edited by J.J. Rawls and R.J. Orsi, pp. 52–77. University of California, Berkeley, CA.

Larson, D.J. (1996). Historical water-use priorities and public policies. In *Sierra Nevada Ecosystem Project: Final report to Congress, vol. II, Assessments and scientific basis for management options.* University of California, Centers for Water and Wildland Resources, Davis, CA.

McClure, J.D. (1948). *California Landmarks: A Photographic Guide to the State's Historic Spots.* Stanford University Press, Stanford, CA.

May, P.R. (1970). *Origins of Hydraulic Mining in California.* Holmes Book Co., Oakland, CA.

Michaelson, J. (2008). *Geography of California.* Course at UC Santa Barbara, Dept. of Geography. [online] http://www.geog.ucsb.edu/~joel/g148_f08/.

Monaco, J.E. (1986). *The changing ethnic character of a California gold mining community: Butte County, 1848–1880.* California State University Chico, Department of Geography and Planning, Masters Thesis.

Monroy, D. (1993). *Thrown Among Strangers: The Making of Mexican Culture in Frontier California.* University of California Press, Berkeley, CA.

Morrison, P.A. (1971). *The Role of Migration in California's Growth.* Rand Corporation, Santa Monica, CA.

Norris, F. (2008). *The Octopus: A Story of California.* IAP.

Orsi, R.J. (2007). *Sunset Limited: The Southern Pacific Railroad and the Development of the American West, 1850–1930.* University of California Press, Berkeley, CA.

Paul, R.W. (1947). *California Gold: The Beginning of Mining in the Far West.* University of Nebraska Press, Lincoln, NE.

Pisani, D.J. (1999). "I am resolved not to interfere but permit all to work freely": the gold rush and American resource law. In *A Golden State: Mining and Economic Development in Gold Rush California,* edited by J.J. Rawls and R.J. Orsi, pp. 123–148. University of California Press, Berkeley, CA.

Pitt, L. (1970). *The Decline of the Californios: A Social History of the Spanish-Speaking Californians, 1846–1890.* University of California Press, Berkeley, CA.

Pitti, J., Castaneda, A., and C. Cortes (1988). A history of Mexican Americans in California. In: *Five Views: An Ethnic Historic Site Survey for California,* pp. 207–264. California Department of Parks and Recreation, Office of Historic Preservation, Sacramento, CA.

Ramsey, E.M. and J.S. Lewis (1988). A history of Black Americans in California. In: *Five Views: An Ethnic Historic Site Survey for California,* pp. 59–101. California Department of Parks and Recreation, Office of Historic Preservation, Sacramento, CA.

Rawls, J.J. and W. Bean (2008). *California: An Interpretative History.* 9th Edition, McGraw-Hill Publishing Co., New York, NY.

Rice, R., Bullough, W., and R. Orsi (2001). *The Elusive Eden: A New History of California.* 3rd edition, McGraw-Hill.

Robinson, W.W. (1979).Land in California: the story of mission lands, ranchos, squatters, mining claims, railroad grants, land scrip [and] homesteads. University of California Press, Berkeley, CA.

Rohrbough, M. (2000). No boy's play: migration and settlement in early gold rush California. In *Rooted in Barbarous Soil: People, Culture, and Community in Gold Rush California,* edited by K. Starr and R.J. Orsi, pp. 25–43. University of California Press, Berkeley, CA.

Schwartz, S. (1998). *From west to east: California and the making of the American mind.* Free Press.

Social Explorer (2009). *United States Census Demographic Maps: California 1850–2007.* [online] http://www.socialexplorer.com/pub/maps/home.aspx.

Starr, K. (1973). *Americans and the California Dream, 1850–1915.* Oxford University Press, New York, NY.

Starr, K. (2005). California: A History. The Modern Library, New York, NY.

Steiner, S. (1980). *Fusang: The Chinese Who Built America.* Harper & Row Publishers, New York, NY.

Sylva, S.A. (1932). *Foreigners in the California Gold Rush.* Masters Thesis, University of Southern California.

Thurman, A.O. 1945. *The Negro in California before 1890.* Masters Thesis, College of the Pacific.

Trazfer, C.E. and J.R. Hyer (1999). *Exterminate them: written accounts of the murder, rape, and slavery of Native Americans during the California gold rush, 1848–1868.* Michigan State University Press, East Lansing, MI.

Walker, R.A. (2001). California's golden road to riches: natural resources and regional capitalism, 1848–1940. *Annals of the Association of American Geographers*, 91:167–199.

Ward, G.C. (1996). *The West.* Little, Brown and Company, Boston, CA.

Watkins, T.H. (1983). *California: An Illustrated History.* American West Publishing Company, New York, NY.

Wey, N. (1988). A history of Chinese Americans in California. In: *Five Views: An Ethnic Historic Site Survey for California*, pp. 105–158. California Department of Parks and Recreation, Office of Historic Preservation, Sacramento, CA.

Williams, J.H. (1996). *A Great and Shining Road: The Epic Story of the Transcontinental Railroad.* University of Nebraska Press, Lincoln, NE.

CHAPTER 9

The Beginnings of Modern California and the Rise of Southern California

Key terms

- Arroyo Culture
- Arts and Crafts
- Big Four
- California craftsman
- California bungalow
- Californio
- Central Overland California Route
- Central Pacific Railroad
- Charles Fletcher Lummis
- Chinatowns
- Chinese Exclusion Act 1882
- Coinage Act of 1873
- Forest Reserve Act 1891
- Health resorts
- Healthy climate
- Helen Hunt Jackson
- John Muir
- Los Angeles
- Mission revival
- Modoc Indian War
- Nome Cult trail
- Northern California
- Orange citrus
- Overland Mail Service
- Pacific Railroad Act 1862
- Petroleum Oil
- Pony Express
- Progressive era
- Ramona
- Rancheria
- Real estate
- Riverside
- Romance
- San Diego
- San Francisco
- Santa Barbara
- Scott Act 1888
- Southern California
- Southern Pacific Railroad
- Theodore Judah
- Transcontinental Railroad
- Transcontinental Telegraph
- Utopian communities
- Yellow peril
- Yosemite National Park 1890
- Yosemite State Park 1864

Introduction

THE DEVELOPMENT OF TWO URBAN EMPIRES

We can consider California up until 1880 as a split state mentally, culturally and developmentally. Yes, stagecoach bumper stickers, if they were available during this time period, that declared "Norcal" or "Socal" would have made obvious sense to people of the state. To understand these geospatial cultural differences and the divergence that developed it, we must keep in mind the last chapter on the Gold Rush and the decline of the Californios.

The contrast in geographic settings between opposite ends of the state could not have been any starker. In **Northern California** resided the majority of the state's population and wealth. **San Francisco** was the hub of enterprise, society and cultural diversity with a population of ~250,000. The main city core had fully developed into the look and feel of an older East Coast city (e.g., Boston, New York, etc.), on which it was based (Figure 9.1). Northern California in general had the best transportation connections with the rest of the country by ship and established wagon and stagecoach trails. Besides the urban setting there were also substantial populations in agriculture: Sacramento Valley wheat farming, Sacramento-San Joaquin Delta vegetable cropping and San Joaquin Valley cattle ranches. The mining communities in the Sierra Nevada foothills also had substantial populations before the downfall in the 1880s of hydraulic mining.

Southern California had been neglected as an economic and cultural backwater since 1848. After the first foreign miners' tax passed against miners of Latino origin, and Californios started losing their ranchos to squatters and through lack of rancho diseño map tenure acceptance in court, many of this group headed to Southern California for protection and cultural support. Southern California during the 1848–1875 gold mining build out in Northern California became a holding area for those left behind and kicked out of the mining region. In contrast to San Francisco, Los Angeles in 1880 had a total population of approximately 15,000, while the total Southern California population (represented as coastal counties and the deserts to the south of Monterrey) remained at only ~76,000. The infrastructure of the main settlements of Southern California—**San Diego, Los Angeles** and **Santa Barbara**—were primitive at best, still based on adobe building materials and some American style wood frame buildings. It was clear that not much of the money from mining and other industrial business ventures was finding its way to the southern parts of the state. In fact, it wasn't until a railroad connection came down from the north and a competing railroad connection arrived from the southwest that Southern California appeared on the settlement radar of Midwestern and East Coast immigrants. These immigrants, however, and the motivations that brought them to Southern California were quite different than those of the firecracker beginnings of Northern California. Romance and lifestyle are the keys to understanding the southern half of the state.

The last chapter clearly showed the shining moments and brutal realities (i.e., treatment afforded ethnic groups other than White Americans) in California's early American beginnings, but more importantly it showed the geospatial development of Northern California brought on by the riches of the Gold Rush and the state's overall settlement control via

FIGURE 9.1   *View of San Francisco from north beach, 1880s. (Howard DeWitt image)*

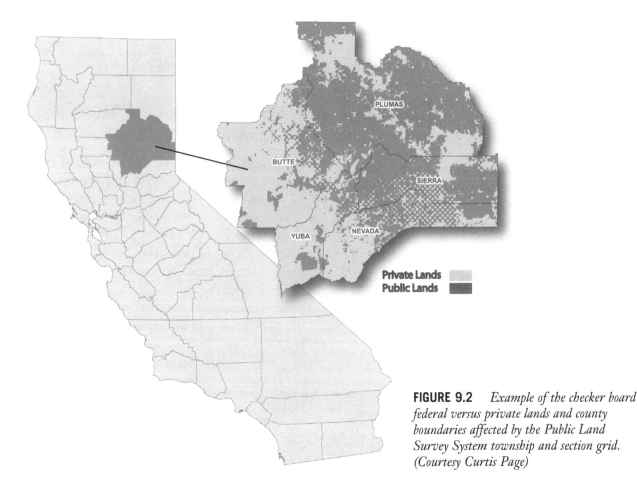

FIGURE 9.2 *Example of the checker board federal versus private lands and county boundaries affected by the Public Land Survey System township and section grid. (Courtesy Curtis Page)*

the Public Land Survey System and the Homestead Act. The geometric rectangularity imparted to the landscape was extraordinary. Property lines, roads, fields and public boundaries (e.g., county borders, parks, forests, military bases, etc.) all exhibited the checkerboard right angles of the cadastral system (Figure 9.2). The settlement infrastructure that conformed to the rigid sectional system soon intensified with the development of the **Southern Pacific Railroad**, which was granted full sections in great numbers that it later used to develop towns and sell larger land units than the simple 160 acres restricted to the Homestead Act.

This chapter provides the geographical context for the modern beginnings of the entire state, and it more exclusively lays out an argument for Southern California's introduction to the development stage. Developments in transportation, promotion of a romantic landscape memory, prospects for health and a different kind of geological "gold" were necessary for this to happen. Now, roll all this into a booster package with the old real estate adage "location, location, location!" and the explanation for the rapid expansion can be complete.

9.1 Early Transportation and Communication

As the last chapter noted, there were three options that miners and settlers used as transportation to get to California during the initial Gold Rush: two routes by ship and one by wagon trail. Communication with the rest of the U.S., however, remained a problem until the mid-1850s when the federal government subsidized the **Overland Mail Service** to run mail stagecoaches via the American Southwest deserts to San Francisco, thus avoiding any snowbound mountains (i.e., Sierra Nevada and Rocky Mountains). This service originated in St.

Louis, Missouri and took three or four weeks to make the run to San Francisco, which was far better than the six months by ship it used to take for mail to arrive from Washington D.C. to California. The federal government realized that California was acting more like an isolated colony than an integrated state in the union. With the outbreak of the Civil War, it abandoned the southern route, and the Wells-Fargo Overland Mail and Stagecoach Company took over and rerouted the mail via the **Central Overland California Route**. This went through Utah and Nevada to California on what is now Highway 50 through South Lake Tahoe and Placerville (i.e., a original Native American trading trail). Its competitor was the infamous **Pony Express** (1860), whose mail service proved that riders on horseback could go from St. Joseph, Missouri to Sacramento in ten days. It was not to last though, as nineteen months later the **Transcontinental Telegraph** connected Sacramento to Omaha, Nebraska, and thus the rest of the Eastern seaboard with California using Morse code. Within the next couple of years the telegraph lines connected other towns and cities along the West Coast.

9.1.1 THE COMING OF THE CENTRAL (SOUTHERN) PACIFIC RAILROAD

Despite clipper ship service via Panama or around the Horn and stagecoach service for the transportation of people and goods, California was still very isolated from the rest of the nation. The snowbound mountains and formidable deserts were a challenge for transportation. The completion of the **Transcontinental Railroad** in 1869 solved this problem; it was a feat of momentous portions that again changed California's landscapes and cultural mix in important geographical ways.

Companies had built smaller local railroads as early as 1854. People used the first one in Humboldt County

FIGURE 9.3 *From left to right and top to bottom: Theodore Judah, Leland Stanford, Collis Huntington, Mark Hopkins, and Charles Crocker. (Courtesy of The Bancroft Library)*

to move giant redwood trees with twenty-foot diameters from the lumber camps in the hills surrounding Eureka to the mills on Humboldt Bay in order to keep up with the demand for lumber products in fast-growing San Francisco. **Theodore Judah** (Figure 9.3) built the first local railroad for transporting passengers in 1854 for the Sacramento Valley Railroad, which went from Sacramento to the Folsom gold fields (twenty-two miles). Judah was the visionary behind building the Transcontinental Railroad, and he became the chief planner for the **Central Pacific Railroad**, which conducted the first surveys in the 1850s. In 1860, he surveyed a safe route over the Sierra Nevada (along what is now Interstate 80). He gained financial backing from Sacramento hardware store merchants—the **"Big Four"**: Leland Stanford, Collis P. Huntington, Mark Hopkins and Charles Crocker, who actually secured the funds and built the Central Pacific (Figure 9.3). After lobbying the U.S. government, Congress passed the **Pacific Railroad Act of 1862** to build the Transcontinental Railroad, which allowed the Central Pacific Railroad to meet up with the Union Pacific Railroad in Utah in 1869 (Figure 9.4). The "Big Four" pushed Judah out of the company at the start of construction in 1863, and he died the same year without realizing his dream.

The Pacific Railroad Act authorized both the making of extensive land grants of public lands and the issuance of bonds to the Central Pacific Railroad (CPRR). As the track was laid out the CPRR received ten square miles (26 km^2) of public land on each side of the tracks, every other section (square mile), for every mile laid except where railroads ran through previously established cities and crossed rivers. The low interest loans were issued at the rate of \$16,000 per mile of tracked grade completed west of the designated base of the Sierra Nevada Mountains. It also provided \$48,000 for tracked grade completed over and within the two mountain ranges (i.e., the Sierra Nevada and the Rocky Mountains), and \$32,000 per mile of completed grade between the two mountain ranges. In California, once

FIGURE 9.4 *Meeting of the Central Pacific and Union Pacific railroads at Promontory, Utah 1869. (Courtesy of The Bancroft Library)*

the Transcontinental Railroad was built the CPRR became the Southern Pacific Railroad (SPRR), which went on to expand and provide new avenues of migration in the state, especially to Southern California. The total acreage eventually owned by the SPRR was 11,588,000 or 11.5 percent of the entire state. The railroad profoundly changed the development of the economic, social and geographical character of the state (see pages 14 and 15 in the *California Atlas*). It solely drove both Northern and especially Southern California into the modern industrial age. The SPRR established towns (i.e., Lancaster, Palmdale, Livermore, Tracy, Mojave, Coachella, Modesto, Merced, Fresno, Tulare, and Hanford to name a few), developed land, forestry and water resources and impacted agriculture and tourism. The SPRR directors were about making huge profits, and they did by having business ventures and the desire for control in every industry (Figure 9.5). As they encouraged settlers to come west to occupy lands and towns closest to the tracks, they drew the highest land prices. Later this chapter will address other detailed aspects of the railroad's impact on California's geography, and the next chapter will focus especially on its impact to agriculture development.

One of the most important aspects of building the western section of the Transcontinental Railroad was the cultural group that did most of the construction: the Chinese. Early on they had proved their hard work and fidelity in building the California Central Line in 1860, which picked up at Folsom (i.e. where the Sacramento Valley Railroad ended) and went north along the Sierra Nevada foothills, crossing many mining camps, to Marysville. After construction problems developed on the Transcontinental Railroad as it left Sacramento in 1863 under White labor, the promise of earning higher wages than they would in the cities or in the mines lured the Chinese. Largely Chinese labor (around 10,000 people) built the section of the railroad through the foothills, over the Sierra Nevada, and across the Basin and Range to Utah (Figure 9.6). Their work was hard and dangerous over the Sierra Nevada, especially through Donner Summit, as they set explosive charges at precarious heights along the way through heavy snow or on hot summer days.

9.1.1.1 Chinese as Scapegoats: Racist Exclusion under the Yellow Peril

One cannot complete the story of the railroad and other mass construction projects in California without discussing the issue of racism, in particular the **"yellow peril"** as viewed among Whites. Early during the Gold Rush Whites stigmatized the Chinese as undesirable aliens of inferior culture and morality. It is unfortunate that the lack of Chinese women and families in California added to an already debased stereotype. Of course, this is strange because the young Chinese men who came to California were no different than the countless other young men of many other cultures that came for the same goal: to find ones riches and bring or send it back to loved ones back home. The Chinese men's hard work and industrious nature in the mining camps made others envious enough to convince the state government to pass a second foreign miners tax against them in 1852, as discussed in Chapter 8.

When the Transcontinental Railroad was completed in mid-1869, the SPRR released the majority of

FIGURE 9.5 *An anti-railroad cartoon. (Courtesy of The Bancroft Library)*

FIGURE 9.6 *Chinese building the Transcontinental railroad near Truckee. (Courtesy of The Bancroft Library)*

its workers (i.e., Chinese labor). This action led to an oversupply of labor on the market, which subsequently suppressed wages. At nearly the same time in the 1870s, a national economic recession set in that was particularly hard hitting in California. While the national problems were large bank failures (similar to the 2008–2012 economic crisis), at the local end in California two issues occurred: economic decline due to the economic impact of the railroad and the national passage of the **Coinage Act of 1873** that demonetized silver. The first issue is the most important as the "Big Four" at the SPRR were hoping that the completion of the Transcontinental Railroad would allow California goods manufacturers to get a jump on trade with the Midwest and East Coast by immediately shipping goods eastward. This was not to happen as the major industrial manufacturers on the East Coast got the jump on California first by immediately dumping low cost goods in California (i.e., similar to our trade with China in the present era). This had the effect of suppressing manufacturing in California and thus further suppressing wages and the need for labor. The castigation was swift as unemployed White workers accused Chinese workers of causing California's economic demise (Figure 9.7). The Chinese were subjected to a series of anti-Chinese riots (many turning violent with public lynchings) between 1871 and 1877 that occurred in Auburn, Chico, Los Angeles, Petaluma, San Francisco, Santa Barbara, Weaverville, and Yreka. By 1879 a new state constitution was ratified with strongly worded anti-Chinese provisions forbidding corporations from employing Chinese workers. It was also during this time period that the authorities officially established many **Chinatowns** (Figure 9.8) or "Chinese Quarters" to police and control their population in many cities (Figure 9.9) and towns across the state (i.e., San Francisco, Oakland, Sacramento, Eureka, Weaverville, Chico, Oroville, San Jose, San Luis Obispo, Los Angeles, and San Diego).

The final blows to the Chinese and Chinese-American community in California came with the federal passage of the **Chinese Exclusion Act of 1882**, which prohibited Chinese laborers, unskilled and skilled, from entering the United States for ten years

162 **CHAPTER 9** *The Beginnings of Modern California and the Rise of Southern California*

FIGURE 9.7 *The Illustrated WASP, 1877. (Howard DeWitt image)*

FIGURE 9.8 *San Francisco's Chinatown in the 1880s. (Courtesy of The Bancroft Library)*

(a)

(b)

FIGURE 9.9 *Modern Chinatown districts of: (a) San Francisco (© Andy Z., 2012. Used under license from Shutterstock, Inc.) and; (b) Los Angeles. (© Shawn Hempel, 2012. Used under license from Shutterstock, Inc.)*

FIGURE 9.10 *The Chinese must go. Anti-Chinese cartoon from the 1870s. (Howard DeWitt image)*

(Figure 9.10). Then in 1888 the **Scott Act** barred re-entry of Chinese laborers to the U.S. if they had left the country temporarily. Finally, the Chinese Exclusion Act of 1882 was renewed in 1892 and 1902, and then extended indefinitely in 1904. The Chinese Exclusion Act was not repealed until 1943 during World War II, but then Chinese immigration was tied to a strict quota system until 1965.

9.2 SEARCH FOR THE IDEAL LANDSCAPE

California boasts one of the most diverse sets of landscapes and the most diverse state and federal parks systems in the country (see pages 16–19 in *California Atlas*). During 1860 to 1900 many environmental events occurred in California, as this was the national era and the rise of the conservationist and preservationist movement. The first focus on land preservation in the state was upon Yosemite Valley and Mariposa Grove (i.e., giant sequoias), which at the time, in the 1850s and early 1860s, travelers were starting to exploit much to the demise of the resident Miwok tribelet group (Figure 9.11). Tourism developers were also starting to build rustic hotels in these areas (i.e., Wawona Hotel). The President and Congress stepped in and deeded this federal land to the state who then created **Yosemite State Park in 1864** in order to preserve the stupendous grandeur of its geomorphologic landscape and giant sequoias. In 1868 the preacher of the Sierra Nevada, **John Muir** (Figure 9.12), arrived from Wisconsin and was spellbound by the beautiful valley and the entire mountain range. He devoted the rest of his life to establishing conservation policies for the region, which culminated in 1890 with Yosemite becoming a national park—**Yosemite National Park** (Figure 9.13).

The Southern Pacific Railroad was behind many of the conservation projects throughout California's forested ecosystems. While on one hand the railroads promoted and spread lumber extraction throughout the Sierra Nevada and other mountainous regions in the state, they were quick to embrace forest conservation for watershed protection, which would affect their agricultural development efforts downstream and especially town development in the Central Valley. The SPRR knew that visitors wanted to see the majestic natural

FIGURE 9.11 *California Native Americans (Miwoks) before being removed from Yosemite Valley. (Painting by Albert Bierstadt, 1872)*

FIGURE 9.12 *John Muir, founder of the Sierra Club and promoter for the preservation of the Sierra Nevada mountain range. (Courtesy of The Bancroft Library)*

landscapes that California contained, and that these visiting tourists (potential passengers) could become future settlers buying land from the numerous SPRR holdings. Therefore they played a significant political role in the establishment of national forests and national parks (see pages 18 and 19 in *California Atlas*). SPRR was one of the key players helping to establish Sequoia, Kings Canyon and Yosemite National Parks in 1890. In 1891, the federal government passed the **Forest Reserve Act**, which helped establish the first forest reserve in the state, the San Gabriel Timberland Reserve (now part of the Angeles National Forest). Over the next fifteen years eighteen national forests were proclaimed with the help of the SPRR.

The biggest booster of California's scenic natural beauty was the SPRR. Everywhere they ran track in northern and down into Southern California they tried to search and retain "ideal landscapes," ones their Eastern tourism passengers would want to see, ones that would encourage them to stay and settle on SPRR's government granted lands." In the 1880s, Southern California experienced tremendous growth stimulated

FIGURE 9.13 *Yosemite Valley in Yosemite National Park.*

by the railroad as the SPRR extolled the region's natural beauty, romantic mission and rancho heritage, its healthy climate and its citrus orange production potential for health and wealth.

9.2.1 FINAL DESTRUCTION OF THE CALIFORNIA NATIVE AMERICAN CULTURAL LANDSCAPE

By 1850 approximately ~50,000 Native Americans remained in California, and they faced an unusually hostile American attitude because most local citizens and miners believed that the Native Americans should have been removed from the state. The Mexican land grants originally protected many of the "tribes" and tribelets as many had clauses protecting the Native Americans on the lands they occupied. Under American occupation, however, people ignored the Native American claims to these lands, which led to mass dispossessions. Further decline and denigration caused by disease, new unworkable social institutions, forced labor and open murder through bounty by the American settlers and miners caused many to move into remote areas (i.e., mountains and deserts) of California and led to ill-fated Native American wars. In 1851, in an effort to persuade California Native Americans to move out of the Sierra Nevada and Klamath Mountains gold belts, the government negotiated a series of eighteen treaties with 130 "tribes" and tribelets for a promised reservation system (Figure 9.14). The proposed reserved land totaled 8.5 million acres (7.5 percent of the state) of land scattered in parcels never greater than 25,000 acres in size located in the Peninsular ranges, the San Bernardino Mountains, the western (i.e., rain shadow) Great Central Valley, the Mendocino Mountains, northeastern Sacramento Valley, the western Klamath Mountains and the Siskiyou Mountains. The remaining Native Americans lost the coast, and they considered much of the proposed land to be set aside undesirable. The U.S. Senate, however, never ratified those treaties and considered them "lost" (actually found in Washington D.C. fifty years later in 1905), and so the Native Americans never received the proposed reservation lands.

Instead, the government established a makeshift reservation system in 1852 at U.S. Army forts around the state (i.e., Fort Tejon, Fort Humboldt, Fort Wright, Round Valley, etc.). The conditions were terrible, and

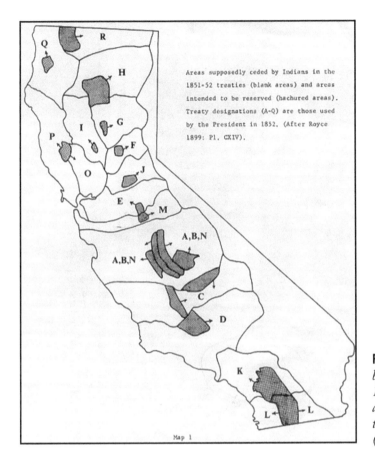

FIGURE 9.14 *Areas supposedly ceded by California Native Americans in the 1851–52 treaties (blank areas) and areas intended to be reserved for them by the federal government (dark areas). (Adapted from Royce 1899)*

several tribal groups became rebellious towards the settlers and militant against the U.S. Army's forced removal policies. This all led to several uprisings in the 1860s in the remote areas of California. First, there were revolts against settlers in the western Klamath mountain region. Second, there was a tragic death march to force the removal of the Konkow tribal group (see page 9 in *California Atlas*, polygon 23) known as the **Nome Cult Trail** (from Chico via Paskenta, over the Mendocino mountains to Round Valley reservation in Covelo, Mendocino county—California's own "trail of tears" (Figure 9.15). Finally, a culminating event transpired in the far northeastern section of California in the **Modoc Indian War** (the most expensive military operation in American history), in which the Modocs held out in the lava beds and lava tubes in the Modoc Plateau for almost a year against the U.S. Army.

Throughout this period the disintegration of the California Native American continued. The government eventually developed an unremarkable and unique reservation system called **"rancherias"** on tiny patches of undesirable federal lands and purchased private lands never greater than a few hundred acres scattered across California (Figure 9.15). By 1900, approximately 15,000 Native Americans remained in California.

9.3 SOUTHERN CALIFORNIA DEVELOPMENT: 1880–1900— DEVELOPMENT OF THE MYTH

The development of Southern California is really the development of a new social memory for the landscape brought on by "boosterist" manipulation of a false past. On the one hand the Southern Pacific railroad was a major booster with elaborate marketing to bring people to settle in Southern California, and on the other

168 **CHAPTER 9** *The Beginnings of Modern California and the Rise of Southern California*

FIGURE 9.15 *California Native American rancherias as of 2011. (Data source: U.S. Geological Survey; Courtesy Curtis Page)*

hand the Spanish mission and Mexican rancho era past was twisted into one of fantasy and romance involving the fading Californios, thus making a literary tourist trap of the landscape. Despite the use of a revisionist fantasy history to promote the region, Southern California also was ideal for other concerns involved with agriculture tied to the missions—citrus oranges. In addition, physicians considered the region a healthy climate, and it became involved in a new kind of geologic "gold" discovery—petroleum oil.

9.3.1 THE MISSION MYTH: ROMANCE TOURISM

In 1884 **Helen Hunt Jackson**, a novelist, wrote an important novel called ***Ramona***, which characterized Southern California as one of Mexican colonial romance (Figure 9.16). Jackson had arrived in California from New York in 1881 to take a respite in the dry climate for her tuberculosis, and she was interested in the area's missions and the ex-mission neophytes. She traveled extensively around Southern California (e.g., Santa Barbara to San Diego Counties) conducting research on the plight of the Native Americans and became very interested in the Spanish-Mexican culture, Roman Catholic heritage and the old ranchos. During this time she became friends with Don Antonio Coronel, a well-known authority on **Californio** life, who was also a former inspector of missions for the Mexican government, a resident since 1830 and the former mayor of Los Angeles from 1853 to 1854. Coronel described to Jackson the plight of the mission Native Americans after secularization led to the giving away of mission lands to create ranchos and how the American occupation destroyed the rancho system and caused mass dispossessions. There were clauses in many of the rancho land grants that protected the Native Americans on the lands they occupied, which the Americans subsequently ignored, thus dispersing their Native American residents. Coronel was very interested in preserving rancho history and tradition, as well as conveying the plight of the ex-mission/ex-rancho Native Americans under American occupation. Jackson became interested and sympathetic to their plight, which allowed her to receive a commission from the Bureau of Indian Affairs to study the condition of the Native Americans in Southern California.

Jackson traveled extensively throughout the region visiting many places that she later used as scenes for incidents in *Ramona*. These included Rancho Camulos in Ventura County, the Santa Barbara mission (the only mission still being run by the Franciscans at this time), Temecula in Riverside county, and so forth. Her extensive field report called for many improvements for the

FIGURE 9.16 *Late nineteenth century Mexican-American fiesta in Los Angeles. (Howard DeWitt image)*

Native Americans, including the purchase of reservation lands and more schools for Native children. She was convinced that the bad treatment of the Native Americans was the complete result of the American takeover of Mexican Alta California (completely disregarding the appalling conditions they endured under Spanish and Mexican rule). The lack of federal response to her report motivated her to write a novel that would have the same effect as *Uncle Tom's Cabin* did for slavery, but in this case the focus would be on ex-mission/rancho Native Americans. When *Ramona* published in 1884 it was an instant success among East Coast society. However, her intention for it to be a work of social protest was overwhelmed by its **romance**-based evocation of a mythical past that Alta California never experienced during the mission or rancho eras. Coincidently, the next year in 1885, another important California novel came out by Maria Amparo Ruiz de Burton, *The Squatter and the Don*. It was the first fictional narrative written from the perspective of the conquered Californio/Mexican population in Southern California. Its story showed that even though this cultural group had full rights of citizenship under the Treaty of Guadalupe-Hidalgo, they had become a subordinated and marginalized minority.

The growth of the *Ramona* legend became a cultural and historical focus for the rapidly growing Southern California region. The Southern Pacific railroad had much to do with connecting Southern California earlier when it ran a rail line down from Sacramento to Los Angeles in 1876. Then in 1887 the Santa Fe railroad arrived from the southwest into Los Angeles. This led by 1887 to the visits of more than 120,000 tourists, health seekers, farmers, artisans and businessmen per month. For the tourists using *Ramona* as their tourist guide, to real Southern California locations (Figure 9.17a) attached to the *Ramona* legend, places developed mythical significance. For example, Rancho Camulos (Figure 9.17b) became the "Home of Ramona," and the Southern Pacific railroad made sure that its railroad line that ran past the rancho had a depot at Camulos for the tourists in 1887. In essence, *Ramona* helped define Southern California for the rest of the country, which led the region to become a major tourist draw. The attraction to Spanish/Mexican artistic and architectural trends that were far different than those in Northern California also allowed the region to establish a local identity. East Coast architecture defined Northern California while Southern California remained rooted in Californio/Latino styles. *Ramona* also was linked to the movement to restore the crumbling missions, and it spawned a multitude of *Ramona* tie-ins, especially with the developing citrus industry.

9.3.2 SOUTHERN CALIFORNIA VITALITY: CLIMATE FOR HEALTH, FREEDOM OF EXPRESSION, AND REAL ESTATE BUBBLES

To travel and medical professionals on the East Coast, California became known as the state of two states. As early as 1857 a book entitled *Climatology of the United States* ranked Southern California's climate as exceptionally healthy (**healthy climate**), characterized by ample sunshine, moderate breezes and low relative humidity (see Chapter 3; pages 6 and 7 in *California Atlas*). At this time in the late 1800s, physicians believed that dry climates could cure tuberculosis or other lung ailments, and thus they often prescribed a change of climate to patients. This is similar to modern day American retirees moving to the southwestern Sunbelt (i.e., Phoenix, Las Vegas, etc.) to ease their arthritic conditions.

The arrival of the Southern Pacific railroad line from Sacramento in 1876, along with the completion of the Santa Fe line from the southwest in 1887 (see pages 14 and 15 in *California Atlas*), triggered ticket price wars to get tourists and potential settlers to settle in Southern California on railroad owned lands. These developments were the turning point in starting Southern California's first major land boom, which also destroyed what remained of the old Californio culture. Even forty years after the Mexican-American War, use of Spanish language, adobe architecture, traditional clothing, social customs and recreations showed American tourists that they were in "foreign" territory when visiting Southern California (a state within a state). The big boom of the 1880s put an effective end to the last stand of the Californios.

The first part of this massive impact of people came in the form of tourists who wanted to relive *Ramona* and to enjoy the large **health resorts** (i.e., sanitariums) being built along the coast from Santa Barbara to San Diego. Many of these health hotels are still in use on the landscape today, including the Hotel Coronado (Figure 9.18) near San Diego and the Biltmore Hotel in Santa Barbara. The longer people stayed to enjoy

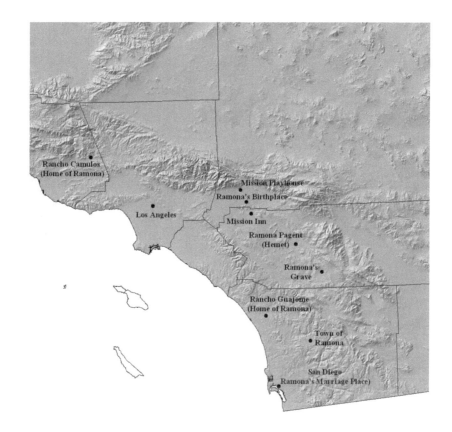

(a)

(b)

FIGURE 9.17 *(a) Tourist map for Ramona (Adapted from Delyser, 2003); and (b) Aerial view of Rancho Camulos (near Piru, Ventura County) in 1888, with vineyards on the left and the Santa Clara River in the background.*

FIGURE 9.18 *Hotel Coronado health resort (image taken c. 1900) located in Coronado Beach near San Diego. (Courtesy of The Library of Congress)*

the "healthy and freedom loving" Southern California lifestyle, the more they wanted to stay. This led to a huge **real estate** rush instigated by the railroads on their government granted lands, and therefore nearly an instant ability to create towns with a railroad station. They promoted Southern California to people from the Midwest as a land of abundant sunshine that allowed for frequent outdoor activities throughout the year (i.e., season was not an issue). The landscapes had natural amenities (e.g., beaches and mountains close by) that created a rich variety of recreational opportunities.

Southern California also became a favored spot for new communities developed around **utopian** (socialism or absolute equality) or spiritualist ideas (Theosophy Society). These communities included Arcadia, Summerland, Pasadena, Point Loma (Krotona) and Anaheim to name a few. During this era, a rejection of urbanism was on the rise in Europe and in the large East Coast cities largely driven by socialist/Marxist literature that was critical of the stifling nature of the urban/industrial complex. Southern California became a natural home to many of these communities as their people could come west to establish a new social pattern based upon a vision of the ideal society that emphasized healthy outdoor living and a self-sustaining agricultural economy for its residents. In Southern California, the new patchwork of purpose driven communities supported by both private real estate and railroad driven land schemes replaced the old Arcadian landscape of the ranchos.

9.3.3 "ORANGES FOR HEALTH, CALIFORNIA FOR WEALTH:" EXPANSION OF THE CITRUS INDUSTRY

One of the most striking landscape changes across the Los Angeles basin during this period is the development of an evergreen treed landscape in place of the rancho coveted grasslands. The **orange citrus** and later lemon agricultural industry played an important role in shaping the image of Southern California by playing up the Mediterranean-type and subtropical climate. Southern Pacific railroad marketing presented Southern California to potential Midwestern farmers as the land of sunshine, orange groves and opportunity. Here one could grow one's own healthy fruit while projecting the healthy, benign climate image as part of the marketing to consumers in the East. The industry played heavily on romantic cultural and natural imagery, which stemmed from the fact that the Spanish originally introduced citrus at the irrigated missions. This was recognized earlier in the 1850s at abandonded and crumbling Southern California Mission San Gabriel where thirty orange trees of the original 400 planted remained and seed was available. By the late 1860s the Los Angeles region hosted 15,000 orange trees and 2,300 lemon trees. Businessmen internally traded the fruit from these trees in Southern California, largely sending the fruit by ship north to San Francisco and on to Sacramento and the miners in the goldfields.

In the 1870s, **Riverside** became the seat of citrus production in Southern California especially after the introduction of the navel orange, which the U.S. Department of Agriculture brought in from Brazil. This variety became very popular, and thus when the railroad arrived from the north and later from the southeast with refrigerated cars, the cultural landscape of the citrus industry boomed into its own "orange rush." The railroads made sure to improve their connections to the citrus growing areas of Southern California and then helped form the California Fruit Growers Exchange (also known as Sunkist) to handle distribution and marketing to eager Midwest and East Coast buyers. Southern Pacific railroad in particular boosted citrus growing opportunities to Midwestern farmers; thus would-be citrus ranch barons came flocking to Southern California. Improved and expanded irrigation methods helped the rapid growth and expansion in the 1880s, with the result that millions of trees covered Los Angeles, Riverside and Orange counties. Southern California alone was able to supply 70 percent of the nation's oranges and 90 percent of its lemons by 1900. The era of "king citrus" came to an end though by World War II when urban encroachment on the orchard lands turned Los Angeles and Orange counties into suburbs (i.e., "housing orchards"), and citrus growing only remained in parts of Riverside County, Ventura County (largely lemons) and moved to the eastern San Joaquin valley (largely oranges).

One of the lasting images of the citrus growing era that helped propel Southern California into its boom time was the citrus crate labels (Figure 9.19). People all over the country saw these labels, which promoted a romantic image of the region. Idyllic landscapes, a postcard view of Southern California stereotypes, orange groves lined by palm trees and backed by beautiful snow capped mountains, Mediterranean-style ranch buildings, romanticized images of the missions and Mexican ranchos, idyllic utopian settlements shown in green

(a)

(b)

FIGURE 9.19 *Citrus crate labels from the late 1800s: (a) Ramona brand; and (b) Mission memories. (Courtesy of the California Historical Society)*

valleys surrounded by mountains and relaxing settings at the beach—all with oranges. All of these things magnetized visitors, potential orange-barons lured from the Midwestern states, land speculators, and other newcomers before 1900 and up to 1930.

9.3.4 THE OIL BOOM: SOUTHERN CALIFORNIA'S "BLACK GOLD RUSH"

Despite all that has been explained to this point, it was the discovery of **petroleum oil** that made Los Angeles. At the start of the industrial revolution, coal deposits were the catalyst for energy required by industry for mass production. In 1859, oil was discovered in Pennsylvania, and soon government geologists that surveyed parts of California predicted great oil wealth but said that the geology of the state (i.e., the land of plate tectonic warping) would make it difficult to drill. In Southern California, from Santa Barbara County south, Native Americans and the Spanish/Californios knew of oil and tar deposits along the coastal beaches and towards the interior, like at La Brea tar pits near Los Angeles. The Chumash Native Americans had used the tar to caulk their tomol plank boats to make them watertight.

The first major oil find occurred in Ventura County by Union Oil in 1890. Soon after that, drillers discovered oil at Summerland beach (1890, home to a utopian group of spiritualists; Figure 9.20), at Los Angeles (1892, near Dodger stadium; Figure 9.21), in Kern County (1899, a massive field west of Bakersfield), then at Huntington Beach and the massive find at Long Beach (in the 1920s at Signal Hill).

Oil brought as much wealth to Southern California as gold did to Northern California. Today companies still pump oil in Southern California, the southern San Joaquin valley (Kern oil field) and at offshore oil platforms from Santa Barbara to Orange Counties. California is the third largest terrestrial oil producer in the U.S., not counting the offshore oil rigs, or fourth if they are included. The coming of the automobile intensified the oil industry in Southern California, prompting such Western-centric magazines such as *Sunset* to run images of the automobile with smiling people in outdoor places declaring "Life's Best Mixture: Sun and Air and Gasoline!"

FIGURE 9.20 *A field of oil wells just offshore, at the former utopian spiritualist colony of Summerland, California (Santa Barbara County), c. 1915. (Courtesy National Oceanic and Atmospheric Administration)*

FIGURE 9.21   *A complex of oil derricks in 1896 near Los Angeles.*

9.3.5 THE CULTURAL CHARACTER OF THE 1880s BOOM

In summary, by 1900 over sixty new towns covering more than 80,000 acres had developed in Southern California with the help of the Southern Pacific Railroad (and its booster publication *Sunset*—"the magazine for Western living"). The city of Los Angeles is an example: its population growth in ten year increments from 1880 to 1930 was as follows: approximately 15,000; 50,000; 102,000; 320,000; 576,000 and 1,200,000. This is an 8,000 percent increase in population within fifty years. A figure for 1890 puts the overall region at 200,000 people. Who were these people that came to Southern California to make new lives for themselves in the various industries for which the region became famous? The migrants to Southern California created a different society because they were a different mix of people than the people who built Northern California during the Gold Rush.

There were fewer foreign born residents, and the Californio/Mexican populations were no longer substantial as those cultures deteriorated. In all, Southern California had a mixture of 15% Europeans, 2% Asians, 2% Californios/Mexicans, 1% Native Americans and Blacks, 33% other Californians, and 47% of Midwestern/ East Coast origin. An apt cultural description of the last group would be conservative farming stock, bankers, merchants and professionals, of Protestant households and guided by Republican politics (defined by the old definition prior to the 1960s as fiscal conservatives), more often than not citizens of considerable affluence.

9.4   THE PROGRESSIVE ERA: ARTS AND CRAFTS MOVEMENT IN SOUTHERN CALIFORNIA

Historians call the timeframe of 1890 to 1930 California's **progressive era**, and the focus is especially upon Southern California. All of the discussion up to this point really lends support to why Southern California supported a different cultural landscape compared to Northern California. The look and feel of the region became etched further into the landscape during this period. In the quest to understand this rethinking of the landscape, one must revisit the utopian mood sweeping the Western world starting in the 1860s that fully moved to the U.S. by the 1890s. It began in England with a strong negative reaction to industrialism, as it separated people from their work and from each other. The goals of the movement were to return to a simpler life that focused on craft guilds and emphasized skilled handicrafts and the material arts. To make this happen there was a further focus on communities living in small villages close to nature.

As with many cultural themes, California had its own version of how to approach the arts community. The immigrants to Southern California who participated came from the East and the Midwest and were enthralled with the romanticized Spanish-Mexican and Native American past. After the publication of *Ramona*, another newcomer from the East Coast named **Charles Fletcher Lummis** arrived in 1885; he was a great admirer of Helen Hunt Jackson and he helped boost the *Ramona* legend. He traveled the Southwest working as a reporter for the *Los Angeles Times* and started his own magazine called *Land of Sunshine/Out West*, which focused on the Spanish-Mexican and Native American past especially through the missions. The Los Angeles Chamber of Commerce subsidized the magazine and sent it to the Midwest and the East Coast as promotional material to sell the southland to potential migrants. As a mission revivalist, Lummis was successful in establishing a society to restore the missions in 1888.

The center of what became known as the **Arts and Crafts** movement out West was Pasadena (twenty-five miles inland, up against the San Gabriel Mountains) and in particular the Arroyo Seco, an intermittent waterway following a westward path from the Sierra Madre to the ocean. This region was relatively rural and natural in the 1890s, and it became the home area of Lummis and a loose grouping of artists and artisans who looked to the Spanish-Mexican and Native American past for regional identity, lifestyles, building styles and materials. Their colony was known as the **Arroyo Culture** which marks the beginning of the **California craftsman** era which looked to nature for guidance for a simple, meaningful life and emphasizing craftsmanship over industrial building styles.

9.4.1 CRAFTSMAN BUNGALOW ARCHITECTURE IN SOUTHERN CALIFORNIA

One of the lasting characteristic buildings to emerge from the California craftsman era was the **California bungalow** single family home (Figure 9.22). The bungalow was originally from India and was a modification of a style of British colonial summer houses built in the Himalayan foothills. It embodied a craftsman style from back East and really represented democratic ideals: It was inexpensive to build, and one could build it by oneself, based on simple plans; it rejected the pretentiousness of Eastern styled and Victorian homes; and it showed the open connectedness of the home with the outdoors. The bungalow worked in Southern California because the benign climate fit its form, i.e., low sweeping eaves, broad porches and patios, and tidy front and backyards. This was the home for the people; it provided respectability in an age that popularized that concept. It provided a fulfillment of the American Dream (i.e., ownership of an individual house on a separate lot) but set in the California style of connecting with the natural and rural through gardens and the artistic use of local stone, rough bricks and wood. Through the magazine Lummis edited and through other national magazines like *The Craftsman*, the California bungalow design spread across the country from California around 1900 to 1930.

After World War I the bungalow court type (Figure 9.23) of housing evolved to provide dwellings for those who wanted a house and garden but who could neither afford a detached house nor bother with the upkeep. The war drove cultural change so that people longed for an

FIGURE 9.22 *Example of a California bungalow. (Image © Diana Lundin, 2009. Used under license from Shutterstock, Inc.)*

FIGURE 9.23 *Spanish style bungalow court near Balboa Park in San Diego.(Courtesy Journal of San Diego History)*

FIGURE 9.24 *Victorian row homes next to Alamo Park, San Francisco. (Image © Aaron Wood, 2009. Used under license from Shutterstock, Inc.)*

independent lifestyle but with a strong sense of community and security. The bungalow court was born by grouping attached bungalows around a central common area and making that area a walking area unencumbered by the automobile (i.e., similar to the modern cul-de-sac housing arrangement). They provided an alternative to cramped multistory apartments and appear as an early example of planned development similar to smart growth or new urbanism promoted in our current modern trend towards sustainable community development.

While the Southern California bungalow eventually made its way into building styles in Northern California, there were still differences in the general look of the built environment. Northern California owes its style to the Victorian age of highly ornate buildings. These were the classic "gingerbread" houses one sees in San Francisco (Figure 9.24), usually painted with a minimum of eight different colors to bring out the ornate accents of woodwork in which these homes excel. This is a style of home imported from the Eastern U.S., and it made sense in Northern California with its vast supply of lumber from the productive coastal and Sierra Nevada forests.

9.4.2 MISSION REVIVAL DESIGN IN SOUTHERN CALIFORNIA

The last phase of romantic myth via cultural building design, which began in the 1890s, was the **mission revival** movement. This was an organized effort to express the Spanish-Mexican past in building architectural styles by using the missions as inspiration. With Lummis' group restoring the missions and the subsequent promotion of the period in romantic novels, marketing literature and magazines, the California contingent to the Chicago World's Fair in 1893 (the World's Columbian Exposition noting the four hundredth anniversary of Columbus' New World discovery) crafted the California building at the fair to look like a blend of mission and southern European Mediterranean architecture. Because over half the country visited the fair during its one year of operation, the mission revival in architecture became deeply ingrained in visitor's minds as a representation of what California's houses and buildings should look like when they visited or migrated. The common building elements included plain stucco walls, red tile roofs, arches over doors and windows and local rock accents—essentially replica mission characterizations (Figure 9.25). This style also readily adapted to the bungalow form and bungalow courts (Figure 9.23).

To this day this style of housing dominates the character of tract home building in Southern California, whether they're condos or homes. The use of stucco and Spanish red tile roofs, along with a patio for connection to the outdoors, and possibly an outdoor oven/BBQ and a fountain or pool, has become synonymous not just with Southern California, but with California architectural style in general—the California dream home.

(a)

(b)

FIGURE 9.25 *(a) Mission Spanish revival design for the famous Mission Inn in Riverside built in 1902. (b) An aspect of the California bungalow as built in a Mission Spanish revival design in 1928 in Northern California's Sacramento Valley (Courtesy Dean Fairbanks).*

FIGURE 9.25 (cont.)   *(c) Santa Barbara Courthouse built in 1926 in Spanish Mission revival style after the 1925 earthquake (Courtesy Tom McFaul).*

9.5 A Development Pattern Snapshot of California: 1880–1930

Modern California took shape after the initial jolt of the Gold Rush to the northern half of the state and the major boosterism that promoted the revisionist myth of the Spanish-Mexican so-called easy life in Southern California. If we start to look at the state regionally we can find the following activities evolved based on resource availabilities and cultural characteristics:

- **North Coast**
  - Redwood logging
  - Fishing industry (salmon)
- **Great Central Valley**
  - Sacramento Valley
    - Cattle ranching
    - Wheat production
    - Irrigation agriculture—fruit tree orchards
  - Delta region
    - Sacramento and Stockton are industrial inland ports
    - Irrigation agriculture—vegetables
  - San Joaquin Valley
    - Cattle ranching
    - Irrigated agriculture—vegetables and citrus
- **Sierra Nevada/Cascade foothills**
  - Minimal hard rock lode gold mining
  - Logging
  - Largely deserted after the end of hydraulic gold mining
- **San Francisco Bay Area**
  - Manufacturing and commercial activities
  - Banking and investment

- Redwood logging
- Irrigated agriculture consisting of vineyards and fruit tree orchards
- Fishing industry
- University higher education

- **Central Coast**
  - Limited population
  - Cattle ranching
  - Limited irrigation agriculture
  - Fishing industry (sardines)
  - Mercury mining

- **Southern California**
  - Service industries—tourism and health spas (sanitariums)
  - Retail trade
  - Real estate speculation
  - Oil supply
  - Irrigated citrus agriculture
  - Early movie production center
  - Limited mining in the Mojave desert

The Western frontier was now considered closed with the inhabitable lands in the Sacramento and San Joaquin Valleys settled and San Francisco represented the social and economic hub of the state. However, 1900 became the seminal shift in the growing modern trend California was facing, represented by a southward shift in population (i.e., approximately 30 percent of the state's population).

When we consider immigration and ethnic patterns, the Irish and Chinese had the rank of the first or second largest foreign-born groups in California. As we have already discussed, however, the 1882 Chinese Exclusion Act started to strongly curtail their immigration by 1900. Irish immigration also witnessed a dwindling immigration independently. Other top ethnic groups were Germans, English, and Canadians. After 1900, the ethnic population patterns changed yet again with a major push of Southern Europeans (i.e., Italians, Portuguese, Basques, etc.) to California because of political and economic turmoil in their home countries. California was a pull for these cultural groups as the climate was similar to their homelands and the state represented far superior opportunities to build new lives. The cultural patterns also changed based on the expansion of irrigation agriculture, and the labor required to maintain the growing specialization of crops that California farmers decided to grow (e.g., vineyards, nut/fruit orchards).

California was on its way to becoming the great state it always wanted to be. It achieved this through two major, highly entwined avenues: agriculture and water. The rise of irrigation agriculture and agribusiness, which was foreshadowed on the horizon with the development of the wheat growing industry, along with the precious resource of water were the major builders of California's overall wealth—north or south. Water became a major pawn for California's two urban empires: San Francisco and Los Angeles. As the saying goes in California, "Where water flows, food and housing grows." Water thus became more important than the cheap labor upon which California agriculture relied.

Bibliography

Almaguer, T. (1994). *Racial Fault Lines: The Historical Origins of White Supremacy in California*. University of California Press, Berkeley, CA.

Baur, J.E. (1951). The Health Seekers and Early Southern California Agriculture. *The Pacific Historical Review* 20(4): 347-363.

Beck, W. (1974). *Historical Atlas of California*. University of Oklahoma Press, Norman, OK.

Boone, C.G. (1998). Real estate promotion and the shaping of Los Angeles. *Cities* 15(3): 155-163.

Booth, S. (2008). *California Geography*. Course taught at Sierra College. [online] http://geography.sierra.cc.ca.us/booth/California/cal_index.htm.

Calisphere (2008a). California Cultures. University of California [online] http://www.calisphere.universityofcalifornia.edu/calcultures/

Calisphere (2008b). A world of California Primary Sources. University of California [online] http://www.calisphere.universityofcalifornia.edu/

Chase, L. (1981). Eden in the orange groves: bungalows and courtyard houses of Los Angeles. *Landscape* 25: 29–36.

Curtis, J.R. and L. Ford (1988). Bungalow Courts in San Diego: Monitoring a Sense of Place *Journal of San Diego History* 34(2). [online] http://www.sandiegohistory.org/journal/88spring/bungalow.htm

Davis, C. and D. Igler (2002). *The Human Tradition in California*. Scholarly Resources, Wilmington, DEL.

Delyser, D. (2003). Ramona memories: fiction, tourist practices, and placing the past in Southern California. *Annals of Association of American Geographers* 93: 886–908.

Deverell, W. and D. Igler (2008). *A Companion to California History.* Wiley-Blackwell, New York, NY.

DeWitt, H.A. (1999). *The Fragmented Dream: Multicultural California.* Kendall/Hunt Publishers.

DeWitt, H.A. (1999). *The California Dream.* 2nd edition. Kendall/Hunt Publishers.

Dilsaver, L.M., Wyckoff, W., and W. Preston (2000). Fifteen events that have shaped California's human landscape. *The California Geographer,* 40: 3–78.

Donley, M.W., Allan, S., Caro, P., and C.P. Patton (1979). *Atlas of California.* Pacific Book Center, Culver City, CA.

Dumke, G.S. (1942). The real estate boom of 1887 in Southern California. *The Pacific Historical Review* 11(4): 425–438.

Durrenberger, R.W. and R.B. Johnson (1976). *California Patterns on the Land.* 5th edition, Mayfield Publishing Company, Mountain View, CA.

Faragher, J.M. (2001). Bungalow and ranch house: the architectural backwash of California. *Western Historical Quarterly* 32: 149–173.

Fogelson, R.M. (1993). *The Fragmented Metropolis: Los Angeles, 1850–1930.* University of California Press, Berkeley, CA.

Franks, K.A. and P.F. Lambert (1985). *Early California oil: a photographic history, 1865–1940.* Texas A&M University Press, College Station, TX.

Garcia, M. (2001). *A World of Its Own: Race, Labor, and Citrus in the Making of Greater Los Angeles, 1900–1970.* University of North Carolina Press.

Gebhard, D. (1967). The Spanish colonial revival in Southern California (1895-1930). *Journal of the Society of Architectural Historians* 26(2): 131–147

Haas, L. (1996). *Conquests and Historical Identities in California, 1769–1936.* University of California Press, Berkeley, CA.

Henderson, G. (1992). *Regions and Realism: Social Space, Regional Transformation, and the Novel in California, 1882–1924.* UC Berkeley, Dept. of Geography, PhD Dissertation.

Hine, R.V. (1983). *California's Utopian Colonies.* University of California Press, Berkeley, CA.

Hornbeck, D. (1983). *California Patterns: A Geographical and Historical Atlas.* Mayfield Publishing Company, Mountain View, CA.

King, A. (1984). *The Bungalow—The Production of a Global Culture.* Routledge & Keegan Paul, London, UK.

Kurtuz, G. (2007). Southern California for Health, Wealth, and Sunshine: The Role of Boomers and Boosters. *South Pasadena Public Library, Volunteers Recognition Day,* March 23.

Lancaster, C. (1985). *The American Bungalow, 1880s–1920s.* Abbeville Press, New York, NY.

Larson, D.J. (1996). Historical water-use priorities and public policies. In *Sierra Nevada Ecosystem Project: Final report to Congress,* vol. II, Assessments and scientific basis for management options. University of California, Centers for Water and Wildland Resources, Davis, CA.

Lindley, W. and J.P. Widney (1888). *California of the South: its Physical Geography, Climate, Resources, Routes of Travel, and Health-Resorts.* D. Appleton and Co., New York, NY.

McClelland, G.T. and J.T. Last (1995). *California Orange Box Labels: An Illustrated History.* Hillcrest Press, Santa Ana, CA.

McWilliams, C. (1973). *California, the Great Exception.* Peregrine Smith, Santa Barbara, CA.

McWilliams, C. (1995). *Southern California: An Island on the Land.* Peregrine Smith Books, Salt Lake City, UT.

Michaelson, J. (2008). *Geography of California.* Course at UC Santa Barbara, Dept. of Geography. [online] http://www.geog.ucsb.edu/~joel/g148_f08/.

Morrison, P. A. *The Role of Migration in California's Growth.* Santa Monica, Calif.: Rand Corporation, 1971.

Nelson, H.J. (1959). The spread of an artificial landscape over Southern California. *Annals of the Association of American Geographers* 49: 80–100, supplement.

Norris, F. (2008). *The Octopus: A Story of California.* IAP.

Orsi, R.J. (2007). *Sunset Limited: The Southern Pacific Railroad and the Development of the American West, 1850–1930.* University of California Press, Berkeley, CA.

Rawls, J.J. and W. Bean. 2008. *California: An Interpretative History.* 9th Edition, McGraw-Hill, Boston.

Rice, R., Bullough, W., and R. Orsi (2001). *The Elusive Eden: A New History of California.* 3rd edition, McGraw-Hill.

Royce, C.C. (1899). *Indian Land Cessions in the United States.* Eighteenth Annual Report of the Bureau of American Ethnology, Washington D.C.

Salkin, J. and L. Salkin (1976). *Orange Crate Art: the Story of the Labels That Launched a Golden Era.* Warner Books.

Social Explorer (2009). *United States Census Demographic Maps: California 1850–2007.* [online] http://www.social-explorer.com/pub/maps/home.aspx.

Starr, K. (1973). *Americans and the California Dream, 1850–1915.* Oxford University Press, New York, NY.

Starr, K. (1985). *Inventing the Dream: California Through the Progressive Era.* Oxford University Press, New York, NY.

Starr, K. (2005). *California: A History.* The Modern Library, New York, NY.

Streatfield, D.C. (1976). The evolution of the Southern California landscape, 1: Settling into arcadia. *Landscape Architecture* 66: 39–46

Streatfield, D.C. (1977). The evolution of the California landscape, 2: Arcadia compromised. *Landscape Architecture* 67: 229–239.

Streatfield, D.C. (1985). Where pine and palm meet: the California garden as a regional expression. *Landscape Journal* 4(2): 61–74.

Thompson, W.S. (1955). *Growth and Changes in California's Population.* The Haynes Foundation, Los Angeles.

Vance, J.E. (1972). California and the search for the ideal. *Annals of the Association of American Geographers* 62(2): 185–210.

Webber, H.J. (1967). History and Development of the Citrus Industry. In *The Citrus Industry, Vol I: History, World Distribution, Botany, and Varieties,* edited by W. Reuther, H.J. Webber and L.D. Batchelor. Unviersity of California Regents. [online] http://lib.ucr.edu/agnic/webber/Vol1/Chapter1.htm

White, G.T. (1962). *Formative Years in the Far West: A History of Standard Oil Company of California and Predecessors Through 1919.* Appleton-Century-Crofts, New York, NY.

Winter, R. (1980). *The California Bungalow.* Hennessey & Ingalls, Inc., Los Angeles, CA.

Wright, D.M. (1940). The making of cosmopolitan California. *California Historical Society Quarterly* 19: 323–343.

Wright, G. (1981). *Building the Dream: A Social History of Housing in America.* Pantheon, New York, NY.

CHAPTER 10

California Agriculture and Irrigation Agri-"Cultures"

Key terms

Agricultural Adjustment Act 1933
Alien Land Law of 1913
Apology Act 2005
Barrios
Ben Haggin
Bracero program
Bureau of Reclamation
California cuisine
California Doctrine
Chinese
Colonias
Dust Bowl
Exodus
Factories in the field
Farm labor
Filipino
Filipino Repatriation Act 1935
Gentlemen's Agreement 1907
George Chaffe
Great Depression 1930s
Hoovervilles
Immigration Act of 1924 (National Origins Act)
Indigent Act 1937
Irrigation
Issei
Japanese
Pensionados
Mexican
Mexican Repatriation (Operation Deportation 1929)
Miller-Lux
Mormon
National Labor Relations Act 1935
National Reclamation Act 1902
Nihonmachi
Nisei
Okies
Paper farms
Picture brides
Prior appropriation rights
Punjabi
Riparian rights
Route 66
The Grapes of Wrath
Total Engagement
Unionization
Wheat
Wright Irrigation Act 1887

Introduction

"The world of California agriculture is formidable . . . whose envoys and practitioners probe every corner of the state's working landscapes. . . . Bold in ambition and vast in extent, California agriculture advances where growers experiment with novel crops rarely before seen, yet soon to be on display in markets across the U.S."

—Field Guide to California Agriculture, 2010

183

If one disregards the urbanization that has occurred in California over the last 150 years, the most remarkable mark on the cultural landscape is farms and farming practice. Since the Spanish era with its mission agriculture to modern industrial production, specialized farming techniques have always been a part of the California agricultural landscape. The physical environment of the state represented by the interplay of soils, topography and climate create a diverse set of environmental conditions from which agriculture has found potential growing niches for a large variety of crops. The environment determines largely what farmers can grow, and whether they can harness technology to push an area's environmental potential (possibilism) for new crop types or crop expansion. Then they can assess the crop type and area for its comparative economic advantage to other states production capacities or other regions in the world for export or import. This combination of natural environment and technology, especially with regards to irrigation development (Figure 10.1), has made many California crops economically advantageous to grow in the state rather than in other parts of the U.S.

California's great drawback as a productive landscape, however, is the timing and amount of the rainfall it receives each year. In a Mediterranean-type climate system rainfall occurs during the winter months, the very opposite of when enough radiation (longer day lengths) and thus warmer air temperatures are available to stimulate plant growth (see Chapter 3 for review; see page 7 in *California Atlas*). In addition, the majority of this orographically driven rainfall falls in northern California and along the Sierra Nevada (see Chapter 3 to review details; see page 6 in *California Atlas*). Meanwhile in the southern San Joaquin Valley

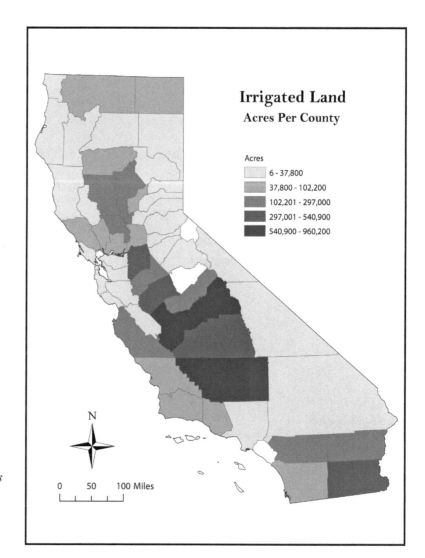

FIGURE 10.1 *Acres of irrigated land as percent of land in farms acreage across California in 2007. (Data source: U.S. Department of Agriculture Census; Courtesy Kirstyn Pittman)*

and Southern California there is insufficient rainfall during the winter to stimulate even winter sown crops. This is why spring wheat was grown so successfully in the Sacramento Valley and cattle ranching was conducted successfully in the San Joaquin Valley during 1860–1900. In compensation for the geographical and temporal precipitation situation that California has always found itself in, the Spanish era and the American era have relied on the **irrigation** of their crops by moving water from regions with surplus amounts to regions of scarcity. Irrigation, realized as far back as the Spanish era in mission design and development, makes California's farming system and agricultural patterns massive in size, diverse in location and very profitable.

10.1 CALIFORNIA AGRICULTURE AND RECLAMATION

In the early years following the Gold Rush, would-be American farmers trying to work with farming conditions in California opted for cattle grazing, thus following the footsteps of the Californios. Eventually, as discussed in Chapter 8, spring wheat became a popular and very successfully grown dryland crop. Indeed, the unusual environment presented to American settlers coming from the humid Midwest and East Coast combined with the isolated nature of California made for confusing and tough conditions. A review of the Homestead Act of 1862 helps to understand how its passage in many ways made it worse for potential farmers in the arid West. Recall from Chapter 8, that the act created a series of townships broken down into one square mile sections, which were further broken down into quarter sections of 160 acres as a means to develop public lands. The sectional title had one clear disadvantage in California: The owner of one could only conduct dryland (rain-fed) farming or grazing on their land; irrigation farming was not allowed. Therefore, for decades following the Gold Rush and especially for those participating in the Homestead Act, California farming was largely a failure for those many farmers who tried to apply the agricultural methods, techniques and crops (i.e., corn, pumpkins, squash, cotton, etc.) that they brought with them from the humid Midwest and the East Coast. They dryland farmed many of these crops in those other parts of the country under conditions of summer rainfall. Thus in those other regions, available heat (warmer temperatures and longer day lengths) coincided with available precipitation. California farmers soon learned what the majority of California Native Americans faced with their lack of classical agriculture (see Chapter 5 for review) and what the Spanish understood upon their arrival. To be successful in a Mediterranean-type climate there are two options:

1. One adopts and grows crops that are indigenous to the local Mediterranean-type system or brings them from another Mediterranean-type system (i.e., wheat, olives, grapes, figs, deciduous fruits, etc.); or

2. One augments the lack of precipitation during the long drought-controlled summers through irrigation, which then allows the growing of both humid-condition and non-humid-condition crops practically anywhere there is a sufficient soil base and appropriate temperatures.

The Spanish missions started the original irrigation agriculture system in California, cropping a wide variety of Mediterranean basin-based crops and New World humid climate plants. We can consider the missions—in many respects due to their large latitudinal extent along both the immediate coast and inland valleys—California's first agricultural experiment stations.

The secularization of the missions and the rise of the cattle rancho put an end to the prospects of irrigation agriculture in California until a small group of **Mormon** African-American farmers settled in San Bernardino County (near Redlands) in 1852 and started irrigation farming. This was the first inkling of farming by group effort, which is what it took to make the control of water (cultural, legal and technical) for irrigation work in California. This group later had to abandon its effort after polygamy problems with the federal government. Other than this first early attempt at post-mission irrigation, farmers in the 1860s who were both under the Homestead Act and those who were not adopted the growing of **wheat**, a Mediterranean basin-based small grain annual grass, as a means to survive and because of demands for wheat in Europe. Spring wheat requires sowing during the winter rain period in January, and farmers can then harvest it in June. If a farmer was lucky and California did not witness cyclical drought during the year, then one could generate a reasonable to substantial harvest. As stated in Chapter 8, wheat rapidly became the crop to grow in Northern California's Sacramento Valley (where sufficient winter rainfall could occur), so that by 1884 California had become the nation's largest wheat supplier and second in global production. Wheat production virtually disappeared by 1910, but its experience left California farmers with national and global trade markets, large farms, absentee ownership, easy financing and hooked on mechanization—which led to the rise of agribusiness in California's future. Why was there a shift away from cattle and annual small grain grasses in California's agricultural landscape? The answer lies in the development of reliable transportation through the railroad and the command and control of water—irrigation water to be precise.

On non-homestead lands (i.e., especially in the Sacramento-San Joaquin Delta region), farmers were experimenting with a variety of fruit, nut and vegetable crops during the cattle-grain era of the 1860s, but California's isolation from national markets and comparatively rather small urban areas made them unprofitable. This started to turn around by the 1880s when irrigation technology developed during the rise of Southern California's citrus orange production boom instigated by **George Chaffe** in Riverside. His was the first technical expansion of irrigation via piped water in the Western U.S. to support a "colony development" he created. It is also important to note that his water system was the first in the Western U.S. to develop hydroelectric power, which later became important for expanding irrigation agriculture to run groundwater pumps in the Central Valley (see page 23 in *California Atlas* for modern hydroelectric generation locations).

10.1.1 WATER DOCTRINE IN CALIFORNIA: PRIOR APPROPRIATION VS. RIPARIAN RIGHTS

The key to understanding the rise of irrigation agriculture involves exploring the water rights issues that plagued the last half of the nineteenth century. Keep in mind that these discussions about water rights of any kind involve "use" rights, not a right to the body of water itself. In the mining areas, the established practice of **prior appropriation rights** or "first in time, first in right" became the standard for water law throughout much of the West. This resembles Spanish water law regarding use since the mission-presidio-pueblo days, which California adopted during the Gold Rush. It allowed mining companies during the hydraulic mining era to amass rights to water across the Sierra Nevada landscape in order to impound and divert it for hydraulic mining purposes. This practice also allowed the establishment of water companies, which grabbed the right to impound and divert water for other uses, both municipal and agriculture. Under the prior appropriation doctrine the first to use water established the primary right to continued use, where the right was conditional on continued use (i.e., use it or lose it). In the era of mining, the evolution of natural resource law in California witnessed the absolute priority of rights established. The corollary to all this was that latecomers lose all water during shortages before users with higher rights lose any. For the first time in the U.S. the right to use water could be bought and sold like property. This encouraged individuals to swiftly divert water and coincided with a rapid development of uses.

In direct contrast to prior appropriation were **riparian rights**, which were based on common water law from England and operated in the Eastern U.S. from colonial times. Both England and the Eastern U.S. are humid, wet environments where the functioning of this type of right is well adapted. Because it was common water law in the rest of the U.S. the California legislature adopted riparian rights as well. Under riparian rights, the rights to water are attached to the ownership of land adjacent to a water source (i.e., a stream, river, lake, etc.). The land owner receives a guaranteed right to the use of full natural flow less any upstream usage by neighbors. The water usage is limited to domestic needs, including livestock, but not irrigation (e.g., it fits the same rules laid out for the national Homestead Act, and it is adapted to areas not requiring irrigation due to humid summer rainfall conditions). At the same time the riparian right is not conditional on usage (i.e., use when you need or not) and both the rights and the shortages are shared equally among users. California's Mediterranean-type climate with its long drought summers and long term drought cycles confused early homestead farmers from humid areas of the country which led to farm failure and subsequently to their abandonment and sale to create large land holdings by cattle barons, especially in the San Joaquin Valley.

Because the mining companies were so powerful, forcing prior appropriate rights, and the Homestead

Act and the California legislature were led to obey riparian rights, both conflicting legal systems co-existed and became incorporated into state water resource law in 1872. This obviously led to increasing political agitation, as large land owners (with riparian rights) grabbed up failed homestead farm quarter sections, while the California government increasingly favored irrigated agriculture on small farms. Many people realized early on that the rules of the homestead system were not sufficient in size or in water rights to survive California's climate situation.

10.1.1.1 Miller-Lux vs. Haggin: A Battle between Two Land Monopolists

In the last half of the nineteenth century, **Miller-Lux** Land Company (cattle ranching) was one of the largest holders of land in California, having acquired 1.25 million acres. The company was a vast network of ranches, hay farms and meat-packing plants developed to supply meat to the San Francisco urban area. Miller-Lux ruthlessly played the lands alienation game by preying on struggling homestead cattle ranchers and failed dryland farmers in the San Joaquin Valley. As David Igler (Historian, U.C. Irvine) importantly states, these gentlemen are the clue to understanding the West's industrial agriculture transformation as they "created a vast machine to engineer the natural landscape and regulate the geography of Western meat production."

In 1879 there was a major drought in the southern half of the already dry San Joaquin Valley, and the Kern River (near Bakersfield) dried up. Miller-Lux raised cattle on 150,000 acres (with riparian rights) on the lower half of the Kern River. Higher up the Kern River was a gentleman named **Ben Haggin** who started the Kern Valley Land and Water Company, which had assembled 400,000 acres (with prior appropriation rights). His water company was diverting all the water from the Kern to his irrigated alfalfa fields. Due to the lack of water for their cattle, the cattle died, and Miller-Lux decided to sue Haggin for taking all the water.

The ensuing battle led to the legal water control hybrid known as the **California Doctrine**, a duel riparian-appropriation system decided in this most famous California Supreme Court case in 1886. Essentially the conflict came down to ranching under riparian law and irrigated farming under appropriation. The court ruled that both sides had rights to the water. It reaffirmed riparian rights for owners of land along a river, but prior appropriation rights trumped riparian rights if the owner was the first user. With both laws declared valid, it came down to timing, i.e., when the river was first *used* (appropriation use date) versus when the land adjacent to the river was purchased (purchase date) determined who had first right to the water during a conflict. Therefore, appropriation prevails over riparian if the appropriator is the first user; otherwise, if the riparian rights holder is the first user then riparian prevails over appropriation rights user. In the case of Miller-Lux vs. Haggin, Miller-Lux had prior riparian rights (an earlier purchase date), which subsequently trumped Haggin's appropriation (late coming appropriation use date). Subsequently, appropriations cannot injure pre-existing riparian rights. Eight other Western states went on to adopt the California Doctrine.

10.1.2 WRIGHT IRRIGATION ACT 1887

After the Miller-Lux vs. Haggin court case exposed how large land monopolists were destroying the original intention of the Homestead Act and generally abusing the use of water in the state, the California legislature passed the **Wright Irrigation Act** of 1887. This single piece of enabling legislation allowed for the creation of independent local irrigation districts. Its purpose was to provide water infrastructure as a public good. A district received the power of eminent domain by having the authority to condemn water rights and irrigation facilities for public use. The districts were generally the owners of water rights and users' contracts. Irrigation districts also had the power to levy taxes and sell bonds to build canals and reservoirs to service their users.

At first it did not work well, as many California farmers were independent and conservative, in contrast to the Mormon group irrigation effort in San Bernardino discussed earlier. Irrigation districts represented group thinking, which was a cultural shift from the dominating rule of individuals making it alone on the landscape. Farmers soon reached the conclusion, however, that to survive and be successful in California, they required access to irrigation. It was the dawning of the golden age of California agriculture. Therefore, by 1890, 30 percent of the state's crops had become irrigated in a transition from cattle and wheat to fruit, nut and vegetable crops, all hooked to a transcontinental railroad network for Eastern markets. After 1900, acreage, production and profitability of these specialty crops increased sharply, thus obscuring family farming as originally a means of subsistence.

The Wright Irrigation Act was a model all the other Western states copied. Finally, in 1902, the federal government passed the **National Reclamation Act**, which changed the homestead rules to allow irrigation agriculture (also increasing the homestead size to 320 acres if a married couple) and provided infrastructure support from the government to develop and supply water to the homestead lands. This landmark legislation also created the **Bureau of Reclamation**, tasked with the construction of large federally funded irrigation systems to supply water to support these "family farms" that the new homestead rules aimed at. By the early 1900s the modern foundations of contemporary agricultural patterns, based on irrigation farming, had become established. In Chapter 11, the command and control of water in the state will be explored as it has pertained to the establishment of the coastal urban empire growth and interior agricultural dominance.

The use of irrigation water again set into motion a transformation of the California landscape to one of ordered trees (Figure 10.2; nuts and fruits) and a mosaic patchwork of flat fields representing a diverse set of vegetables and fruits. The process leading to this mass land transformation began with the Spanish unknowingly bringing in exotic annual grass species from the Mediterranean basin that slowly changed the native perennial bunch grassland. Along with the existence of large cattle ranches, this change helped turn the landscape into a grass-meat production landscape under the Mexican era. The Americans adopted the cattle ranch-

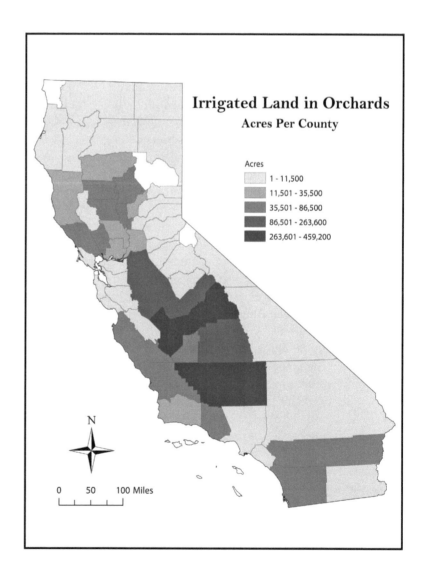

FIGURE 10.2 *Irrigated land in orchards across California in 2007 (Data source: U.S. Department of Agriculture Census; Courtesy Kirstyn Pittman)*

ing ways of the rancho days and learned to crop the interior Sacramento Valley and Delta region with another productive Mediterranean grass: wheat.

Meanwhile, early irrigation and re-examination of the missions' early irrigation production by farmers in Southern California's drier Mediterranean-type climate led to the development of both citrus tree production and vineyards (which were more plentiful in the Los Angeles basin than in Northern California at the turn of the century). Once farmers and/or irrigation districts could develop irrigation projects anywhere in California, the exotic grasslands and wheat fields became transformed into tree production landscapes and vegetables. At the turn of the century in 1900, the reclaiming of the desert "wastelands" or the greening of the deserts began. A generous soil base, ample temperatures and especially lots of water were the ingredients for success—"where water flows, food grows."

10.2 Do You Know What We Grow? California Agriculture and Cuisine

Agriculture is one of the state's longtime leading industries, outside of the high technology sector. In 2010, the total agricultural production value reached $37.5 billion, which is almost twice as large as that of the next largest agricultural state, Iowa. In addition, nine of the nation's top 10 producing counties are in California. The production occurred on 81,700 farms with an average farm size of 311 acres (while the U.S. average farm size is 418 acres) covering 25.4 million acres. The number of farms versus the average farm size varied significantly until 1990 when they both stabilized (Figure 10.3).

The size of the export from California is $9.8 billion with 24 percent of the agricultural production going to foreign markets, especially the European Union, Canada, Japan, China and Hong Kong, and Mexico. By rank, California is the world's fifth largest supplier of food and agricultural commodities.

The production range in California is very wide, considering that the state produces more than four hundred plant and animal commodities. This means everything from general cattle production to such specialty crops as artichokes, almonds, and strawberries. In fact,

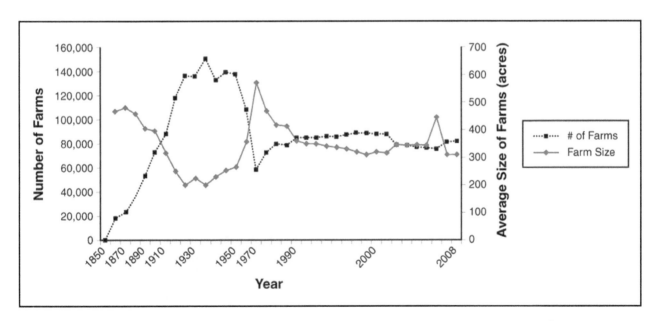

FIGURE 10.3 *Time series of farm size versus total acreage changes since 1850. Note that the average farm size prior to 1850 was on average 4465 acres (not charted) due to their association to the Mexican era ranchos. (Data source: U.S. Department of Agriculture)*

many of the crops cannot grow elsewhere in the U.S., or if they do, the farmers cannot grow them in large quantities. Crops, where California is the sole or top producer, range in origin from subtropical and Mediterranean to temperate:

- Avocados, dates, kiwifruit
- Almonds, artichokes, table & wine grapes, raisins, persimmons, pomegranates, olives, prunes, walnuts
- Sweet rice (Calrose medium grain), broccoli, lettuces, seed/ladino clover

Many of these crops helped California gain its own cultural representation in cuisine—**California Cuisine**—started by celebrity chefs in the 1960s and '70s.

Some final general points to say about California's agricultural landscape are that, besides being almost totally irrigated, the cropping occurs on flat fields (rarely terraced, with the exception of vineyards) in and around California's river valleys (Figure 10.4). Management of the modern California farm is by heavy industrialization in production and processing equipment, but even more important is its strong dependence upon a steady supply of cheap farm labor. Later in this chapter a discussion of the many cultural groups (agri-"cultures") that over time have been a part of the California agricultural machine or what a famous chronicler of California, Carey McWilliams, termed the "Factories in the Field" will be examined.

The following subsections provide a basic production ranking of the top twenty agricultural products (based on 2010 statistics) and where the state grows them.

10.2.1 FRUITS, TREE NUTS, AND BERRIES

California produces 46 percent of the total U.S. fruit and nut crops (Figure 10.5), with grapes being the top valued crop in this sector.

- Grapes—wine, table, and raisin—(Production rank: 2nd)
  - Table grapes and raisin production occurs in the hotter climate of the San Joaquin Valley. Wine grape production predominantly occurs in coastal valleys and on hills from Santa Barbara to Mendocino Counties, although there is a young vineyard region developing in Riverside and northern San Diego Counties, the central Sierra Nevada foothills and the Sacramento Valley (Figure 10.5).
  - A further differentiation among the wine growing areas comes in the American Viticultural Area (AVA) designation, which represents a region distinguishable by geographic features. An AVA specifies a region from which ≥ 85 percent of the grapes used to make wine it must have grown. Additional requirements are name recognition locally or nationally, that the boundaries are legitimate historically and that the growing conditions (e.g., the terroir) consisting of climate, soil, elevation and other physical features are distinctive to the area (Figure 10.6).
  - The following AVAs cover California with the number of smaller AVAs inside larger AVAs, showing the variety of California's viticultural areas.
    - Central Coast and Santa Cruz Mountains—29 AVAs
    - Central Valley—17 AVAs
    - Klamath Mountains—3 AVAs
    - North Coast—45 AVAs
    - Sierra Foothills—6 AVAs
    - South Coast—8 AVAs
- Table grapes are California's leading export to foreign markets.
- Strawberries—fresh and processed—(Production rank: 6th)
  - Coastal valleys from San Diego to Santa Cruz (Figure 10.5): more specifically the Oxnard Plain, Pajaro Valley, Salinas Valley, Santa Maria Plain, Lompoc Valley and Orange County.
- Almonds and Walnuts—(Production ranks: 3rd and 10th)
  - California produces all of the U.S. supply of almonds and walnuts. Almonds represent the second leading export to foreign markets. The production area covers the entire Central Valley (Figure 10.5). The San Joaquin Valley produces more almonds, and the Sacramento Valley produces more walnuts (see page 28 in *California Atlas*).
- Pistachios (Production rank: 9th)
  - San Joaquin Valley

County	Top Crops (2010)	(Millions)
Fresno	grapes, almonds, tomatoes, poultry, cattle and calves	$4,843
Tulare	milk, oranges, cattle and calves, grapes, alfalfa hay and silage	$3,871
Monterey	leaf and head lettuce, strawberries, nursery, broccoli, artichokes	$3,490
Kern	almonds, grapes, milk, carrots and citrus	$3,477
Merced	milk, chickens, almonds, cattle and calves, tomatoes	$2,284
Stanislaus	milk, almonds, cattle and calves, chicken, walnuts	$2,148
San Joaquin	milk, grapes, tomatoes, almonds, walnuts	$1,685
Ventura	strawberries, lemons, celery, woody ornamentals, tomatoes	$1,506
San Diego	flower and foliage plants, trees and shrubs, bedding plants, avocadoes, tomatoes	$1,461
Imperial	cattle, alfalfa, carrots, leaf and head lettuce, cotton	$1,308

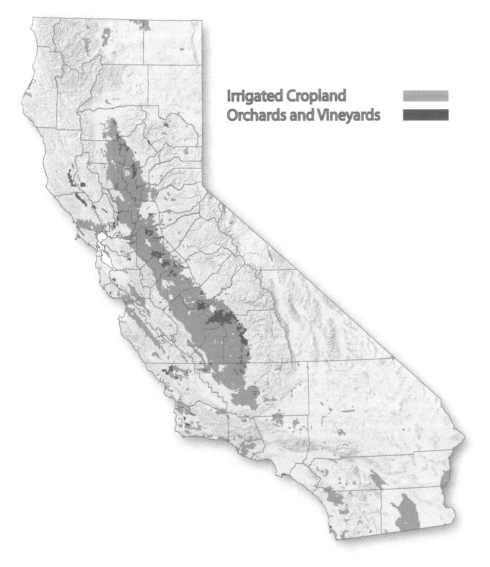

FIGURE 10.4 *Generalized agriculture map of California. (Data source: U.S. Geological Survey; Courtesy Curtis Page)*

192 **CHAPTER 10** *California Agriculture and Irrigation Agri-"Cultures"*

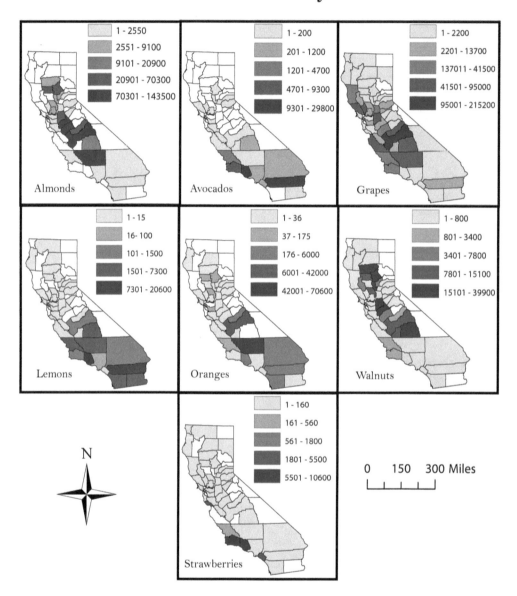

FIGURE 10.5 *Distributions in acres from top to bottom starting left to right for: Almonds, Avocados, Grapes, Lemons, Oranges, Walnuts, and Strawberries. (Data source: U.S. Department of Agriculture, 2007; Courtesy Kirstyn Pittman)*

- Oranges (non-Valencia)/Lemons—(Production rank: 15th and 20th)
  - Orange production takes place only in the southeastern San Joaquin Valley thermal belt just below the Sierra Nevada foothills (Figures 10.5 and 10.7). Some lemon production also occurs in this area. Remnants of the large Southern California lemon industry, however, remain in Ventura, Riverside, San Diego Counties and the Imperial Valley as well as the Colorado River around Blythe (Figure 10.5).

- Avocados—(Production rank: 19th)
  - Frost-protected areas in coastal region ranging from San Luis Obispo to San Diego Counties, with a focus in Santa Barbara, Ventura, and Riverside Counties (Figure 10.5).

FIGURE 10.6 *California cultural vineyard-scapes amongst the rolling oak woodland draped hills and mountains of the Coast Ranges. (Image © Galina Barskaya, 2009. Used under license from Shutterstock, Inc.)*

FIGURE 10.7 *Orange groves among the Sierra Nevada foothills in the San Joaquin Valley. (Image © Richard Thornton, 2009. Used under license from Shutterstock, Inc.)*

10.2.2 VEGETABLES—TRUCK CROPS

California is the leader in fresh market vegetables (Figures 10.8 and 10.9), which account for 48 percent of U.S. production. All species variations of lettuce make it the leading vegetable crop in terms of value.

- Tomatoes—fresh and processed—(Production rank: 8th)
  - Produced in the hotter Central Valley region: western side of the Sacramento Valley—Colusa to Sacramento Counties—and the entire San Joaquin Valley (see page 29 in *California Atlas*).
- Broccoli (Production rank: 17th)
  - Coastal valleys: Santa Cruz to Ventura Counties, including Imperial Valley and parts of the San Joaquin Valley. It dominates in the Salinas River Valley (see page 28 in *California Atlas*).
- Lettuce—(Production rank: 7th)
  - Coastal valleys: Santa Cruz to Ventura counties, including Imperial Valley, Colorado River, and parts of the San Joaquin Valley. It is dominant in the Salinas River Valley (see page 29 in *California Atlas*).
- Carrots (Production rank: 18th)
  - Produced in the hotter inland valley regions of the state, i.e., the San Joaquin Valley and the Imperial Valley.

10.2.3 FIELD PASTURE, SEED CROPS, AND NURSERY/FLORICULTURE

- Hay and Alfalfa—(Production rank: 12th)
  - Imperial and San Joaquin Valleys (especially Tulare County), northwestern California (Siskiyou, Del Norte and Humboldt Counties), and the Basin and Range counties representing Modoc, Lassen, Sierra, Alpine, Mono, and Inyo Counties.
- Cotton (Pima)—(Production rank: 16th)
  - Cotton production (Figure 10.10a) only takes place on the hotter and drier (e.g., rain shadow) lower west side of the San Joaquin Valley (see page 29 in *California Atlas*). Pima cotton requires ample irrigation. The luxury and upscale clothing brands favor Pima cotton, which they typically market as Egyptian cotton or Sea Island cotton. Pima cotton

194 **CHAPTER 10** *California Agriculture and Irrigation Agri-"Cultures"*

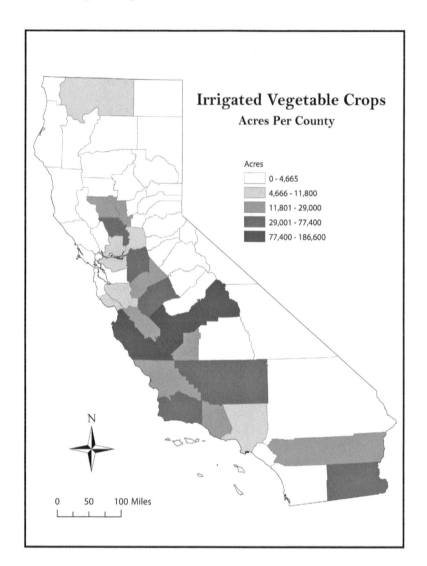

FIGURE 10.8 *Distribution of irrigated vegetable crops in 2007. (Data source: U.S. Department of Agriculture Census; Courtesy Kirstyn Pittman)*

FIGURE 10.9 *The rich green "truck crop" agricultural fields before the harvest. (Image © David P. Smith, 2009. Used under license from Shutterstock, Inc.)*

CHAPTER 10 *California Agriculture and Irrigation Agri-"Cultures"* 195

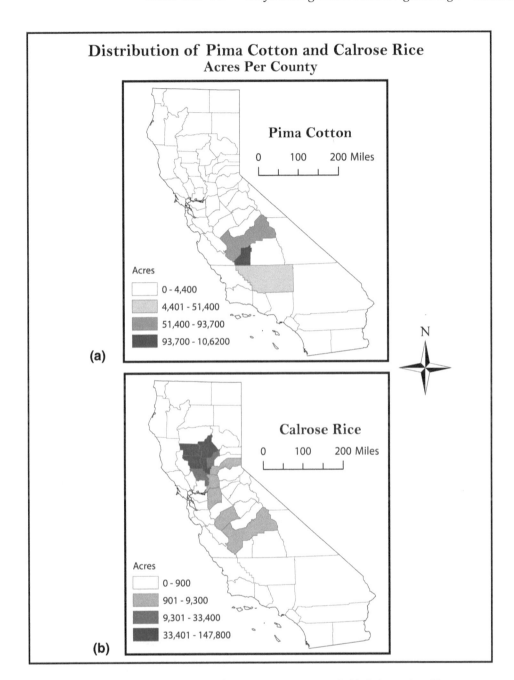

FIGURE 10.10 *2007 distributions for (a) Pima cotton; and (b) Calrose rice. (Data source: U.S. Department of Agriculture Census; Courtesy Kirstyn Pittman)*

represents the state's third leading export to foreign markets.

- Rice (Calrose medium grain)—(Production rank: 13th)
  - Medium grain rice (Calrose variety sushi style rice) dominates in California (Figure 10.10b); all other rice production grown in the U.S. is the long grain variety produced in the lower Mississippi River Valley areas of Arkansas, Mississippi, and Louisiana, as well as Texas. Rice production dominantly occurs from the Delta north to Butte (Figure 10.11) and Glenn Counties, following the Sacramento River flood channel that has deposited heavier clay soils able to hold water to form rice paddies. The Calrose variety of rice represents one of California's leading exports to Japan. This crop

FIGURE 10.11 *Rice paddies in the Sacramento Valley with the Sutter Buttes in the distance. (Image © Bryan Cole, 2009. Used under license from Shutterstock, Inc.)*

is both highly water intensive and requires a particular temperature of water in order to germinate and grow. The irrigation water supporting rice in the Sacramento River Valley becomes warmed via the Thermolito reservoir (e.g., large and very shallow in order to solar heat quickly, which is part of the Oroville Dam and State Water Project system).

- Nursery products and cut flowers—(Production rank: 4th and 11th)
  - Frost protected coastal areas from Sonoma to San Diego Counties. San Diego and Ventura Counties represent the dominant areas for these commodities (see page 28 in *California Atlas*).

10.2.4 DAIRY, CATTLE, AND POULTRY

Since the Mexican rancho era, livestock continues to be the state's most valuable farming commodity. The production of milk and beef makes it among the top five earners in the state. Milk is number one (Figure 10.12), which makes California far surpass Wisconsin as the number one dairy state. Yes, happy cows (and wealthy farmers) do come from California. California's large population is one of the main reasons for the massive supply of these products, and isolation still plays a role in limiting imports of especially milk and eggs from the Midwest or East Coast. The production of grains and hays in almost all the counties and the development of water supplies make the livestock industry a success, despite possible downsides from harsh rangeland environments (e.g., hot climate). It's no wonder that the Imperial Valley and Tulare Counties not only rank number one in alfalfa and hay production, but that they also hold these ranks in cattle production via the concentrated feedlot system. The concentration in dairy farming to only a couple of counties is also linked to the production of feed. San Bernardino and Riverside Counties are the "milk shed" that supplies dairy products to the South Coast metropolitan areas of Los Angeles to San Diego, while Marin and Sonoma Counties along with San Joaquin County are the suppliers to the greater San Francisco Bay Area and the Sacramento metro region. Poultry and eggs production follow a similar theme of being on the urban fringe. Therefore, they are nearby population centers because they are highly perishable.

CHAPTER 10 *California Agriculture and Irrigation Agri-"Cultures"* **197**

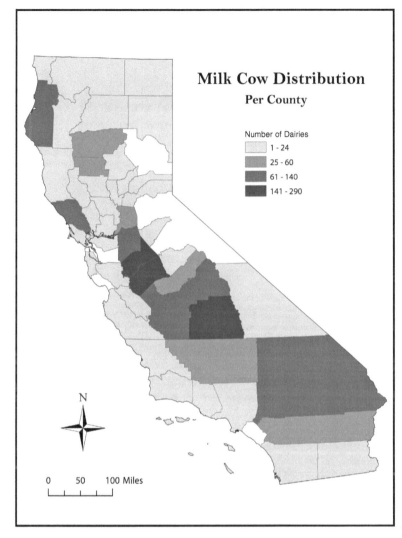

FIGURE 10.12 *Milk cow distribution in 2007. (Data source: U.S. Department of Agriculture Census; Courtesy Kirstyn Pittman)*

- Dairy cows—(Production rank: 1st)
  - This labor and industrial heavy commodity exists heavily in the following "milk sheds": San Joaquin Valley (especially Tulare and Merced Counties), Riverside, San Bernardino, Marin and Sonoma Counties.
- Cattle and calves—Largely feedlot raised—(Production rank: 5th)
  - San Joaquin and Imperial Valleys dominate in cattle production, as well as Riverside and San Bernardino Counties. However, other areas to note in the state are Alpine, Del Norte and Marin Counties with rather large cattle ranching densities.
- Poultry and Eggs—(Production rank: 14th)
  - Poultry farming exists only in the central San Joaquin Valley between Fresno and Sacramento Counties and Sonoma County. While industrial scale egg production exists only in Sonoma, San Diego, Riverside, San Bernardino, Kern and the Central San Joaquin Valley between San Joaquin and Merced Counties (centered on Modesto).

10.3 Irrigation Agri-"cultures": Labor for the Factory Fields

Throughout the span of California's agricultural prosperity and especially after American occupation, the need for **farm labor** has always existed. In the early years much of the farm labor came from the Californios working on cattle ranches. Late in the nineteenth century, when the fruit and specialty-crop farming sector became the most productive and profitable, the industry required a large number of farm workers to support it. Many of these crops, such as citrus, deciduous fruit, berries, vegetables, melons and cotton, required high levels of organization from farmers. This led to the development of marketing cooperatives like Sunkist for the citrus growers. These cooperatives became powerful complex organizations in the period from 1900 to the1930s, with large political leverage at both the state and federal levels with respect to influencing policy for the immigration of farm laborers. One of the key issues with specialty crops is their high demand for labor to harvest the crops on time, with further specialized packaging or processing in the production chain. The cooperatives made sure they always had an oversupply of labor during these times, no matter what they were harvesting or where, in order to keep the labor costs down. To this end the cooperatives led the California growers to organize farm labor bureaus that controlled the input and flux of labor on California's large and complex production landscapes.

From the standpoint of innovation, these agri-"culture" immigrants introduced many of the crops and cultivation methods now used in California; not necessarily American farmers moving in from the Midwest or Eastern seaboard. Some of these immigrants brought their indigenous knowledge from places that had similar growing regions to California, and therefore they could readily transfer their experiential learning to the fertile beginnings of the California agricultural landscape. For example, in viticulture the California industry owes a debt to the early Spanish Franciscan fathers at the missions, then to the French, Hungarians, Italians and Portuguese for bringing in rootstocks from their home countries to try in the California soil and climate. French and German immigrants initially brought in many of the deciduous fruits farmed in the state. Finally, one of the groups that contributed the most was the Chinese, who brought in the knowledge of irrigation and the skills to develop the vegetable ("truck garden") industry.

10.3.1 Early Chinese Labor and Agriculture: A Glimpse of the Future

Recall the origin of the **Chinese** during the Gold Rush in Chapter 8. They were largely farmers that came from the Pearl River Delta region in Guandong Province (southeast coast of China). It is important to know that China by the mid-1800s had already developed some of the most advanced technology in irrigation, crop rotation, field layout (plots with diverse adjacent crops) and fertilization. Their farming method could be termed bio-intensive, which focuses on an organic system and maximum yields from the minimum area of land while improving the soil base.

They came to California to seek a better life, first in the gold mining regions, later through their association with the Transcontinental railroad, and building river levees in the Central Valley and finally in the beginnings of irrigated agriculture in the Central Valley. Towards the end of the first decade of gold mining in California, many Chinese miners had left the gold fields to open laundries and farm small patches of vegetables, sweet potatoes and fruits to provide a more steady living income. The Chinese were the only ones growing vegetables near or around every mining camp situated along major transportation routes of the Sacramento and San Joaquin Rivers, which they used for irrigation. Soon after the Chinese began working for the Central Pacific Railroad in 1864, other groups found their dedication to work valuable for the reclamation of land in the Sacramento-San Joaquin Delta (e.g., marshy swamp lands with cyclical flooding). This consisted in building the levee system to control flooding, which by 1880 had reclaimed 88,000 acres of marshlands for irrigation agriculture production (Figure 10.13). Fruit tree orchards, vegetable and vineyards became the dominant crops in the Delta region in contrast to the wheat dominance at this time that

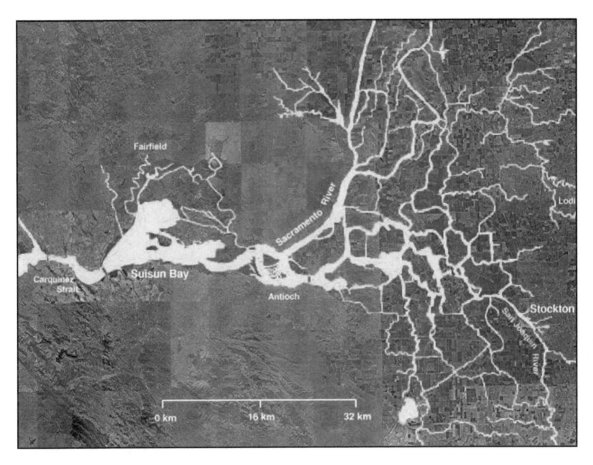

FIGURE 10.13 *Sacramento-San Joaquin Delta. (Courtesy U.S. Geological Survey)*

covered the Sacramento Valley. Therefore, from 1860 to 1910 the Chinese were the dominant farm workers contributing to the irrigated agricultural development of vegetables especially in the Central Valley, Santa Clara Valley, and San Francisco districts. In 1890, eight years after the passage of the Chinese Exclusion Act, they were forced to vacate their agricultural jobs for White workers. The White farm labor, who later found the work unattractive, forced the farmers to fill the farm labor hole with Japanese workers. However, the Chinese provided the trend in California's agricultural future: irrigated flat fields laid out in large plots with a variety of crops adjacent to each other, crop rotation and the strong need for a steady supply of labor. One only has to drive down the Salinas Valley or through the Delta region to see the Chinese legacy displayed through beautifully green fields in a mosaic of dozens of different crop types. In contrast, a drive through the Midwest will display miles of homogenous dryland crop types, such as corn, winter wheat, and soybeans. Truly, California's agricultural mosaic owes a lot to the early Chinese community.

10.3.2 JAPANESE FARM LABOR

Historians really know little about the first **Japanese** immigrants to California, who started arriving in the late nineteenth century as their country did not legally allow them to leave until after 1884. While we know that the Chinese were both "pulled" by gold and "pushed" from China as it became victim to economic depression and British imperialism, Japan was quite the opposite case. Japan was on the ascendancy as a modernizing nation and a growing imperial power. It appears though that this modernization caused social upheaval in the rural areas of Japan and that this "push" in conjunction with the "help" of the Chinese Exclusion Act started a flow of Japanese immigrants to California. The first flows of immigration after 1884 occurred when the Japanese government signed an agreement with Hawaiian sugar plantations to allow labor immigration. From Hawaii, many Japanese continued on to California via San Francisco.

Not unlike the Chinese before them, the Japanese were also agricultural innovators and conducted specialized irrigation agricultural production back in Japan. The majority of the Japanese came from Hiroshima, Kumamoto, Yamaguchi, and Fukushima prefectures with an expertise in agriculture, a tradition of hard work and a willingness to travel. The Japanese immigrants also shared a similarity with their early Chinese counterparts; they were largely young and male. In addition, as an ethnic minority, the Japanese immigrants faced adversity, hard work and defeats, and they experienced triumphs and community initiative as they looked with hope towards the future.

After 1890, farmers used the Japanese to fill the void left by the removal of the Chinese in the Delta agricultural region. After 1900, the Japanese were the dominant farm labor force in California, with unrestricted immigration to supply the farm labor bureaus until 1907. Therefore, they largely lived in rural northern California agricultural areas. In 1900, however, a large population node had begun in Los Angeles County most likely lured by the citrus industry. By 1910 a distinct change in the geospatial pattern occurred, illustrating a rapid expansion of Japanese to Southern California and making Los Angeles County the most populous Japanese settlement in the state. This stimulus for internal movement ("pull factor") was the Southern California boom period, which had ample opportunities. It was also the time of the San Francisco earthquake of 1906, a major push factor.

In many communities the Japanese farm laborers built houses for their families in an area known as a **nihonmachi**, or a Japanese section of town. As a further social differentiation, first generation Japanese farm laborers and newly established farmers were known as **issei** while their children (second generation born in the U.S.) were known as **nisei**.

Just like the Chinese before them, the Japanese faced adversity from the White supremacist organizations, the White farm owners in the form of the farm cooperatives, and the politicians. In 1907, California's moneyed farm interests and White labor organizations forced the federal government to pass a piece of racist policy halting all further Japanese labor immigration to the U.S. through the passage of the **Gentlemen's Agreement** with Japan. There was a provision, however, in the agreement that permitted laborers' wives and children to continue to enter and laborers already in the U.S. to remain. This led to the cultural phenomenon of **"picture brides,"** which occurred from 1908 to 1924. Since arranged marriages in Japan were the rule, the nihonmachi's opened marriage arrangement offices where a Japanese male could look through books of photos of potential brides, select one and then have the marriage arranger write a letter of introduction to the bride's family to arrange marriage via entering the bride's name in the groom's family registry. This legally constituted a marriage in Japan and thus passed the tenets of the Gentlemen's Agreement. The 1900 census illustrates the effect of this policy: At that time there were 410 Japanese females and 23,916 Japanese males in California. By the 1920s there were nearly equal numbers of females to males. What this policy also did is to increase the Japanese population rather than decrease it, as the families produced many nisei children.

Anti-Japanese organizations felt that the "picture bride" network was violating the Gentlemen's Agreement. They therefore pressured the federal government to totally exclude Japanese immigrants through the passage of the **Immigration Act of 1924 (National Origins Act)**. This act imposed restrictive quotas upon some cultural groups, while excluding Japanese, Chinese and other Asians. The Japanese thus had no immigration rights to the U.S. until 1952. The act created a change in the character of the Japanese population, making it 66 percent nisei by birth (though not allowed the right of naturalization) by the start of World War II.

While Japanese laborers were an important element in California agriculture, by 1900 they began to buy property to establish their own farms, vineyards, and orchards. Theirs was a community that did not want to remain simple farm laborers, as they were accomplished farmers in their own right. Thus they made a rapid change from being farm laborers to farm owners. In the Central Valley, several all-Japanese communities developed in Sacramento, Fresno and Merced Counties. They were important producers and growers of truck vegetable crops, in the Central Valley and in Southern California along the coast. They grew grapes and tree fruits in the Central Valley and Southern California, strawberries along the coast in Southern California and rice in the Sacramento River Valley. In fact, it was the Japanese farmers that helped introduce rice growing culture to California. Rice is now the thirteenth largest commodity produced in the state—with the majority of the production being exported to Japan.

After the Gentlemen's Agreement, but before the 1924 immigration act, anti-Japanese state laws were developed. One particular law involved land ownership. As noted above, by 1904 Japanese farmers owned 50,000 acres in California, and this figure continued to grow. The White farm cooperatives did not want the competition from the Japanese; rather, they wanted them to be only farm laborers for *their* farms and processing facilities (e.g., packing and canning plants). They convinced the California legislature to pass the Webb-Hartley Law (known as the **Alien Land Law of 1913**), which limited land leases by "aliens ineligible to citizenship" to three years and barred land purchases. By 1919, however, the Japanese were farming 450,000 acres, as the law had loopholes that remained until 1922. The biggest hole was not to acknowledge the nisei. Because they were second generation Japanese born in the U.S., they were considered citizens (even though they were not allowed to be) and thus could buy or lease land in their name for use by their issei parents or themselves. This ended in 1922 when the U.S. Supreme Court declared that all Asians were ineligible for citizenship under the U.S. Constitution and that only "free White persons" or "Africans by birth or descent" are allowed.

10.3.3 FILIPINO FARM LABOR

As the Japanese continued to transition from farm laborers to farm owners, the farm labor bureaus prior to World War I, and more so after the war, started to bring **Filipinos** in to fill the farm labor void. The U.S. had taken over the Philippines from Spain as part of the Spanish-American War of 1898. In the early 1900s, the immigration of young Filipino males to California began as part of a U.S. government-sponsored territorial scholarship system known as **pensionados**. As part of the U.S. government's territorial system Filipino women were forbidden from entering the U.S. After World War I, the farm bureaus started importing male Filipino farm workers in large groups with the idea that they could keep their expenses down by employing a work force of single men. It is also important to note that the Immigration Act of 1924 did not specify Filipinos as people of Asian descent, which gave them free mobility in the U.S. via territorial status and law. By the time of the Great Depression in the early 1930s, these male workers had organized into associations that led to some powerful strikes for better wages and living conditions. By 1930, Filipinos composed 30 percent of the farm labor in California.

Their powerful convictions to force a living wage from the White farm cooperatives and the racist backlash from Whites who felt that the Filipinos and other "ethnic" groups not like them had taken their jobs led to the U.S. government passing the **Filipino Repatriation Act of 1935**. This act applied pressure on the Filipinos to return to the Philippines by offering free transportation. It did not work very well and was later declared unconstitutional in 1940.

10.3.4 INDIAN PUNJABIS IN THE SACRAMENTO VALLEY: A UNIQUELY CONCENTRATED LABOR GROUP

Another unique category of Asian ethnic group came from India. In the early part of the 1900s **Punjabi** males immigrated to California to find work in the agricultural sector; like the Chinese and Japanese many were farmers in their own right. When the 1924 Immigration Act closed the immigration path, it also created a situation where the Punjabi men did not go back to India. They instead pooled their resources and leased out lands to farm for themselves largely in the Sacramento Valley, centered in Sutter and Colusa Counties with Yuba City regarded as their cultural center. The Punjabi immigrants found the physical landscape of the Sacramento Valley, Sutter Buttes and Sierra Nevada Foothills similar to that of the Punjab province (western India's border with Pakistan) where they originated. Between the years closed to immigration (1924–1965) many of the Punjabis married Mexican women, creating a bi-racial society uniquely evident in the Sutter and Colusa County region with restaurants that serve both Indian curries and rotis, and Mexican enchiladas, as well as religious sites representing Sikh temples and Muslim mosques among Catholic churches.

10.3.5 MEXICAN FARM LABOR

By the early 1900s the stage was set for the growth of **Mexican** immigration to the U.S. because of social and economic change inside Mexico. These changes in Mexico had become hostile to small landowners who were losing their holdings to larger farming businesses and being forced to become farm workers under a system of peonage and accumulating debt. Finally in 1910, the Mexican Revolution broke out across the country, with its deep social, economic and political ramifications. The chaos from this event drove many Mexicans north over the border where jobs were available in industry—mining, railroad construction and agriculture—and at better wages than they could receive in Mexico. Their pace of immigration increased after World War I as the farm labor bureaus started to use Mexican labor for harvest along with the Filipino men. The Immigration Act of 1924 did not affect Mexicans or any Latin American group; it only imposed the quota system on Southern and Eastern Europeans.

The post-World War I boom in manufacturing, food processing, garment manufacture and construction industries in Southern California, coupled with the expansion of irrigated croplands, pulled more than 30 percent of the Mexican-born U.S. residents to California by 1930. In the urban areas they settled in large communities called **barrios**, the one in East Los Angeles (East LA) being one of the most famous southwestern barrios. The cities of Santa Barbara, San Diego and Santa Ana are also good representatives having barrios. The largest in East Los Angeles was created because of White discrimination forcing the Mexican families to not be allowed to settle in Los Angeles proper. It kept the core city 90 percent White during the 1920s by making sure that the Mexicans were at a distance. In the rural areas of the state the Mexicans settled in **colonias** on the fringes of agricultural towns and cities, i.e., Fresno.

The Mexicans also became the first "mobilized" labor force due to the fact that they lived in centers of settlement's and began to use cars and trucks that were becoming popular in California during the 1920s with the expansion of roads and highways to travel to the fields to stay temporarily (Figure 10.14). They represented mobile harvest labor, which continues today in California's agribusiness sector where the men leave the women behind in the barrios to travel around the state and follow the harvest schedules of the many crops. Additionally, other industries at this time recruited the Mexican women for work in the food processing and garment manufacturing sectors.

Racist attitudes and adverse conditions were also foisted onto the Mexican communities, but they were able to develop stability because complete families immigrated into California. The tenets of the Roman Catholic religion, strong family ties and social networks, historical connections to California and a strong work ethic helped this cultural group avoid the major problems that afflicted the Asian groups. By 1930, Los Angeles County had the largest concentration of Mexicans outside of Mexico City. In addition, Mexicans carried out 70 percent of the farm labor.

Just like what happened to the Filipinos, anti-Mexican attitudes flared during the start of the Great

FIGURE 10.14 *Mexican workers in a San Joaquin Valley tent city. (Courtesy of The Bancroft Library)*

Depression. These led to the movement known as **Mexican Repatriation (Operation Deportation 1929)**, where a Mexican was called a *repatriado*. This was both a voluntary and an involuntary migration of Mexicans back to Mexico. Between 400,000 and 500,000 Mexicans left the U.S. due to unemployment, fear of deportation by the Immigration and Naturalization Service (INS), encouragement by welfare agencies and the lure of the Mexican government for their own farmers to help the country's development. Mexicans became an easy target because of the close border. Easily distinguishable physically, and geographically, their barrios and colonias gave them away to the authorities who sought to repatriate them.

> "Now I go to my country where although at times they make war [Mexican Revolution] they will not run us from there. Goodbye, my dear friends, you are all witness of the bad payment they give." Mexican Folk ballad or *corrido* from this era of the *repatriado*.

It is important to note that the state of California passed the **Apology Act of 2005** officially recognizing the "unconstitutional removal and coerced emigration of United States citizens and legal residents of Mexican descent during the 1930s" and apologizing to residents of California "for the fundamental violations of their basic civil liberties and constitutional rights committed during the period of illegal deportation and coerced emigration."

Despite the repatriation schemes, Mexicans and Mexican-Americans have continued to be the largest ethnic group working in California agriculture since the 1930s. Their ranks became augmented during World War II when the farm bureaus required more farm laborers to support harvests during the war years. This culminated in the **Bracero Program** (1942–1964), which was an agreement between the U.S. and Mexican governments to allow the issuing of work permits to Mexicans wanting to leave Mexico for a set period of time to work in U.S. agriculture. Braceros (Figure 10.15) were principally experienced farmers in their own right who hailed from crucial agricultural regions in Mexico.

The development of farm labor contracts were controlled by independent farmers associations and the Farm Bureau. The system was, however, flawed from the beginning as the contracts were in English and

FIGURE 10.15 *Mexican migrant farm worker (Bracero) in the Salinas Valley 1960s. (Courtesy of The Bancroft Library)*

signed without the Mexican laborers understanding their full rights and the conditions of their employment. Under the contract they could work in the U.S. for seven years. Then after expiration they had to turn in their work permits and return to Mexico. During the contract period Braceros could return to their native lands in case of a family emergency, but only with written permission from their farm labor boss. In the agreement the U.S. agreed to: provide transportation to the U.S. and the return to Mexico; for health care; housing; a minimum wage; a savings account; and unemployment pay. Unfortunately, during the Bracero program many of these promises were not met (i.e., some were never paid; there was a loss of saving accounts, and no health care, and so called "decent" housing usually was a tent). Furthermore, these farm workers were often exploited by the farm labor bosses. In October 2008, a successful lawsuit was settled in the U.S. District Court of Northern California that allowed many living Braceros or their families to recover losses from this time period.

10.4 Factories in the Fields: Great Depression— Dust Bowl—Okies

Many scholars consider the 1920s a boom time in the California agriculture, manufacturing and entertainment industries. The dark side, however, was low wages and the production of too many unaffordable goods the masses could not purchase. Then the U.S. stock market crashed in 1929, which rippled out to a worldwide economic collapse—the Great Depression of the 1930s had begun. By 1933 California's unemployment rate had reached 25 percent. California, however, remained an oddity in people's minds across the country. Many could only see the myth of California: a balmy paradise of orange groves and Hollywood movie stars. The reality, however, was that California was more vulnerable to economic disruptions then any other state because of the industries that required people to spend money: specialty agricultural crops, tourism, and entertainment.

The geographical picture of the state during the Great Depression could not have been more different between Southern and Northern California. The financial calamities of the times basically devastated Southern California. There are several reasons why this region was the hardest hit. First, there was an unusually high number of service jobs tied to the tourism and entertainment industries. Second, connected with this was the large number of lower-middle-class, white-collar workers involved in the real estate, tourism, oil and entertainment industries. The third strike against the area's economy was due to the fact that the region had the highest proportion of elderly people in the nation. Southern California was a magnet for Midwestern and East Coast migrants going there to retire and live in a mild climate linked to the area's projection as a healthy recreation mecca. These are the same people that were living on their savings and investments; however, with the stock market crash and the bank failures they lost nearly everything.

Northern California was a different story altogether in at least three different geographical ways. First, the San Francisco Bay Area survived fairly well as it was the home of the state's largest, oldest and strongest corporations (e.g., manufacturing and financial). There were a high percentage of salaried white-collar employees able to hold on to their jobs and thus invest their incomes back into the local economy. In the Sacramento area, the state capital was swamped with relief requests from the government, and tent cities sprang up around the city. At the same time, government employment largely drove the local Sacramento economy, which survived if not expanded during this time. More interestingly, there was an increase in the population in the Sierra Nevada foothills as gold mining became popular again, with large groups of unemployed men staking claims and conducting placer mining. Thus gold output from California spiked during the Great Depression even though it had been a dead industry since the late 1880s. Many small mining groups were able to make a living since the gold standard still backed the dollar. In summary, there was massive unemployment throughout the whole state, which meant more discrimination and racism against Hispanics, Asians and African-Americans. This is the same story that occurred during similar economic downtimes in the 1870s.

10.4.1 THE DUST BOWL: DISASTER FROM A MAJOR DROUGHT AND POOR LAND MANAGEMENT

The **Dust Bowl** (1930–1936) was a period of severe dust storms that led to major agricultural damage to the Midwestern prairie region (Figures 10.16 and 10.17). The event was both natural and human induced—severe drought coupled with decades of poor farming practices, i.e., no crop rotation or erosion controls. The Dust Bowl affected millions of acres of farmland centered on Texas and Oklahoma, but it included parts of New Mexico, Colorado, Kansas, Missouri and Arkansas. These farms became useless, forcing hundreds of thousands of people into foreclosure and into leaving their homes. In addition, the process of farm foreclosure and loss of leased land increased with the passage of the **Agricultural Adjustment Act** of 1933 which stopped farmers from farming to raise food prices as a way to spur inflation, since the Depression was a deflationary cycle. In the end it made the large corporate farms stronger and those who leased land as sharecroppers were forced off the land.

CHAPTER 10 *California Agriculture and Irrigation Agri-"Cultures"* 205

FIGURE 10.16 *Farmer walking in dust storm, Cimarron County, Okalahoma. (Courtesy of The Library of Congress)*

Many of these poor Dust Bowl stricken families, known as **Okies** since so many came from Oklahoma, began the great American **exodus** to California to find better economic conditions in the golden state. Historians still consider this event one of the largest internal mass migrations to occur in the U.S. The migrants traveled along old **Route 66**—"66 is the mother road, the road of flight"—to California only to find economic conditions little better than those they had left behind (e.g., the agricultural act had cut the acerage and thus decreased the demand for farm labor). Little did they realize that the prosperous California farms of the 1920s were struggling from their concentration on producing specialty crops (i.e., strawberries, oranges, almonds, etc.) that were irrelevant and extravagant in a time when people required just the basics to survive, i.e., a sack of flour, corn meal, milk, etc. They came into California to become farm laborers, traveling from farm to farm to pick the crops, often at starvation wages. John Steinbeck chronicled the plight of these people in

FIGURE 10.17 *Dust storm approaching Spearman, Texas 1935. (Courtesy National Oceanic and Atmospheric Administration)*

California's agricultural jungle in novels such as *In Dubious Battle* and **The Grapes of Wrath**.

10.4.2 THE "OKIES" IN AGRICULTURE, AN AGRI-"SUBCULTURE"

The Okies came at an interesting time to the California agricultural scene, for as was described earlier, Mexican and Filipino laborers largely staffed the farms. In addition, the farming system in California was creating fewer farmers. Under a system of farm aggregation ("agribusiness corporate farms"), many so called farmers were living in the cities and had never seen the farms they owned (also known as **"paper farms"**). Instead they relied on underpaid "slave wage" immigrant labor to make them operate. The farm bosses hated the Okies, as did the Mexican and Filipino workers whose jobs they threatened. It was a difficult situation as the Okies provided an important distinction in the traditional California harvest labor used by the farm labor bureaus up until this time. They were White Anglo-Saxon Protestants (WASPs) who sought permanence in the Central Valley towns, and they were U.S. citizens, which meant they could register to vote, perform jury duty, send their children to the local schools and have their men drafted into the military. What made the farm bosses uneasy and angry was that the Okies were no different from themselves, and they could not ignore them in their poverty. The Okies life of squalor in mobile tent camps known as **Hoovervilles** (named after President Hoover, whom the Depression was blamed on; Figures 10.18 and 10.19) in the rich Central Valley agricultural lands exposed the farm labor exploitation that California farmers hoped to carry on to create healthy profits for themselves.

> "Pray God some day kind people won't all be poor. Pray God some day a kid can eat. And the association of [farm] owners knew that some day the praying would stop." John Steinbeck, *The Grapes of Wrath*, 1939.

Anti-Okie hysteria was made policy in California's **Indigent Act of 1937**, which made it a crime to knowingly bring an indigent person into the state. This was an attempt to stop the flow of Okies at the border

FIGURE 10.18 *Tented Hooverville behind Southern Pacific railroad advertisement. (Courtesy of The Library of Congress; photographer, Dorthy Lange)*

FIGURE 10.19 *Okie family living in a squalid Hooverville, San Joaquin Valley. (Courtesy of The Library of Congress; photographer, Dorthy Lange)*

(a migration hinder), but it was later struck down by the U.S. Supreme Court as unconstitutional under the 14th Amendment.

The Mexican, Filipino and Okie farm laborers, as well as other day laborers in the industry played into the struggle between labor and capital—it was the era of **"Total Engagement."** This was a difficult time as agricultural workers suffered the peculiar agony of watching food rot in the fields because the farmers could not sell enough crops to pay the costs for harvesting and marketing. Therefore, in an attempt to keep prices higher on the market by making the commodity rare, the farm bosses let food rot rather than use it to feed starving people. The Mexican and especially the Filipino farm workers attempted to form new organizations to fight for improvements in wages and working conditions, and at some stages these efforts resulted in strike actions (i.e., the San Joaquin cotton strike of 1938). This was the era of **unionization** (Figures 10.20 and 10.21) that was often linked to the Communist party. California was on center stage for the saga of industrial capitalism's transformation of would-be farmers into a low-paid, multi-racial army of farm workers toiling on huge factory farms, i.e., **"Factories in the Fields."** This distinction was necessary as there really was no difference between factory work and farm work. Both involved a highly industrial process, and therefore union organization was the first step in guaranteeing farm workers the right to strike (Figure 10.22 and 10.23) and collective bargaining for better wages.

FIGURE 10.20 *Early ethnic labor unionists demand change. (Courtesy of The Bancroft Library)*

FIGURE 10.21 *Filipino labor union, 1930s. (Howard DeWitt image)*

FIGURE 10.23 *Strikers calling for scab labor to leave the fields, 1933. (Courtesy of The Bancroft Library)*

The **National Labor Relations Act of 1935** guaranteed a worker's right to join a union and a union's right to bargain collectively. It also outlawed business practices deemed unfair to labor. Unfortunately, no such transformation occurred in California agriculture. While workers started several unions, i.e., the Cannery and Agricultural Workers Industrial Union or the United Cannery, Agricultural, Packing and Allied Workers of America (of the Congress of Industrial Organizations), they were essentially failures. The fundamental nature of California farming, the composition of its labor force (multi-racial, immigrants with no rights, etc.) and the very powerful farmers' organizations resisted the process of unionization and halted recognition of the "Factories in the Fields."

FIGURE 10.22 *Women strikers demanding food, 1933. (Courtesy of The Bancroft Library)*

In fact, it was largely the Okie subculture, along with the repatriation pressures described earlier for the Filipino and Mexican labor, that stopped the transformation of the unfair system. The Okies were strong, conservative individuals who were very patriotic and had antiradical tendencies. They alone impeded union movement in agriculture, as they viewed it as the evil of communism. If we explore the Okie subculture and its impact on the cultural landscapes of California and especially the Central Valley, one can see their influence even today. The term Dust Bowl encompasses all Depression refugees from the southern Midwest and Southwest, which totaled approximately 300,000 migrants between 1935 and 1940. The San Joaquin Valley has preserved their rural Midwestern "plain-folk American" values and folkways through their evangelical Protestantism, distinctive accents, food preferences (i.e., fried) and love of country music. Their cultural capital is Bakersfield, also known as "Nashville West." Baptist and Pentecostal churches thickly populate the San Joaquin Valley, which also represents the state's highest number of Christian radio stations. Since the 1940s this cultural group has produced and dominated the country music industry. Such Bakersfield stars as Gene Autry, Bob Wills, Buck Owens, Merle Haggard, and Dwight Yoakum have all called this region home.

> "California is a garden of Eden, a paradise to live in or see. But believe it or not, you won't find it so hot. If you ain't got the do re mi." Woody Guthrie's Okies ballad, *Do Re Mi*, 1937.

Eventually, it was the Okies call to military duty as patriots and citizens in World War II that forced the farm organizations to rely again on Mexican immigrant and Filipino labor. Because the young Okie men and the Mexican-Americans left for war, the agricultural industry moved towards less labor intensive agriculture whose labor fluctuations did not have such major effects on the industry. In the end though, as described earlier, the Bracero program developed, which increased the demand for Mexican farm labor throughout the rest of the century.

Bibliography

Almaguer, T. (1994). *Racial Fault Lines: The Historical Origins of White Supremacy in California*. University of California Press, Berkeley, CA.

Balderrama, Francisco. 1982. *In Defense of La Raza: The Los Angeles Mexican Consulate and the Mexican Community, 1929–1936*. Tucson: University of Arizona Press.

Beck, W. (1974). *Historical Atlas of California*. University of Oklahoma Press, Norman, OK.

Beiser, V. (2004). The salads of wrath. *Chico News and Review*, January 15, pp. 14–17.

Booth, S. (2008). *California Geography*. Course taught at Sierra College. [online] http://geography.sierra.cc.ca.us/booth/California/cal_index.htm.

California Department of Food and Agriculture (2009). California Agricultural Resource Directory 2008–2009. [online] http://www.cdfa.ca.gov/statistics/.

Calisphere (2008a). California Cultures. University of California [online] http://www.calisphere.universityofcalifornia.edu/calcultures/

Calisphere (2008b). A world of California Primary Sources. University of California [online] http://www.calisphere.universityofcalifornia.edu/

Camarillo, Albert. 1984. *Chicanos in California: A History of Mexican Americans*. Sparks, NV: Materials for Today's Learning, Inc.

Cardoso, Lawrence. 1980. *Mexican Emigration to the United States 1897–1931*. Tucson: University of Arizona Press.

Cardellino, J. (1984). *Industrial Location: A Case Study of the California Fruit and Vegetable Canning Industry, 1860–1984*. UC Berkeley, Dept. of Geography, Masters Thesis.

Carosso, V.P. (1951). *The California wine industry, 1830–1895: a study of the formative years*. University of California Press, Berkeley, CA.

Chen, J. (1984). The contributions of the Chinese. In *The Book of California Wine*, edited by D. Muscatine, M.A. Amerine and B. Thompson, pp. 22–24. University of California Press, Berkeley.

Chinn, T.W., ed. (1969). *A History of Chinese in California*. Chinese Historical Society of America, San Francisco, CA.

Cox, C.J. (2008). *California Geography*. Course taught at Sierra College. [online] http://faculty.sierracollege.edu/ccox/california_geography/index.html.

Davis, C. and D. Igler (2002). *The Human Tradition in California*. Scholarly Resources, Wilmington, DEL.

Deverell, W. and D. Igler (2008). *A Companion to California History*. Wiley-Blackwell, New York, NY.

DeWitt, H.A. (1999). *The Fragmented Dream: Multicultural California*. Kendall/Hunt Publishers.

DeWitt, H.A. (1999). *The California Dream*. 2nd edition. Kendall/Hunt Publishers.

Dilsaver, L.M., Wyckoff, W., and W. Preston (2000). Fifteen events that have shaped California's Human Landscape. *The California Geographer*, 40: 3–78.

Donley, M.W., Allan, S., Caro, P., and C.P. Patton (1979). *Atlas of California*. Pacific Book Center, Culver City, CA.

Durrenberger, R.W. and R.B. Johnson (1976). *California Patterns on the Land*. 5th edition, Mayfield Publishing Company, Mountain View, CA.

Garcia, M. (2001). *A World of Its Own: Race, Labor, and Citrus in the Making of Greater Los Angeles, 1900–1970*. University of North Carolina Press.

Gonzalez, G. (1994). *Labor and Community: Mexican Citrus Worker Villages in a Southern California County, 1900–1950*. University of Illinois Press.

Gregory, J.N. 1989. *American Exodus: The Dust Bowl Migration and Okie Culture in California*. Oxford University Press.

Haas, L. (1996). *Conquests and Historical Identities in California, 1769–1936*. University of California Press, Berkeley, CA.

Hayes, S.E. (1984). Those who worked the land. In *The Book of California Wine*, edited by D. Muscatine, M.A. Amerine and B. Thompson, pp. 25–29. University of California Press, Berkeley.

Hoffman, Abraham. 1974. *Unwanted Mexican Americans in the Great Depression: Repatriation Pressures, 1929–1939*. Tucson: University of Arizona Press.

Hornbeck, D. (1983). *California Patterns: A Geographical and Historical Atlas*. Mayfield Publishing Company, Mountain View, CA.

Kelly, R. (1959). *Battling the inland sea: floods, public policy, and the Sacramento Valley*. University of California Press, Berkeley, CA.

Kiser, G.C. and D. Silverman. (1979). Mexican Repatriation During the Great Depression. In *Mexican Workers in the United States*, G.C. Kiser and M.W. Kiser, editors. University of New Mexico Press, Alberqurque.

Jelinek, L.J. (1979). *Harvest Empire: A History of California Agriculture*. Boyd & Fraser, San Francisco.

Larson, D.J. (1996). Historical water-use priorities and public policies. In *Sierra Nevada Ecosystem Project: Final report to Congress, vol. II, Assessments and scientific basis for management options*. University of California, Centers for Water and Wildland Resources, Davis, CA.

Leonard, K. (1989). California's Punjabi Mexican Americans: Ethnic choices made by the descendants of Punjabi pioneers and their Mexican wives. *The World & I*, vol. 4(5): 612–623. [online] http://www.sikhpioneers.org/cpma.html.

Liebman, E. (1981). *The Evolution of Large Agricultural Landholdings in California*. UC Berkeley, Dept. of Geography, PhD Dissertation.

McWilliams, C. (1973). *California, the Great Exception*. Peregrine Smith, Santa Barbara, CA.

McWilliams, C. (1995). *Southern California: An Island on the Land*. Peregrine Smith Books, Salt Lake City, UT.

McWilliams, C. (2000). *Factories in the Field: The Story of Migratory Farm Labor in California*. University of California Press, Berkeley, CA.

Michaelson, J. (2008). *Geography of California*. Course at UC Santa Barbara, Dept. of Geography. [online] http://www.geog.ucsb.edu/~joel/g148_f08/.

Miller, C.S. and R.S. Hyslop (2000). *California the Geography of Diversity*, 2nd edition. Mayfield Publishing Company, Mountain View, CA.

Mitchell, D. (1996). *The Lie of the Land: Migrant Workers and the California Landscape*. University of Minnesota Press, Minneapolis, MN.

Morrison, P. A (1971). *The Role of Migration in California's Growth*. Rand Corporation, Santa Monica, CA.

Orsi, R.J. (2007). *Sunset Limited: The Southern Pacific Railroad and the Development of the American West, 1850–1930*. University of California Press, Berkeley, CA.

Rice, R., Bullough, W., and R. Orsi (2001). *The Elusive Eden: A New History of California*. 3rd edition, McGraw-Hill.

Sackman, D.C. (2005). *Orange Empire: California and the Fruits of Eden*. University of California Press, Berkeley, CA.

Starrs, P. and P. Goin (2010). *Field Guide to California Agriculture*. California Natural History Guide Series; no. 98, University of California Press, Berkeley, CA.

Steiner, S. (1980). *Fusang: The Chinese Who Built America*. Harper & Row Publishers, New York, NY.

Stoll, S. (1998). *The Fruits of Natural Advantage: Making the Industrial Countryside in California*. University of California Press, Berkeley, CA.

U.S. Department of Agriculture (2009). The Census of Agriculture: Historical years. National Agricultural Statistics Service. [online] http://www.agcensus.usda.gov/

U.S. Department of Agriculture (2009). 2007 Agricultural Census Atlas of the U.S. National Agricultural Statistics Service. [online] http://www.agcensus.usda.gov/Publications/2007/Online_Highlights/Ag_Atlas_Maps/index.asp

U.S. Department of Agriculture (2009). 2007 Agricultural Census U.S. County Profiles. National Agricultural Statistics Service. [online] http://www.agcensus.usda.gov/Publications/2007/Online_Highlights/County_Profiles/California/cp99006.pdf

University of California (2009). Agricultural Issues Center. [online] http://aic.ucdavis.edu/pub/exports.html.

Wagner, R.L. (2004). *Sleeping Giant: An Illustrated History of Southern California's Inland Empire*. Stephens Press.

Walker, R. (2004). *The Conquest of Bread: 150 Years of California Agribusiness*. The New Press, New York, NY.

Weber, D. (1996). *Dark Sweat, White Gold: California Farm Workers, Cotton, and the New Deal*. University of California Press, Berkeley, CA.

CHAPTER 11

Urban-Rural California
Water, Land, and Design

"Water is the true wealth in dry land; without it, land is worthless or nearly so. And if you control water, you control the land that depends on it."
—Wallace Stegner, *Beyond the 100th Meridian*, 1992.

"Water is the foundation upon which California's ecosystems and economic vitality rise."
—Arthur Guy Baggett, Jr. Chair, California State Water Resources Control Board, from *Introduction to Water in California*, David Carle, 2009.

Key terms

- All American Canal
- Boulder dam
- California Aqueduct
- California Endangered Species Act
- California Water Doctrine
- Canal
- Central Coast Branch
- Central Valley Project
- Chinook salmon
- Clare Engle lake
- Climate change
- Coachella canal
- Colorado River
- Colorado River Aqueduct
- Colorado River compact
- Contra Costa canal
- County-of-origin law 1931
- Delta Cross channel
- Delta-Mendota canal
- Delta smelt
- Department of Water and Power
- Department of Water Resources
- Endangered Species Act
- Family farming
- Feather River
- Folsom dam
- Friant Dam
- Friant-Kern canal
- Groundwater
- Hetch Hetchy Aqueduct
- Imperial Valley
- Lake Castaic
- Lake Perris
- Los Angeles
- Los Angeles Aqueduct
- Los Angeles River
- Mokelumne Aqueduct
- New Melones dam
- Oroville Reservoir
- Owens River Valley
- Paper farms
- Pueblo rights
- Reclamation Act
- Reclamation Reform Act 1982
- Resources
- Sacramento River
- Sacramento-San Joaquin Delta
- Salton Sea
- San Francisco
- San Joaquin River
- San Joaquin Valley
- Shasta Reservoir
- State Water Project
- Tehama-Colusa canal
- Tulare lake
- Water Resources Development Act
- Westlands Water District
- William Mullholland
- Zanjero

Introduction

EARLY CALIFORNIA WATER LAW—A BASIC PRIMER

Under the Spanish and Mexican governments the missions and pueblos had rights to control water ownership and delivery. The missions had built their own waterworks to support human consumption but more importantly for irrigating agriculture. These water systems, some of which still operate today, remained confined to the mission buildings and immediate surrounding agricultural lands. In the pueblo, however, the **zanjero** (or water master) was a powerful person who controlled the system that supported settlement development in a dry landscape. This person controlled the development, operation and maintenance of the municipal and irrigation canals with regards to allotment of water, and he made sure the livestock stayed out of the series of ditches (zanjas). A requirement of the water master was to bring equality to the division of water for both the fields and the homes within the pueblos. Therefore, the Spanish and Mexican eras considered and promoted a communal use of water resources for the national interest. The Spanish *Law of the Indes* (see Chapter 6) stated the communal decree for water, where water in a pueblo (municipality) could not belong to separate individuals, but rather passed from the Spanish Crown to the entire community as an incorporated body. When there was not enough water within the common lands granted to a pueblo to sustain the people and agriculture, it could lead to failure.

The attitude regarding a communal and fair interest in water changed with the Gold Rush. Recall from Chapter 10 that the **California Water Doctrine** became established under the U.S. era of California's development as a hybrid between prior appropriate and riparian water rights. Under prior appropriation, an unbridled individualism was dominant, allowing the creation of water companies that grabbed rights to impound and divert water. In the mid-1850s the courts upheld the right of companies to retail water not directly connected to hydraulic mining, and individualism led to corporations. The right to use water could be bought and sold like property; therefore it encouraged the swift diversion of water and the subsequent rapid development of land uses. Riparian rights, on the other hand, are attached to the ownership of land adjacent to a stream or a river with the land owner being guaranteed the right to use the full natural flow less any upstream usage. Because the right of usage was limited to domestic needs, including livestock but not irrigation, it encouraged the acquisition of large land holdings.

The co-evolution of California's inhabitants and the waterscape is both dynamic and legendary in the American West, if not the world. California's spectacular growth is deeply linked to the creation and maintenance of the world's largest engineered water works. In 2010, four of the 20 most populous cities in the U.S. were in California—two in Southern California (Los Angeles and San Diego) and two in Northern California (San Jose and San Francisco). The availability of water propelled urban expansion in this semi-desert in disguise. What will become clear to the reader is that the California story is one of constant water seeking, with support from the growing populations of urban dwellers, corporate agriculture, industrial use, and strongly shaped by government policy and use directives.

> *"... water issues are so closely intertwined with the core elements of California's (and the American West's) political, economic, legal and cultural evolution."*
>
> —Norris Hundley Jr., *The Great Thirst: Californians and Water: 1770s–1990s*, 2001.

11.1 Urban Water Imperialism: Stories from the South and North

The rapid spread of California's cities since the Gold Rush and their connection via the Transcontinental railroad during the first fifty years of the American era is unheralded. In these early years local sources of water for the major urban nodes of **Los Angeles** and **San Francisco** could sustain their growing populations, but by 1900 their booming populations and diminishing local supply met with the reality that a long-distance transfer of water would become a necessity.

11.1.1 Los Angeles: The Owens River Valley Saga

"Few of the great cities of the land have had such humble founders as Los Angeles. Of the forty-four pobladores (settlers) who built their huts of poles and tule thatch around the plaza vieja . . . not one could read or write. Not one could boast of an unmixed ancestry. (Their only concern was water [sic] . . .)" J. M. Guinn, *Historical and Biographical Record of Los Angeles and Vicinity*, 1901.

The Spanish *Law of the Indes* granted Los Angeles **pueblo rights** to owning and thus controlling water under communal law (see Chapter 6 for review), which also transferred into the Mexican era. The Treaty of Guadalupe Hidalgo with the U.S. preserved these pueblo rights, which the state upheld under the town's charter. Since the town council directly controlled the community's water system, a full time zanjero position was put in place to issue water-use permits, regulate water distribution, collect special water taxes and maintain and expand the system of aqueducts, canals and wells.

At the founding of Los Angeles, the main source of water was the **Los Angeles River**. Townspeople constructed a series of diversion canals (zanjas) to divert the waters. By the mid-1880s there were canals covering 93 miles working to support the water needs of Los Angeles, a far cry more extensive from its small pueblo days.

As early as the 1860s, however, the growing population required additional sources of water and methods of delivery. The town council authorized a thirty-year contract with a private water company (Los Angeles Water Company; LAWC) to construct the city's first reservoir and brought the first use of above ground wooden pipes into the system for human supply. In addition, the city asserted its authority by making sure it fought against any water diversions along the Los Angeles River, which the town council believed the city fully owned under Spanish pueblo rights, supported by the Treaty of Guadalupe Hidalgo and upheld by the city charter. Fights with upstream landowners over the rights to the full flow of the Los Angeles River (including groundwater) went back and forth until the courts finally sided with the city in 1895. The legal authorities supported Los Angeles' right to become a "great city." The courts extended the pueblo rights once more in 1899 by allowing any annexed land by the city of Los Angeles to include ownership of the water rights as well. Annexation used to secure water rights became the Los Angeles survival strategy. By 1902 a series of ten canals and a newly installed system of underground pipes replacing the wooden pipes served the city.

During the Southern California boom period the population continued to increase at a rapid pace. Therefore the city continually sought new sources of water in the most imaginative of ways. In 1898 the city finally changed the city charter to require full public control, development, and maintenance of the water system (thus eliminating the use of private water companies), by establishing the board of water commissioners to take a proactive and strategic course in the city's water source searches. Irish immigrant, **William Mulholland**, was appointed as the first superintendent of the water system (he was a former deputy zanjero and superintendent of the LAWC), head of the **Department of Water and Power (DWP)** in 1902.

Guided by the former mayor of Los Angeles (who was also the former superintendent of the former LAWC), the DWP conceived the idea of bringing water from the Eastern Sierra Nevada—the fertile **Owens River Valley**. Mulholland considered the Owens River plentiful enough to provide enough water for all the city's future needs. He set out to build the **Los Angeles Aqueduct** from the Owens Valley across the Mojave Desert over the San Gabriel Mountains and ending in the San Fernando Valley (Figure 11.1; see page 22 in *California Atlas*).

The construction of the Los Angeles Aqueduct was completed in 1913 with the support of the Bureau of

214 CHAPTER 11 Urban-Rural California

(a)

(b)

(c)

FIGURE 11.1 *(a) channel of diverted Owens River to Los Angeles Aqueduct from headgate intake bridge looking southeast; (b) cast iron pipe that forms the Los Angeles Aqueduct as it runs across the Mojave Desert; and (c) view south/southeast looking down the aqueduct cascades towards the filtration plant and Los Angeles Reservoir in the eastern San Fernando Valley with Interstate 5 in the background (Courtesy of The Library of Congress).*

Reclamation, the use of stealth in secretly buying options on land with riparian water rights in the Owens Valley, bond measures passed by Los Angeles voters and the grant by Congress of the aqueduct's right of way over federal land. City officials annexed land in the Owens Valley and the San Fernando Valley to Los Angeles, which extended its pueblo rights and thus water ownership rights to these new landscapes.

Once the ranchers and orchard farmers in the Owens Valley discovered what the City of Los Angeles was doing, resentment for their imperialistic ways built. The publication of *Land of Little Rain* by Mary Austin even generated national sympathy. It described in vibrant tones the landscape beauty and mystique of the Owens Valley and Mojave Desert—its flora, fauna and coexistence with people. While the project went through and was completed, the people of Owens Valley had an uneasy relationship with Mulholland's DWP. Animosity escalated in the early 1920s when a drought led to increased water diversions by Los Angeles. The continuing loss of ranchers' livelihoods led to California's first "eco-terrorist" events when the ranchers dynamited the aqueduct in 1924. In the ensuing standoff, superintendent Mulholland refused to deal with the ranchers. The ranchers and their families then continued their battles by seizing the Alabama water gates, shutting the water off from the aqueduct and diverting the water back into the Owens River. At the start of the troubles Mulholland, worried about the continuation of water supplies to Los Angeles, had the St. Francis Dam (1924–1926) constructed in San Francisquito Canyon (just north of Saugus and Santa Clarita in Los Angeles County) to store water in case of more vandalism problems along the aqueduct. The dynamiting of the aqueduct and economic "battles" with the city continued. Then in 1928 a disaster that rivaled the 1906 San Francisco earthquake took place. The St. Francis Dam broke (they unknowingly built it on an earthquake fault and a water-swelling metamorphic rock type) killing 450 people and destroying over 900 buildings. Still reeling from the physical and public relations disaster of the dam failure, a year later the stock market crash of 1929 occurred, which caused the collapse of Owens Valley banks that held properties and water rights the city needed to further secure the water and squeeze out the ranchers. This forced Los Angeles to buy much more of the Owens Valley, which entailed passing another bond to expand the aqueduct into the Mono Basin. By 1941 the extension was complete and the extended water diversions began.

The extension into Mono Basin started to have adverse ecological effects on the ancient Pleistocene era Mono lake. By the 1970s researchers documented severe problems with drops in the lake levels on wildlife due to the diversion of its tributary source waters into the Los Angeles aqueduct by the DWP. Through the 1980s the DWP fought to maintain their rights to the waters against enormous environmental pressures and public trust law doctrines (communal property laws). In 1994, the California Water Resources Control Board ordered the protection of Mono Lake and any of its source streams from diversion by the DWP for the dedicated use by the public for future generations.

11.1.2 SAN FRANCISCO: HETCH HETCHY FOR A CITY WITH NO WATER RIGHTS

San Francisco represents a far different starting story than Los Angeles but with a similar trend and long distance solution to its water problems. The location of the City of San Francisco was on a site that lacked substantial water sources and represented rather inhospitable geography. There was enough water in nearby springs and small lakes when the only population was the soldiers at the Presidio of San Francisco and the priests and neophytes at Mission San Francisco de Asis (Mission Dolores) during the Spanish era. After the securing of Mexican ownership and secularization closed the mission, American settlers arrived in small numbers and founded the settlement of Yerba Buena (around Portsmouth Square). A year after the Americans took California from Mexico, they renamed the settlement of Yerba Buena to San Francisco. The biggest legal difference between San Francisco and Los Angeles is that San Francisco never started as a pueblo chartered under the Spanish—therefore no pueblo water rights.

The Gold Rush brought in a flood of people who used San Francisco as their arrival and staging town before going to seek their riches. The population went from 1,000 in 1848 to 25,000 in 1849, which set off severe water shortages. During this period private entrepreneurs handled much of the development and distribution of water, as the social environment was virtually impossible for any local government authority to build a water strategy, let alone a system. Many groups competed to control the meager local sources that came from springs on Twin Peaks or were shipped in kegs from Marin County by barge and then sold door to door. The numerous fires that destroyed the city from 1850 to 1852 (see Chapter 8) further exacerbated its water problems. Finally in 1857 a group of investors formed the San Francisco Water Works to take water from Lobos Creek, which was a free flowing stream on the western boundary of the presidio's property. They constructed five miles of redwood flume to reach the city center. Several years later another company, the Spring Valley Water Works, absorbed the San Francisco Water Works. Spring Valley Water Works had created a monopoly of water sources along the peninsula (bought water rights) and received an exclusive franchise to provide water to the city in exchange for allowing its use for fighting fires.

Rather than San Francisco owning its own public water supply system, since it did not have pueblo rights, it was under the yoke of the Spring Valley Water Works to supply the city residents with water. By 1874 city officials convinced the state to approve its creation of a public water system, but it was unable to do so since Spring Valley Water Works owned the majority share of the water rights throughout the peninsula. The city tried to purchase the company, but the owners wanted more than the city would offer, and these were powerful people in their own right—represented by the Southern Pacific Railroads' "Big Four" and Ben Haggin (i.e., *Miller-Lux vs. Haggin* fame). Fighting continued until 1900 when the city charter was changed to mandate a city-owned water system.

The city drew up a plan to circumvent the Spring Valley Water Works monopoly by obtaining water from a distant source outside the monopoly's domain. It considered a distant source on the Tuolumne River in the Hetch Hetchy Valley of Yosemite National Park (Figure 11.2), a promising area for a reservoir, plus it was accessible across the Central Valley via aqueduct. In 1906, the great earthquake struck, and the city blamed the Spring Valley Water Works for San Francisco's burning down by not supplying enough water to put out all the fires. The city pushed for a reservoir at Hetch Hetchy and got the U.S. Secretary of the Interior to grant access rights to build the reservoir in Yosemite National Park. Local and national politics delayed the process, with fierce opposition from Spring Valley Water Works, the San Joaquin Valley farmers and John Muir. San Francisco succeeded in receiving final approval in 1913. Despite further political and engineering problems delaying the construction, the city completed the **Hetch Hetchy Reservoir and Aqueduct** by 1934 (see page 22 in *California Atlas*). In addition, San Francisco was able to buy the Spring Valley Water Works in the late 1920s. In a parallel process occurring across the San Francisco Bay, the East Bay Municipal Utility District built the **Mokelumne Aqueduct** by 1929, which also took water from the Sierra Nevada (Mokelumne River) to support Oakland, Alameda, and Berkeley.

(a)

(b)

FIGURE 11.2 (a) Hetch Hetchy Valley (1914) as it appeared before it was transformed into a reservoir. (Courtesy of the Geological Survey); and (b) Hetch Hetchy Valley filled (2001) by O'Shaughnessy dam (Courtesy of The Library of Congress).

11.2 WATER PROJECTS: COLORADO RIVER, THE CENTRAL VALLEY PROJECT AND THE STATE WATER PROJECT

While the municipal projects conducted by Los Angeles and San Francisco were marvels and engineers and politicians admired them, an earlier project conducted on the **Colorado River** in conjunction with the Bureau of Reclamation showed both the prospects and the folly of large-scale irrigation projects. None of these projects though could match the breadth and sheer vast amounts of water transferred by the **Central Valley Project** that the federal government sponsored during the 1930s New Deal Depression era or the California-built **State Water Project** of the 1960s and 1970s.

In a state where inequality happens among cultures and economies, it is the regional inequality of water as a natural resource that these massive transfer schemes have tried to ameliorate. In the water deficient **San Joaquin Valley** and in Southern California these transfers have supported large scale irrigation agriculture, urbanization, and industrial expansion. In the northern half of the state however, although the water capture and transfer process has provided hydroelectric power (see page 23 in *California Atlas*), flood control and recreation, it has been detrimental to fish and wildlife protection and brings up passionate pleas to protect California's water heartland—the Sacramento River Valley—from further water transfers and loss of productive agricultural lands. Engineering for extending natural carrying capacity has not been an easy road for California to follow, especially in a land of historical long term droughts, high biodiversity and boosterism policies that have brought multitudes of people to the state. The state's hydro-politics has not only pitted a water abundant north versus a water impoverished south, but also urbanization developers and the tax base they create against high powered agribusiness and their lucrative global trade contracts.

11.2.1 CONTROLLING THE COLORADO RIVER AND THE "ACCIDENTAL" CREATION OF THE SALTON SEA

The Colorado River is an exotic river, the "American Nile" according to some views. It is a moderate discharge river flowing through an arid region of the Southwestern U.S., and it creates part of California's southeastern border with Arizona. Its upper basin begins in the states of Colorado, Wyoming, Utah and New Mexico then flows south-southwest into the lower basins of Nevada, Arizona and California and on into Mexico to discharge into the Gulf of California (Figure 11.3).

Hydrologists consider it a river of modest average flow, but it is highly variable throughout the year and over decades. Heavy spring rain and snowmelt in the upper part of the basin can lead to late spring-summer floods in the lower part of the basin, with late summer-fall being a very low flow period. Coupled with long-term drought cycles that punctuate the climatic record of the Western U.S., scientists consider the river to have both a large annual range and a very high inter-annual variability. The river also cuts through sedimentary bedrock formations owing to its steep gradient along its flow route (e.g., upper reaches of the Grand Canyon) resulting in erosion, entrainment, and transport of a very large sediment load (i.e., 160 million tons per year of silt). The river's gradient flattens below the Grand Canyon, finally creating a large depositional delta in the Gulf of California, Mexico.

Upon the passage of the **Reclamation Act** in 1902 the federal government (i.e., newly formed Bureau of Reclamation) wanted to approach the harnessing of the Colorado River to "green" the Southwestern desert "wastelands." The **Imperial Valley** (formerly named the Valley of Death) located in the Salton Basin of California's portion of the Colorado Desert in Southern California had already shown in earlier studies to have great potential as a fertile agricultural landscape (see page 5 in *California Atlas*). The valley consists of very fertile soils and a twelve month growing season, but there is an extreme lack of water. In 1901, a private diversion canal, the Alamo Canal, was cut from the Colorado River and routed through Mexico. It then turned northward and entered the U.S. to support a highly successful irrigation agriculture project in the Imperial Valley. In 1904 the canal silted up due to the high sediment loads of the Colorado River. Two successive bypasses north of the border were cut, but they also became blocked with silt almost immediately.

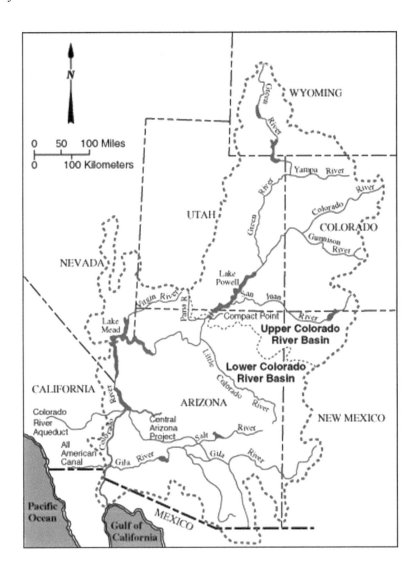

FIGURE 11.3 *Colorado river drainage basin. (Courtesy of the U.S. Geological Survey)*

In late 1904 a private water company received permission from Mexico to cut a new temporary canal. The company intended the construction work to be temporary so they installed flimsy control gates on the canal's entrance to the Colorado River. In 1905, the spring flood arrived early, and high water tore out the gates. Soon the full flow of the Colorado River surged through the diversion canal, swept northward and flooded the entire Salton Basin, transforming it into the **Salton Sea**. It took close to two years (1907) to close the gap, reroute the Colorado River back into its old channel and to restore its regulated flow through the Alamo Canal. By this time, the human-made disaster that is the Salton Sea (525 mi^2/1,360 km^2)had become part of the Colorado Desert landscape, along with the agricultural mosaic to the south of the sea that is the Imperial Valley (Figure 11.4). The Salton Sea has been shrinking with the high evaporative demand of the desert climate and swelling via agricultural runoff for over a hundred years, and therefore it has become highly saline and polluted.

11.2.2 COLORADO RIVER COMPACT AND BOULDER DAM

The Bureau of Reclamation saw the possibilities of the Imperial Valley and the disaster of the Salton Sea as a clear sign that it required a much larger framework to tame the Colorado River for resource use. While the Imperial Valley farmers wanted a new irrigation canal to be on the U.S. side of the border the bureau saw the greater need for a dam on the Colorado River to control the spring floods in order to effectively irrigate the lower basin. To make their plan work, however, required a water allocation agreement—"Law of the

FIGURE 11.4 *Satellite image of the Salton Sea, Imperial Valley and border with Mexico. Notice the differences in agricultural patterns on either side of the border. (Courtesy of the U.S. Geological Survey)*

River"—among the seven states that contained the Colorado River drainage basin. The inequality of such a feat was apparent to most as the water comes from the upper basin states (Colorado, Wyoming, Utah and New Mexico), but the possibilities for agriculture and domestic uses are largely in the lower basin, especially California. Therefore in 1922 the states involved created the **Colorado River Compact**, based on an assumed average annual flow of 17.5 million acre feet, which unfortunately was based on a short record that turned out to be the wettest period in several centuries (i.e., we know this now from Southwestern tree ring analyses and archaeological research). The compact introduced the following allocations: 7.5 million acre feet to Colorado, Wyoming, Utah and New Mexico and 7.5 million acre feet to Arizona, California and Nevada, with the remaining 2.5 million acre feet left in the river for Mexico. Under the compact there was no further attempt to divide the allocations down to a states proportions within a basin. After much disagreement and lack of movement on the process, Congress took action in 1928 and authorized the building of **Boulder Dam** (i.e., also known as Hoover Dam) and the **All American Canal** to supply the Imperial Valley. Congress set lower basin allocations by state as follows:

California received 4.4 million acre feet, Arizona received 2.8 million acre feet and Nevada received 300,000 acre feet. Six states agreed to the deal and signed the compact before construction began, but Arizona refused to sign. There was years of bickering and hostilities despite construction commencing, which eventually lead to a Supreme Court case in 1963 in which the compact was finally ratified.

The federal construction of Boulder Dam (that created Lake Mead reservoir) began during the worst of the Great Depression in 1931with a majority of the labor coming from the devastated economy of Southern California. By the early 1940s the All American Canal to the Imperial Valley and the **Coachella Canal** to the Coachella Valley (i.e., the northern end of the Salton Sea) completed the modern irrigation system to California's Colorado Desert (see page 22 in *California Atlas*).

The final great Southern California tapping of the Colorado River came when William Mulholland convinced the citizens of Los Angeles to pass a bond in order to build the **Colorado River Aqueduct** (see page 22 in *California Atlas*). Construction began with the building of Parker Dam that created Lake Havasu reservoir in 1933, much to the anger of Arizona, which sent state national guardsmen to stop dam construction on their side of the river. The Supreme Court and the U.S. Congress acted in 1935 to allow the completion of Parker Dam, with the Colorado River Aqueduct being completed in 1941. The additional water source for Southern California helped major industrial and housing expansion during World War II.

Over the years since the 1940s Southern California has used more than 1 million acre feet over its original allocation of 4.4 million acre feet. Arizona has had numerous lawsuits against California since the 1930s. Finally, in a landmark decision the Department of Interior in 1998 ordered the Los Angeles Department of Water and Power to come up with a plan to reduce its usage back to the original 4.4 million acre feet allocation. In addition, the federal government has put rules in place regarding allocations during low flow drought years to protect the integrity of the river for endangered species and to honor the treaty with Mexico on their minimum allocation of 2.5 million acre feet. Drought years have allowed the government to cut back on the original allocations to any of the lower basin states, much to their dismay as their populations have boomed.

11.2.3 CENTRAL VALLEY WATER PROJECT: A NEW DEAL IN WATER DELIVERY

The Central Valley consists of three major drainage basins: the **Sacramento River** Valley to the north of the Sacramento-San Joaquin Delta, the **San Joaquin River** Valley to the south of the Delta (but the northern two-thirds of the entire southern Central Valley) and the **Tulare Lake** basin between the San Joaquin basin and the Tehachapi Mountains in the far south (see Figure 2.21b in Chapter 2). Despite numerous rivers flowing through the valley, with sources in the Klamath, Cascade and Sierra Nevada Mountain ranges (Figure 11.5), the usable surface water for irrigation agriculture and domestic needs is quite scarce. Floods in spring and early summer from snow melt create a swampy environment with subsequent loss of crop lands. By late summer and fall the region's extended drought leaves many of the riverine systems with low flow environments. In addition, the water availability is geographically biased towards the entire eastern side of the Central Valley (windward slopes), as the storm track patterns across California leave the western and especially the southwestern portions (the rain shadow slopes) quite dry during the winter precipitation period (see Chapter 3 for review). The dry precipitation pattern and warm interior temperatures make the southwestern portion of the valley nearly similar in climate to the Mojave Desert.

As noted in Chapter 10 the soils in the Central Valley are exceptionally fertile and consist of a balanced texture for agriculture. The type of agriculture available to California, outside of dryland farmed spring wheat crops (see Chapter 8 and 10 for review), is irrigated agriculture. Therefore to make it through the long extended summer drought Central Valley (particularly the San Joaquin Valley) farmers started to tap the valley's enormous **groundwater** supplies. Researchers have estimated that there may be as much as 750,000,000 acre feet of groundwater accumulated over hundreds of thousands of years as the miles-thick alluvial sediments have acted like a gigantic sponge. In the early 1900s farmers dug wells and used inefficient pumps to bring this water to the surface. In some cases the groundwater pressure was so great that they could easily tap flowing artesian wells and make access to the water easier (i.e., no pumping required). After World War I the development of efficient electric turbine pumps allowed irrigated acreage to rapidly expand three times the 1900 level of less than a million acres. The cost of over pumping the ancient aquifer in the San Joaquin Valley, however, led to severe drops in the groundwater level, land subsidence (Figure 11.6) and the death of deep rooted riparian trees (i.e., cottonwoods).

Early state plans to bring water to the Central Valley, while mused about as early as 1856, took place as a proposal in 1919. The proposal called for a large dam on the upper Sacramento River and two aqueducts flowing south through the Central Valley with a spur to provide water to San Francisco Bay cities. The state promoted the idea as providing controlled flows to the Sacramento and San Joaquin Rivers for improved navigation and to meter out the release of water into the Delta to prevent salt water intrusion. The State also proposed a southern component that would divert the Kern River to Los Angeles via a tunnel through the Tehachapi Mountains. This plan, linked to riparian law, failed both in the legislature and with voters all through the 1920s. Recall that riparian users (owners of property next to a river) after 1902 could engage in irrigation as long as they understood that all riparian owners possessed the same right. Thus riparian rights holders had an obligation to use water reasonably, i.e., to share. Riparian rights users had a problem with this obligation to be reasonable when it came to those that had prior appropriation rights on the same river. If they

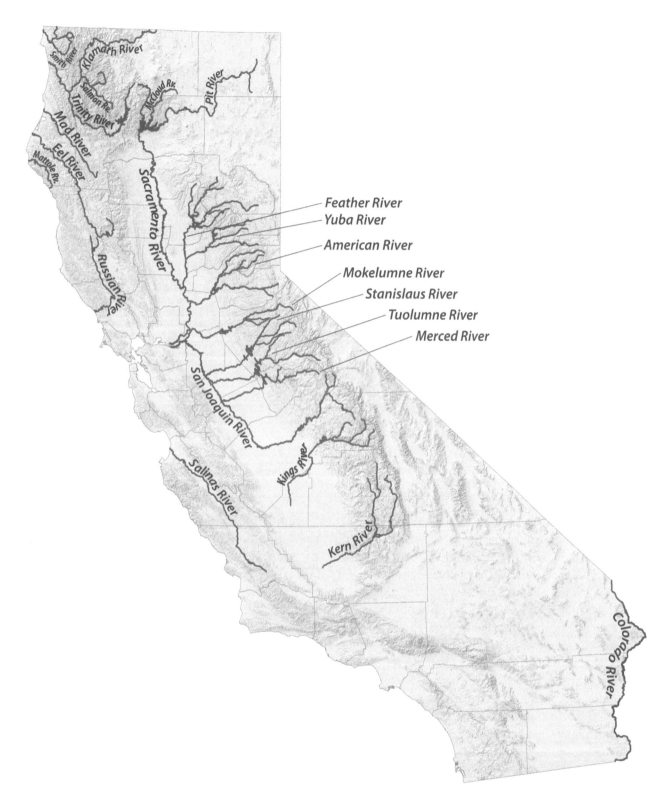

FIGURE 11.5 *Major rivers in California with average annual runoff greater than 0.5 millions of acre feet. For example the range of values includes the Sacramento River (17.0) to the Salinas River (0.6). (Courtesy Curtis Page)*

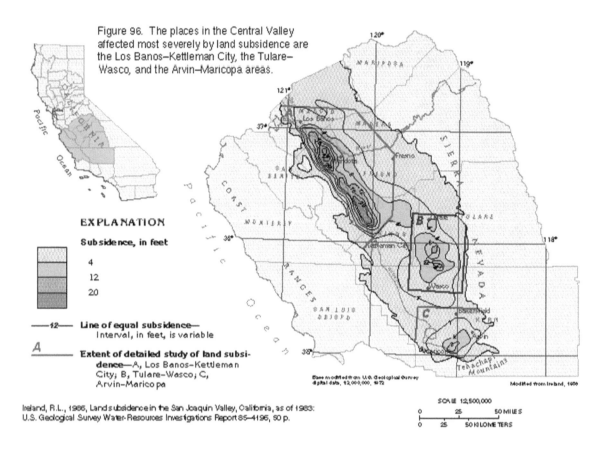

FIGURE 11.6 *The places in the Central Valley affected most severely by land subsidence from intensive groundwater pumping. (Courtesy of U.S. Geological Survey)*

wanted to waste water they felt they should be able to be like the appropriators who ordinarily did not live on the rivers from which they took water. A 1926 State Supreme Court case upheld the riparian users' right to use water any way they saw fit. This decision also prevented appropriators from building dams and therefore trapping the spring flood waters vital to new development. The public was not happy with this ruling and in a separate ballot initiative passed an amendment to the state constitution to prevent the "*waste of water or unreasonable use.*" This amendment removed a significant roadblock to a statewide water plan and remains a central theme of California water rights law.

Because of the tremendous damage that Los Angeles caused to the agricultural communities in the Owens Valley, the Northern California counties wanted reassurance that they had superior water rights in relation to the massive transfer schemes being proposed by the state. Therefore in 1931 the state legislature passed the **county-of-origin law** that stated that counties could regain water rights to water originating within their borders if ever needed. Now the state could proceed with a scaled-down plan that only included the large dam on the upper Sacramento River, an irrigation canal and reservoir system to transport water south to the San Joaquin Valley, as well as regulate releases to improve navigation and prevent saltwater intrusion in the Delta. The plan essentially became a Central Valley water plan.

California suffered a drought beginning in 1929 that continued until 1935. Thus in the depths of the Great Depression the state legislature and voters approved the **Central Valley Project** (CVP). Geographically it was an interesting mix of support. Northern California was in favor (the northern and central parts of the state were hardest hit by the severe drought), Southern California was against it as their municipalities saw no direct benefit for themselves, and the private electrical power companies (i.e., Pacific Gas & Electric (PG&E) and Southern California Edison) were against it because the plan had public power provisions. No one, however, would buy the $170 million in issued bonds needed to make the project happen. Therefore the federal government stepped in and took over the project as its own in 1935 for the Bureau of Reclamation to develop. The bureau began construction in 1937 and completed the system in the late 1950s. The project distributes more than 5,600,000 acre feet of water with 5,000,000 acre feet for farms (enough to irrigate about 33 percent of the agricultural land in California) and 600,000 acre feet for municipal and industrial use. The irrigation water is provided via canals as a subsidy to farmers residing on the western-side of the Sacramento River Valley and the eastern-side of the San Joaquin Valley. The project also generates 5.6 billion kilowatt hours of electricity that it sells to PG&E to supply to urban users.

The entire CVP consists of twenty dams and reservoirs, eleven power plants and five hundred miles of canals, conduits, tunnels, and so forth (see page 22 in *California Atlas*). The main components are Shasta Dam on the upper Sacramento River, which created **Shasta Reservoir** (Figure 11.7); **Clare Engle Lake** on the upper Trinity River (with an inter-basin transfer to Whiskeytown Reservoir); the **Tehama-Colusa canal** (that distributes water to agricultural users on the west side of the Sacramento Valley); the **Delta Cross channel** and the **Contra Costa canal** (keeps the Delta flowing to impede saltwater intrusion); **Folsom Dam** on the American River and **New Melones Dam** on the Stanislaus River (flood control and irrigation water storage); the **Delta-Mendota canal** (that moves water from the Delta to the San Joaquin Valley farmers); and the **Friant Dam** on the Kern River and the **Friant-Kern canal** (that moves water north along the east side of the San Joaquin Valley for farmers).

A final addition proposed for the CVP in the 1950s had envisioned a dam collecting water on the Klamath, Salmon, and Trinity Rivers system (Figure 11.5) with pipes and tunneling sending it south to urban areas but was abandoned as it was too complicated and expensive. The only current expansion proposal for the CVP

FIGURE 11.7 *Shasta dam on the upper Sacramento River creating Shasta reservoir. Note the view of snow covered Mt. Shasta in the distance. (Image © Andy Z., 2009. Used under license from Shutterstock, Inc.)*

is to raise Shasta dam by an additional 7 feet by 2020, however, this is still under review.

11.2.3.1 Irrigated Acreage Limitations: A Big Problem in California

Recall from Chapter 10 that the passage of the Reclamation Act of 1902 allowed homesteads to use water for irrigation, although it still limited irrigated projects to 160 acres per person (or 320 acres if married). In addition, the federal government was supposed to sell off excess land in areas being provided irrigation water through a Bureau of Reclamation project at pre-project prices. Other stipulations included that there was not to be absentee ownership of a farm, also known as "**paper farms**." The government intended for all of these rules to foster the spread of small **family farming** areas in the Western U.S. and especially in California, with the support of federally subsidized water projects. Because the Bureau of Reclamation made no attempts to enforce the acreage limitation until 1943, largely in reaction to the portrayals of Okie farm workers in John Steinbeck's novels, California farmers abused the 160 acre rule. Each member of a family could petition the government for 160 acres, which they cumulatively worked as a large land holding. For example, the number of farms increased from 75,000 in 1900 to 150,000 by 1935 in the Central Valley, while the average farm size dropped from 400 to around 200 acres during the same period (see Chapter 10, Figure 10.3). The percentage of the land in large holdings, however, remained high and continued to rise dramatically. Therefore by the mid-1930s, at the start of the Central Valley Project, two-thirds of the irrigable land was in land holdings larger than 160 acres. (For example, combine the 160 acres each of dad, mom, sister, brother, grandfather, grandmother, aunt and uncle to really farm 1,280 acres under one "farmer"). This exploitation led to 6 percent of the owners controlling 35 percent of the land in the Central Valley, and the Bureau of Reclamation was obligated to support irrigation to these lands under the revised Reclamation Act rules via the Central Valley Project water at a subsidized rate of payment. The idea of a true family farm in California as envisioned by the federal government never worked. It didn't help that the Bureau of Reclamation (by the late 1940s) tried to fix things for these large agri-business farms to continue to receive subsidized water by encouraging the farmers to develop technical compliance schemes. The farmers created corporations and "sold" pieces of land (160 acres or less) to each stockholder, but in reality they deeded the land back to relatives, children, employees and even pets. By 1982, after many decades of abuse, Congress passed the **Reclamation Reform Act** that recognized the reality of modern California farming operations. It was shown that to make the economics work farms had to be bigger than the original 160 acres (even the revised 320 acres). The reality was that two-thirds of California farms were on less than or equal to one hundred acres, but 80 percent of the land was in holdings of over a thousand acres, and 75 percent of the agricultural production in the state came from only 10 percent of the farms. The acreage limitation was raised to 960 acres for subsidized federal water (the price for water was increased however), and the act ended the residency requirement for farmers (i.e., "paper farms" were allowed). The California agri-business farm was here to stay.

11.2.4 STATE WATER PROJECT: WORLD'S LARGEST WATER WORKS

To the large agri-business concerns in the San Joaquin Valley the technical compliance loopholes in reclamation law were ephemeral. Legislation could always be rewritten to truly enforce the acreage limitations that made the large corporate farm both dependent upon and wary of the Bureau of Reclamation. At one point they tried to push the state to buy the Central Valley Project from the Reclamation Bureau, but the price was too high for the state and the Bureau was not politically interested in giving up its expanding empire of controlled waterscapes. Corporate agriculture therefore got together to lobby for the state to underwrite, develop and own its own waterworks for the benefit of the whole state, not just the San Joaquin Valley. The beginning of the gigantic **State Water Project (SWP)** was set, and with the post-World War II population booming across the state its proposal came at a most auspicious time.

The main proponents in the push for a State Water Project were located in the southern end of the San Joaquin Valley—Kings and Kern Counties—and especially the southwestern side of the valley known as the **Westlands Water District** and the ancient **Tulare Lake** basin (see figure 2.21b in Chapter 2). These parts of the San Joaquin Valley are very dry with no useable surface water, very deep groundwater tables that are often brackish (this was an ancient seabed) and no connections to the Central Valley Project canals. Most of the owners held large land holdings ranging from 11,000 to 220,000 acres. These owners varied from the Kern County Land Company (cattle) to Standard Oil, Shell Oil, Richfield Oil, Belridge Oil, Tidewater Oil, Southern Pacific Railroad and Tejon Ranch (whose main owner was the publisher of the *Los Angeles Times*).

Even after the arrival of the Central Valley Project to the eastern side of the San Joaquin Valley the farmers continued to pump a tremendous amount of groundwater (i.e., 6,000,000 acre feet/year). All the Central Valley Project did was to cause the conversion of more lands to agriculture, rather than reduce pumping, and thus the water table continued to drop. The corporate landowners wanted State Water, at all cost, to avoid the acreage restrictions set by the Reclamation Bureau and to reduce some pressure on the groundwater system. To make the State Water Project work, the farmers needed urban buy-in as there was no way that the state could afford the project without this majority block of taxpayers. They had earlier support when the Water Resources Board issued a statement in the early 1950s stating that water was being "*wasted*" as it emptied into the ocean and that "The greatest challenge [facing the state] was redistribution of the water supply from areas of surplus to areas of deficiency." The plan developed in 1951 called for the world's highest earthen dam on the Feather River (near Oroville, Butte County) to apportion out water back into the Feather River, which was a tributary of the Sacramento River. The water was then to be pumped out of the Delta to be carried by an aqueduct along the west side of the San Joaquin Valley to the Tehachapi Mountains where it would then be lifted 3,300 feet over the mountains at a huge pumping station into Southern California.

The plan nearly sold itself when the state's worst flooding occurred in the winter of 1955–56 (i.e., a large pineapple express storm occurred; see Chapter 3 for review) on the Feather and Yuba Rivers inundating 1,000 square miles and killing people. A dam on the Feather River at Oroville could have prevented this, so the state established the **Department of Water Resources** to combine all fifty-two independent state water agencies under one roof for statewide planning. In 1959 the **Water Resources Development Act** passed the legislature and was put to a vote in 1960. It was written to authorize the selling of $1.75 billion in bonds with additional money coming from the state's share of royalties from offshore oil production for a total project cost of $2.5 billion. In economic terms, this bond issue nearly equaled the state's budget at the time, was the largest bond issue ever considered by any state up until this time and represented a record sum to be used for any water project. The true cost was still underestimated.

Gov. Pat Brown Sr. (Gov. Jerry Brown's father) promoted the project, which San Joaquin Valley farmers (agri-businesses) backed heavily, but it had only moderate support from much of Southern California. There was a feeling in Southern California that they really did not need the extra water for urban use at this time, but they were worried that litigation with Arizona in the U.S. Supreme Court over the Colorado River Compact was intensifying. The bond measure still passed by a 0.3 percent margin on a strong north-south split: north being against the project and the San Joaquin Valley and Southern California being for the project . . . barely. As it stands now, the SWP provides approximately 70% to urban users and 30% to agricultural users.

SWP construction ran from 1962 to 1973 with the main components being the Oroville Dam (completed in 1968, it represents one of the tallest and largest dams in the U.S.) on the **Feather River** that created the **Oroville Reservoir** (when filled, contains enough water to supply about 40% of California's urban water needs for one year) and the **California Aqueduct** (Figure 11.8) which travels along the San Joaquin Valley's Westside to the San Luis reservoir in Pachecho pass for secondary storage before being metered back into the aqueduct. Near Bakersfield the aqueduct is lifted (3,300 ft/1005 m) by pipes and pumps over the Tehachapi Mountains and then split into a west branch into Los Angeles County (**Lake Castaic**) and an east branch into Riverside County (**Lake Perris**). The water to supply the California Aqueduct comes from pumping the flows coming down from the Sacramento and Feather Rivers via the Delta. The latest addition to the project was the completion of the **Central Coast branch** that voters of San Luis Obispo and Santa Barbara Counties approved after the drought that occured in the early 1990s. The Central Coast pipeline was completed in 1997. The total water contracted for delivery consists of the following allocations: San Joaquin Valley (mainly Kings and Kern Counties—Westlands district) receives 1.35 million acre feet; Southern California receives 2.5 million acre feet; and the new Central Coast district receives 400,000 acre feet.

FIGURE 11.8 *Aerial view of the California aqueduct carrying water south to Los Angeles and the rest of Southern California. (Image © iofoto, 2009. Used under license from Shutterstock, Inc.)*

11.3 CALIFORNIA'S PERENNIAL WATER WOES: WHAT'S NEXT FOR THE DRY STATE?

Engineers identified the first big problem in California's waterworks in the 1970s—the **Sacramento-San Joaquin Delta**—but voters abandoned a peripheral canal to route around the Delta as a solution in 1982 only to have it come back into focus as California experienced another drought (2007–2010). The Delta's fragile ecosystem is in decline, but, more importantly, the federal government's **Endangered Species Act (ESA)** of 1973 must now dictate its management. The ESA and the **California Endangered Species Act** are strong pieces of policy adopted the same year completion took place for the main components of the State Water Project. The great pumps that pull the fresh water out of the Delta and into the CVP canals and California Aqueduct are severely threatening two state and federally listed endangered fish: the **Delta smelt** and the **Chinook salmon**. The California component of the endangered species legislation is more stringent than the federal as California requires that all losses be fully mitigated (avoid incidental take or fully replace taken species). Therefore the greatest threat to the conveyance of water supplies in California is the Delta. California voters in 1982 voted down the building of a peripheral canal that would have moved the pumps and built a canal farther north and around the eastern edge of the Delta to pump water from the Sacramento River.

In 2007, a federal judge ordered that the southern pumps be shut down to stop the further destruction of the Delta smelt. This scared Southern California, which currently receives 30 to 40 percent of its water supplies via the California Aqueduct. In an act designed to revive the idea of a peripheral canal to solve the environmental problems, the state produced a report at the end of 2008 entitled the "Delta Vision Strategic Plan." The plan acknowledges that the Sacramento-San Joaquin Delta is caught in a toxic cycle of evolving conflicts concerning ecosystem decline, endangered species, flood control, water supply and quality, and drought. The co-evolving system has resulted in failed policies, numerous lawsuits, court orders and a renewed focus on conservation practices within the urban and agricultural systems that rely on the Delta as a critical link in the state's chain of water conveyance.

Second, the basis of the Central Valley Project and the State Water Project as well as with the local projects produced by Los Angeles and San Francisco was the substantially underestimated population projection numbers. The current systems are old and can barely keep up with the demand of 38.5 million people as of the 2010 census. The projection is that the state will exceed 49 million people by 2030, while some estimates place the population at 90 million by 2100. The way California manages its water supplies has to change to a conservation driven and more sustainable path, as predictions indicate that the supplies will shrink rather than expand. Scientists don't consider the ocean desalination technologies touted in the Persian Gulf countries as a possible solution, can produce enough supply to remotely meet future demand.

Finally, the issue of shrinking supplies is the third major state water issue. This brings to the fore the effect global **climate change** will have on California's regional climate and especially upon precipitation and snow pack. The tree ring record for the state has already alluded to large-scale and long-term droughts having occurred in the past. The natural range of variability, coupled with the effects of anthropocentric enhanced greenhouse gas emissions that cause the atmosphere to heat up faster than normal, will have a detrimental impact on California's water supply. These include the impounding of more water during mega-drought years, which deprives free flowing cold water for endangered fish species, combined with the potential rise in sea level of 55 inches by 2100, which will further exacerbate salt water intrusion into the Delta and up the Sacramento River. The prediction for the state's winters is not good as they will be wetter but have less snow pack (70–80 percent loss in the Sierra Nevada alone), thus leading to smaller spring and summer inflows to the rivers and reservoirs. For example, the upper American River has already witnessed a change in the snowline over the last thirty years of up slope movement of nearly five hundred feet.

California has some serious challenges regarding water. The development of reasonable solutions that rely on sustainable development values and serious conservation rather than upon voluntary public action are required, rather than just relying on more dam building. Are the citizens in the three large urban nodes of Los Angeles, San Diego and the San Francisco Bay Area able to come to serious, operational domestic water conservation agreements? Is agriculture, the largest user of water in the state, ready to confront its own wasteful practices to secure a sustainable food producing future? *"Where water flows, food grows"* is the rallying cry in the Central Valley farming communities, but those flows are not for certain. Sustainably designed urban and rural waterscapes for better land management will be necessary in the future if California is to move forward with an environmentally and socially secure future. Many of the ideas discussed in this chapter and more will be forthcoming in the California State Water Plan Update 2013.

Bibliography

Booth, S. (2008). *California Geography*. Course taught at Sierra College. [online] http://geography.sierra.cc.ca.us/booth/California/cal_index.htm.

California Department of Water Resources (2010). California Water Plan Update 2009. Bulletin 160-09 Department of Water Resources, Sacramento, CA.

Calisphere (2008a). *California Cultures*. University of California [online] http://www.calisphere.universityofcalifornia.edu/calcultures/

Calisphere (2008b). *A world of California Primary Sources*. University of California [online] http://www.calisphere.universityofcalifornia.edu/

Carle, D. (2003). *Water and the California Dream: Choices for the New Millennium*. Sierra Club Books, San Francisco, CA.

Carle, D. (2009). *Introduction to Water in California*. 2nd edition. University of California Press, Berkeley, CA.

Chinn, T.W., ed. (1969). *A History of Chinese in California*. Chinese Historical Society of America, San Francisco, CA.

Cox, C.J. (2008). *California Geography*. Course taught at Sierra College. [online] http://faculty.sierracollege.edu/ccox/california_geography/index.html.

Deverell, W. and D. Igler (2008). *A Companion to California History*. Wiley-Blackwell, New York, NY.

DeWitt, H.A. (1999). *The Fragmented Dream: Multicultural California*. Kendall/Hunt Publishers, Dubuque, IA.

DeWitt, H.A. (1999). *The California Dream*. 2nd edition. Kendall/Hunt Publishers, Dubuque, IA.

Donley, M.W., Allan, S., Caro, P., and C.P. Patton (1979). *Atlas of California*. Pacific Book Center, Culver City, CA.

Durrenberger, R.W. and R.B. Johnson (1976). *California Patterns on the Land*. 5th edition, Mayfield Publishing Company, Mountain View, CA.

Fogelson, R.M. (1993). *The Fragmented Metropolis: Los Angeles, 1850–1930*. University of California Press, Berkeley, CA.

Glennon, R. (2010). *Unquenchable: American's Water Crisis and What to Do About It*. Island Press, Covelo, CA.

Guinn, J.M. (1901). *Historical and Biographical Record of Los Angeles and Vicinity*. Chapman Publishing Co., Chicago.

Hornbeck, D. (1983). *California Patterns: A Geographical and Historical Atlas*. Mayfield Publishing Company, Mountain View, CA.

Hundley, N. Jr. (2009). *Water and the West: The Colorado River Compact and the Politics of Water in the American West*. 2nd Edition, University of California Press, Berkeley, CA.

Hundley, N. Jr. (2001). *The Great Thirst: Californians and Water-A History*. 2nd edition, University of California Press, Berkeley, CA.

Kelly, R. (1959). *Battling the inland sea: floods, public policy, and the Sacramento Valley*. University of California Press, Berkeley, CA.

Lantis, D.W., Steiner, R., and A.E. Karinen (1989). *California: The Pacific Connection*. Creekside Press, Chico, CA.

Larson, D.J. (1996). Historical water-use priorities and public policies. In *Sierra Nevada Ecosystem Project: Final report to Congress, vol. II, Assessments and scientific basis for management options*. University of California, Centers for Water and Wildland Resources, Davis, CA.

Leung, P.C.Y. (1984). *One Day, One Dollar: Locke California and the Chinese Farming Experience in the Sacramento Delta*. Chinese American History Project, CA.

McClurg, S. (2000) *Water and the Shaping of California*. Water Education Foundation, Heyday Books, Berkeley, CA.

McWilliams, C. (1973). *California, the Great Exception*. Peregrine Smith, Santa Barbara, CA.

Michaelson, J. (2008). *Geography of California*. Course at UC Santa Barbara, Dept. of Geography. [online] http://www.geog.ucsb.edu/~joel/g148_f08/.

Miller, C.S. and Hyslop, R.S. (1983). *California: The Geography of Diversity*. Mayfield Publishing Company, Mountain View, CA.

Mitchell, D. (1996). *The Lie of the Land: Migrant Workers and the California Landscape*. University of Minnesota Press, Minneapolis, MN.

Mulholland, C. (2002). *William Mulholland and the Rise of Los Angeles*. University of California Press, Berkeley, CA.

Rawls, J.J. and W. Bean (2008). *California: An Interpretative History*. 9th Edition, McGraw-Hill Publishing Co., New York, NY.

Reisner, M. (1993). *Cadillac Desert: The American West and Its Disappearing Water*. 2nd edition, Penguin, New York, NY.

Rice, R., Bullough, W., and R. Orsi (2001). *The Elusive Eden: A New History of California*. 3rd edition, McGraw-Hill.

Righter, R.W. (2005). *The Battle Over Hetch Hetchy: America's Most Controversial Dam and the Birth of Modern Environmentalism*. Oxford University Press.

Starr, K. (2005). *California: A History*. The Modern Library, New York, NY.

State of California Resources Agency (2008). *Delta Vision Strategic Plan*. Blue Ribbon Task Force. Sacramento, CA. [online] http:// www.deltavision.ca.gov.

Stegner, W. (1992). *Beyond the Hundredth Meridian: John Wesley Powell and the Second Opening of the West*. Penquin, New York, NY.

Steiner, S. (1980). *Fusang: The Chinese Who Built America*. Harper & Row Publishers, New York, NY.

Wey, N. (1988). A history of Chinese Americans in California. In: *Five Views: An Ethnic Historic Site Survey for California*, pp. 105–158. California Department of Parks and Recreation, Office of Historic Preservation, Sacramento, CA.

Worster, D. (1992). *Rivers of Empire: Water, Aridity, and the Growth of the American West*. Oxford University Press, New York, NY.

CHAPTER 12

The Rise of the Modern California Landscape

"World War II left an indelible imprint on the economy [and culture] of the American West [sic California]. No other event in the twentieth century had such far-flung influence."
—Gerald Nash, *World War II and the West: Reshaping the Economy*, 1990.

Key terms

African-Americans	Ghettoization	Rumford Fair Housing Act
Aircraft industry	Hippies	1963
Automobiles	Hispanics	Santa Clara Valley
Beat Generation	Hollywood	Sequence of occupation
Black Panthers	Japanese	Ship building
California ranch house	Manzanar	Silicon Valley
Chain migration	Mexican-Americans	Suburban sprawl
Civil Rights Act 1964	Military bases	Summer of Love
Collier-Burns Act of 1947	Military-industrial	Tract homes
Concentration camps	complex	Watts Riots
Dot-com	Newell	White flight
Film industry	Rosie the riveter	Zoot-suit riots

Introduction

WORLD WAR II AS THE DEMOGRAPHIC, ECONOMIC AND CULTURAL INFLECTION POINT

The growth of California's major urban nodes continued to expand during the Great Depression but at a much slower pace for any time since the U.S. takeover of Alta California. Up to World War II the state had witnessed a series of suburban growth spurts largely attributed to the further development of rail transport and to the introduction of the automobile, as well as to further expansion of the power grid network (see pages 14, 15, 21 and 23 in *California Atlas*). The completion of rail lines between cities during the latter half of the 1800s and the early 1900s spawned much of the suburbanization with low density housing that continued, and co-evolved with the automobile, to shape the modern urban cultural landscape up until today. The Depression era projects that built the automobile-focused Golden Gate and Bay Bridges were the beginning of this trend in the San Francisco Bay Area (Figure 12.1). California's early adoption of electrical power also led the way to intra and interurban trolley lines, allowing for the spread of cities or the establishment of new ones. These old trolley lines created many of the key urban and suburban commuting corridors that use of the automobile further developed. The early California appetite for electrical power led to the development of state-regulated public utility companies: Pacific Gas and Electric (PG&E) in Northern California and Southern California Edison (SCE) in the southern section of the state. Before the 1950s the state's large hydroelectric capacity created at reservoirs in the Klamath, Cascades, and Sierra Nevada Mountains generated much of this power (see page 23 in *California Atlas*). The population boom in the 1940s and 1950s changed the energy mix to steam-generated power from imported coal, natural gas and, later in the 1980s, nuclear power (Figure 12.2). In the more recent green or sustainability era of the 21st century, California once more has become the vanguard with the development of wind and solar power energy (Figure 12.3).

At the beginning of World War II several other major enterprises, both private and public, had become established in the state, with some having a strong geographic focus. The **film industry**, arguably one of the most defining symbols of California, had its early start in the 1900s and became fixed in **Hollywood** by 1913 (Figure 12.4). Film production companies left New York and New Jersey for California to escape the imposition of the monopolizing Edison Company and to take advantage of the longer sunny days available to shoot film. Southern California was noted for its lack of overcast weather and its diverse landscapes (i.e., its proximity to deserts, mountains, and coast), with the result that the young semi-rural suburban area of Hollywood became the center of U.S. filmmaking by contagion. One of the most memorable times of filmmaking history occurred between the Great Depression and World War II, known as the "*Golden Age of Hollywood*." Numerous famous films and actors convinced the world that Southern California was the capitol of film making. Some of these films include: *Hell's Angels, King Kong, Dracula, Frankenstein, The Public Enemy, Gone with the Wind, The Wizard of Oz, The Adventures of Robin Hood* and *Snow White and the Seven Dwarfs* to name a few.

The second major enterprise, primarily located again in Southern California, was the **aircraft industry**, which had a strong start with several leading companies, including: Lockheed Aircraft, Northrop Corporation, North American Aviation, Douglas Aircraft and Hughes Aircraft. Southern California offered weather for year-round flying and the ability to build planes both inside and outside the factory hangars. Such a large group of aircraft manufacturing companies in California placed Southern California especially as an important war manufacturing hub in the Pacific and European theater conflicts during World War II. In contrast, Northern California with its focus on the San Francisco Bay Area had a concentration of commercial

FIGURE 12.1 *(a) Golden Gate bridge with San Francisco in the background (© Nagel Photography, 2012, Shutterstock, Inc.); (b) San Francisco-Oakland Bay Bridge with the cities of Oakland, Emeryville and Berkeley in the background (© aspen rock, 2012, Shutterstock, Inc.).*

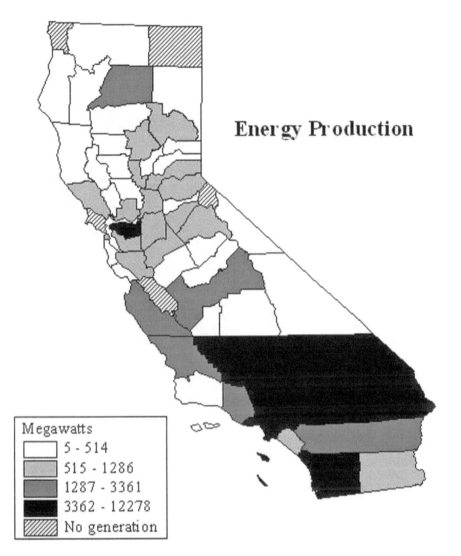

FIGURE 12.2 *Amount of electrical power generated (2009) in California to keep urbanization and industry running. (Data Source: California Department of Energy; Courtesy James A. Bauml)*

and naval **ship building** with large shipyards at Richmond, Vallejo (Mare Island), Alameda, Oakland and Hunters Point (naval shipyard).

Finally, the accelerated establishment of universities and colleges (public and private) around the state (Figure 12.5) had a major impact not only upon the World War II era has maintained a strong focus on economic capacity development in the state. The focus of some of California's first universities in the San Francisco Bay Area upon technological research (i.e., Berkeley's Lawrence Radiation Lab and the Stanford Research Institute) gave them a large role to play during the war. In addition, the concentration of academic researchers and innovative projects later spawned the personal computer revolution and the **Dot-com** Internet revolution.

Throughout this chapter one will recognize that from World War II until the present the various major ethnic groups became more prominent on the landscape. Modern California at the outset of this era was a testament to the cultural landscapes developed since Spanish occupation ("**sequence of occupation**") within the major urban metropolitan areas and the rural landscapes. Each major ethnic group has made its mark in unique ways across the state. They provide the cultural mosaic that California carries across the land and that defines the state to the rest of the world.

232 **CHAPTER 12** *The Rise of the Modern California Landscape*

FIGURE 12.3 *(a) Wind turbines generating electricity in Palm Springs area (© Vlad Ghiea, 2012, Shutterstock, Inc.); (b) Solar panel array in the city of Los Angeles (© egd, 2012, Shutterstock, Inc.).*

FIGURE 12.4 *Hollywood in foreground with the city of Los Angeles in the background (© Andy Z, 2012, Shutterstock, Inc.).*

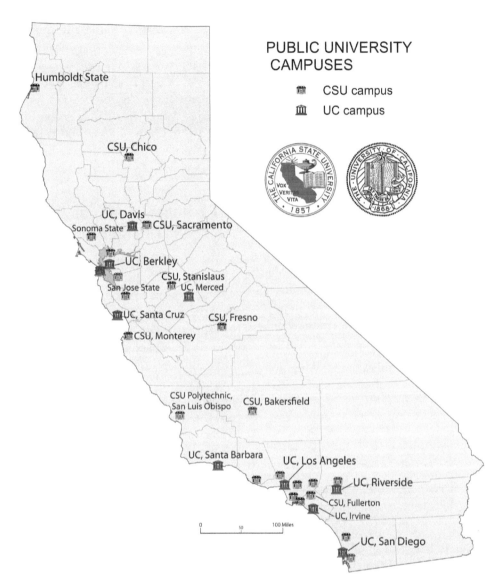

FIGURE 12.5 *Public education for all: Locations of all the California State University and University of California campuses. (Courtesy Curtis Page)*

12.1 WORLD WAR II: INDUSTRY, INFRASTRUCTURE, AND IMMIGRATION

World War II proved an important watershed moment for California, setting the stage for extraordinary growth on several different fronts. Like every other state in the U.S., California was in the doldrums of the Great Depression when aircraft factory orders from the British government signaled the call for wartime buildup before the U.S. had entered either the Pacific or European war theaters. When the declaration of war occurred, it lifted the state from the Depression and set in motion the *"high-technology"* industries and status that California carried to the present. The war also unleashed significant growth rates in new migration, e.g., population 1940—6,900,000 to 1960—15,650,000. Two ethnic groups stand out because of their major migrations to

California during the war and in the postwar period: The **Hispanics** due largely to the Bracero Program with Mexico (see Chapter 10) and **African-Americans**, who followed wartime job availability to the large urban centers. Unfortunately, however, during this time a hundred years of anti-Asian racism in the state came to a head with the placement of largely U.S.-born (66 percent nisei, but not legally naturalized) **Japanese** families in concentration camps located in the Basin and Range high deserts of Owens Valley (**Manzanar**) and Modoc Plateau's Tule Lake (**Newell**).

The war years were the beginning of a new era for the state in many ways. The amount of federal money that came into the state (1940–1947) amounted to approximately 10 percent of the entire federal budget spent in California for wartime manufacturing, military training, shipping and military research. Accordingly, the war brought four fundamental changes to the state:

1. There was a dramatic industrialization and modernization of the San Francisco Bay Area and Southern California (including the rise of San Diego with naval investment).

2. The war provided for a broad set of infrastructure and technology investments to all the urban areas of the state.

3. The military directly controlled a tremendous expansion of California lands, with a focus in Southern California.

4. As mentioned earlier, the war sparked an extraordinary population rush that has persisted until the present.

Under further industrialization and modernization programs the major harbors around the state, including the San Francisco Bay Area (in particular Richmond, Vallejo, Mare Island, Alameda, Oakland and Hunters Point), San Pedro, Long Beach and San Diego Bay, readily got expanded and retrofitted as organizing and collecting points for the military. Approximately 30 percent of the wartime tonnage came through the San Francisco Bay Area alone. In conjunction with that, these military transportation nodes became expanded for navy ship building and retrofitting and as official home ports for the Pacific fleet after the disaster of Pearl Harbor. Los Angeles County was host to over four thousand defense related manufacturing plants producing everything from uniforms to war planes.

The important message here is that all the wartime investments in infrastructure had a direct effect in making trans-Pacific trade via California harbors and airports important today in the global economy (see page 31 in *California Atlas*). The ports of Los Angeles (San Pedro) and Long Beach alone account for more than 40 percent of the nation's imports, from largely Asian countries. When San Pedro is combined with Long Beach (which is the adjacent port) this trade complex represents the sixth busiest container-port in the world (2011 ranking).

12.1.1 NON-MILITARY INDUSTRY AND MILITARY LANDS

World War II was a boon to the Southern California and southern San Joaquin Valley oil industry. The military machine needed oil, and it got it with a 50 percent increase in production in the 1940s. The other business that received handsome contracts from the military was the agricultural industry. After the slump of the Great Depression and its many labor issues, the war effort provided agri-business and its specialty crops a special market for processed fruits and vegetables to feed the troops. Agricultural output soared, and the Bracero farm labor program helped make sure that agri-business was able to make good on its huge contracts with the military. Finally, the funds that the federal budget brought into California focused on building new airports, expanding roads, updating water treatment facilities and supplies, upgrading the communication network between the major urban nodes and allowing the two public utilities (i.e., PG&E and Southern California Edison) to further develop the electricity transmission grid to support the wartime industries and the growing population.

During this time, and into the 1950s, California also witnessed the large-scale takeover of federal lands and the rights of eminent domain proclaimed on private lands to create **military bases** (Figure 12.6). The total acreage under military control reached 3.3 million acres at its height, but starting in the late 1980s the military acreage has decreased from base closures and transfer to civilian or state uses. Many of the original bases, shipyards, supply depots, army air bases and so forth were created or greatly enlarged to take advantage of California's varied terrain and landform regions for training troops and practicing with military machinery. California was the main deployment area for troops sent to the Pacific theater, which led to these expansions and their accompanying infrastructure upgrades that also occurred at civilian sites.

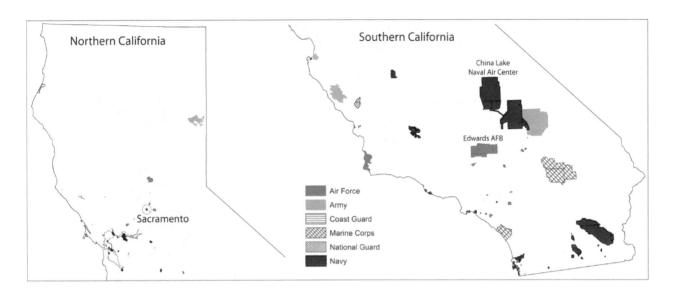

FIGURE 12.6 *Military lands across California. (Data source: U.S. Geological Survey; Courtesy Curtis Page)*

The geographic spatial distribution of these military lands differs between Northern and Southern California (Figure 12.6). In the north they tended to be small and near urban centers, but in the south the military lands were expansive, with some consisting of hundreds of thousands of acres, largely in the Mojave and Colorado Deserts. Large military bases such as Edwards Air Force Base and China Lake Naval Air Warfare Center hosted troops and tanks training for General Patton's North African campaign. The fragile desert soils and vegetation still bear the marks of the tank training maneuvers seventy years later. Several of these bases are on the old Pleistocene playa lake beds discussed in Chapter 2, which are naturally flat and compact and allowed for the landing of NASA's space shuttles from the 1980s until that program ended in 2011.

12.2 MODERN CULTURAL CONFLICT AND MIGRATION

The World War II era brought to the fore two racial issues that had simmered in the golden state for at least the last hundred years—the so called Asian and Mexican "problem." Of course it was only the dominant White population that regarded these groups as socio-economic "problems." The era also witnessed the large migration of African-Americans to the state from the Southern U.S., creating a unique cultural group uniquely tied only to the state's major urban areas.

12.2.1 JAPANESE INTERNMENT

One of the biggest stains on California's cultural past was the forced "evacuation" (a euphemism for forced incarceration) of Japanese-Americans during the war. In 1942, a matter of months after the U.S. entered the war against imperial Japan following the bombing of Pearl Harbor, the government eliminated close to 100,000 Californians of Japanese ancestry (loyal and productive Californians) from the public scene. These people were forced to give up their homes, farms, businesses and jobs for imprisonment in desert lands throughout the Western U.S. on the basis of ancestry alone. The same treatment did not take place against any persons of German or Italian descent. This action against those of Japanese ancestry represented the culmination of anti-Asian sentiment in California. The basis for this forced internment into **concentration camps** was fear, confusion, power, and racial prejudice. Approximately 66 percent of these people were nisei born in America. They would have been eligible for citizenship if the immigration laws had allowed it to happen. Of the eight "relocation centers" built to house the families, two were located in California: **Manzanar** in the Owens Valley (Figure 12.7) and **Newell**, Tule Lake on the Modoc Plateau in northeastern California. In 1988 and again in 1992 the U.S. government issued formal apologies and financial retribution to the Japanese American community for "race prejudice, war hysteria, and a failure of political leadership."

12.2.2 MEXICAN-AMERICANS AND THE ZOOT-SUIT RIOTS

By 1940, more than 35 percent of the Mexican-born U.S. residents lived in California, specifically in Southern California and the San Joaquin Valley. While the Great Depression and the Mexican repatriation movement by the U.S. government had left their mark on this community with respect to a decline in population, devastated communities and broken dreams, World War II brought growth, resurgent communities and new expectations. The war had brought a labor shortage in agriculture and food processing, as many **Mexican-Americans** joined the military along with the Oakies. A further expansion in the numerous military-related industries from garment manufacture to building tanks, opened the doors to Mexican-Americans to find jobs outside the agricultural arena.

Along with the many new opportunities, the Mexican-Americans experienced increased tension with the Southern California police and the U.S. servicemen passing through Southern California on their way to the Pacific front. One of the most famous race riots to occur in Los Angeles was the **"Zoot suit riots."** Zoot suits (e.g., baggy suits) were popular clothing just before and during the war as their development occurred largely during the swing dance craze. Because of the rationing of textile cloth materials during the war, however, it was illegal to produce zoot suits with their generous use of fabric. But because many of these

FIGURE 12.7 *Japanese internment camp at Manzanar in the Owens Valley. (Courtesy of The Library of Congress)*

youths' parents worked in the Los Angeles garment district, they made these suits for their children. Mexican-American youth had developed a subculture that made the zoot suit popular, and they were often acknowledged as *zoot-suiters* or *pachucos*. These second and some third generation youths with their adopted distinctive style of dress were more affluent and more self-confident than their largely Mexico-born parents. In a most unfortunate time of racial hatred, promoted by the Los Angeles newspapers and law enforcement, racial sentiment boiled over in June 1943 when a huge number of servicemen (largely from the U.S. South and Southwest) started attacking the Mexican-American youth with the assistance of the Los Angeles Police Department. Later investigation revealed that racial prejudice, discriminatory police practices and inflammatory newspaper coverage were the principal causes of the seven days of rioting. Despite this event, however, the World War II era proved generally positive for further establishment of the Mexican-American populations.

12.2.3 AFRICAN-AMERICANS: FROM RURAL TO UNIQUELY URBAN

The presence of **African-Americans** in California dates back to the Spanish explorers, the Mexican era and the Gold Rush, though their population numbers were always very low. This all changed from 1940 up until the present. A large migration of African-Americans arrived from rural areas and cities of the Southern states for wartime employment with respect to the expansion of infrastructure and defense related factory jobs. In Northern California the focus was on the railroad and shipyards located in Richmond, Vallejo, Oakland, Alameda and Hunters Point, San Francisco. In Southern California, the focus was on the railroad yards and the aircraft manufacturing industry located around Inglewood, Compton, Carson and south-central Los Angeles. This cultural group was unique in California because it stayed so highly urbanized as a population with no representation in the agricultural businesses—either in farms or in food processing.

After the war this group lost most of the peacetime jobs available to the returning White male veterans. The loss of stable jobs forced African-Americans into poverty and a position of welfare. Matters became worse with the level of housing discrimination leading to the **ghettoization** and creation of low income housing projects in inner city Los Angeles and the San Francisco Bay area.

12.3 HOUSING: SUBURBANIZATION—THE RISE OF THE TRACT HOME

The California post-WWII population increases rivaled the increases of the 1920s. Between 1940 and 1970 the state added another 13 million people. These tremendous increases have made California the most populated state in the union since 1962, with 66 percent of that growth coming in the 1940s and '50s represented by migration from other states. The large proportion of the incoming flux of people settled in Southern California, but as a whole by 1960, close to 90 percent of Californians lived in the large urban metropolitan areas across the state. In a repeat of the past, these new residents came for the same old reasons: the mild climate and the surging economic opportunities in California's productive capacity. These people were largely ex-service men and women who had encountered California during the war while they were stationed, being trained or in transit. The GI bill of rights for funding education swelled the California State University and University of California campuses (Figure 12.5), adding to the capacity development that continued to propel the state in economic productivity.

To house all of the incoming families was quite a problem in the early stages after the war ended; however, builders developed mass production techniques taken from those used in the defense industries for building aircraft, tanks, etc. to make efficient gains in building homes. From the mid-1940s until the present-day, California builders became "tract" homebuilders (e.g., modeled after the planned community development of Levittown, NY). **Tract homes** referred to numerous houses of similar or complementary design constructed on a suburban tract of land with medium density (Figure 12.8). Sometimes referred to as "cookie-cutter homes", this housing style is able to

FIGURE 12.8 *California housing tract sprawl in Riverside.*

rapidly encompass many square miles of suburban land in a limited amount of building time because the limited designs allow for building the homes all at once and labor is segmented in the process, which also brings down costs. The most successful style of home was the **California ranch house** (Figure 12.9), which pushed "California living." It fused modern ideas and styles with the notions of Western period open style ranch homes to create an atmosphere of informal and casual living. The outdoor patio, the built in BBQ, the attached two-car garage and, in Southern California the swimming pool, characterized the California style. There were strong Northern versus Southern California design characteristics. The north constructed homes with wood siding and wood shingle roofs in contrast to the mission/Hispanic look of the Southern California homes characterized by the use of red tile roofs and stucco siding. The universal nature of **suburban sprawl** in connection with the automobile was the catalyst for the major encroachment into agricultural lands in the state's urban centers. In Los Angeles and Orange Counties, this sprawl led to the near complete removal of the citrus orchards for housing, which caused the orange citrus production to shift to the eastern side of the San Joaquin Valley (see page 25 in *California Atlas*). Remnants of the lemon citrus industry, however, remained in Ventura County.

In addition to the suburban sprawl, the return of mainly White servicemen from the war caused a major displacement of women, as well as ethnic minorities, in the industrial employment sector. During the war many women took the place of men on the industrial assembly lines and in the shipyards of Northern California, and aircraft and weapon assembly lines in Southern California. The war had broken the former established social structure where women stayed at home. Instead they filled the gaps left by their husbands, fathers and significant others who joined the armed services to fight in the war. The U.S. War Department quickly built a campaign for getting women into these manufacturing jobs, therefore giving California women a taste of the workplace and of financial freedom (Figure 12.10). This was the era of "**Rosie the Riveter**," which came to a sudden halt for the White, Mexican-American and

CHAPTER 12   *The Rise of the Modern California Landscape*   **239**

FIGURE 12.9   *The California ranch home, 1960s style.*

FIGURE 12.10   *Images of women working to win World War II in California military equipment production plants and ship yards. (Courtesy of The Library of Congress)*

African-American women employees with the return of White male employees. For the cities, this loss of jobs by the minorities coupled with the development of the suburbs meant the beginning of the era of inner city decline. Post-WWII became the era of **"White flight"** where White families migrated to tract home planned suburban "bedroom" communities; the male heads of the homes then commuted into the cities to work. The lack of economic opportunity and job discrimination combined with the 1938 Department of Housing Act, which encouraged racial segregation, left the inner cities in poverty and dominated by ethnic minorities. All the while there was a continued **chain migration** of African-Americans into south-central Los Angeles, Vallejo, Richmond, Oakland, Alameda and Hunters Point; and the Mexican population swelled in the inter-city barrios and the rural colonias. This left California's urban centers with inferior housing, high unemployment, welfare dependent families, school segregation, business discrimination and high crime rates. All this occurred as a link to the postwar housing development and employment changes. One of the saddest marks of this postwar period was the collusion created by real estate operators and developers to discriminate and to limit access to the new tract housing for White families only.

12.3.1 TRANSPORTATION AND ECONOMY

"The East built the cars, but California taught us how to live with them." Donald Meinig, *Symbolic Landscapes: Some Idealizations of American Communities*, 1979.

Californians have always had a love affair with **automobiles**, and they hold tightly to them like a cellphone or wallet. The automobile has etched its identity into the landscape through thousands of miles of roads, highways, freeways and city streets. At the end of 2011 there were 24 million people with driver's licenses and 32 million registered vehicles. This was twice the number of vehicles registered in California than any other state. The automobile's transportation flexibility in conjunction with the suburban sprawl that began towards the end of the war led the state legislature to pass the **Collier-Burns Act of 1947**. This gasoline-driven tax committed the state to a construction program to build some 13,000 miles of freeways and expressways (see page 21 in *California Atlas*). In its co-evolution it was a self-fulfilling creation as it promoted suburbanization farther away from the city centers and therefore committed Californians to lives lived in commuter traffic (Figure 12.11) and shopping at roadside strip malls.

The act's passage also indirectly led to the demise of the electric trolley and the city rail systems in the 1950s that many small towns and major cities had built during the earlier era of electrification. Essentially, it put any remote ideas about mass public transit to rest until the oil supply shocks of the early 1970s. Public transport finally came back to life (barely) in the mid-1970s with the construction of BART (Bay Area Rapid Transit) in the San Francisco Bay Area and later in the 1990s with the Metrolink rail service in Los Angeles County and the red trolley line operating in San Diego that provides service to the Mexican border (Figure 12.12). In this current era of energy conservation and sustainable development, California is now looking to build a hydrogen highway fuel network to reduce the state's dependence on foreign oil for gasoline distillation. In addition, voters passed a bond in 2008 to build a high-speed train service to link Northern and Southern California urban centers. The high-speed rail network will link Sacramento with San Diego through the San Joaquin Valley and have connections with BART for service to the San Francisco Bay Area.

Economically, California has always been a great experiment with its diverse cultures and constant regenerating "newness" that tends to lend itself to a future full of creative energy. Its embracement of being a world community has combined with the expectation that tomorrow means new opportunities for the accumulation of wealth and the pursuit of a quality of life that one cannot find in the other forty-nine states.

". . . people are the common element in California's vital economy. It has been California's good fortune to have citizens possessing the imagination, daring and skill to create new industries and to invent social, political, and economic institutions that facilitated and encouraged growth. The California economy is people believing in themselves and sharing dreams of tomorrow." David Hornbeck, *California Patterns*, 1983.

The attitude described above has led California to develop a distinctly diversified economy that has varied over time but has held onto some broad themes of productivity. Outside of the specialty agricultural and natural resource extraction sectors of the economy (i.e.,

CHAPTER 12   *The Rise of the Modern California Landscape*   **241**

FIGURE 12.11   *Aerial view of complex highway interchange in Los Angeles California. (© ifoto, 2012, Shutterstock, Inc.)*

FIGURE 12.12   *Downtown San Diego's Harbor Drive, including the convention center and red trolley line. (© N Christopher Penier, 2012, Shutterstock, Inc.)*

oil, natural gas, lumber, mining, cement, etc.) the state is the foremost producer and designer of electronic equipment (computers, cellphones, and handheld devices), computer software, biotechnology (San Diego), machinery, and metal products, as well as of motion picture, television film and related entertainment industries. California places much of its economic might in emerging industries of the new computer/Internet-based economy, biased towards design and services, rather than focusing upon the manufacture of durable goods. One industry that used to dominate, especially in Southern California, was aerospace as discussed ealier, which had grown steadily since World War II. The basis of its expansion was the rising tensions of the Cold War with the former Soviet Union (USSR). Defense funding kept many of the wartime plants working from the 1950s to the mid-1980s through the auspices of the **military-industrial complex**. In Southern California, the Intercontinental Ballistic Missile (ICBMs), Patriot missile, jet aircraft, as well as NASA space shuttles were designed and built. By the end of the Cold War in 1989, however, these industries had become shattered, shutdown or bought out and moved to other states.

In Northern California, many of the aerospace companies had diversified into developing and manufacturing electronic circuits and space technology such as communication satellites, and computer technology. These companies concentrated in the **Santa Clara Valley**, which became known as **"Silicon Valley"** (see Chapter 1, Figure 1.3). The region was ideally situated between Stanford University, the Stanford Research Institute, U.C. Berkeley and the Lawrence Livermore National Labs. The war generated major momentum in all these groups. The profound evolution of this valley and subsequently the California landscape came with the invention of the Intel 8080 microprocessor in 1973, which fueled the development of the personal computer, especially in 1977 with the debut of the Apple II and later the Apple Macintosh in 1984). It's hard to imagine today not having the microprocessor in our lives, as it controls everything from television sets and coffeemakers to cellphones and iPods. The computer industry went on to reshape the state's economic landscape and cultural landscapes with the early software development that made the Internet and World Wide Web a required element in our lives. The net has allowed a certain level of decentralization in business that has enabled other neglected areas of the state to grow since the 1980s (i.e., the Sierra Nevada foothills) and to survive based on the **Dot-com** economy.

Despite the positive "golden" nature usually described for California's economy, many of its industries have moved their manufacturing or production bases to other countries or states. The cost to do business in this state or in the U.S., in general, is high, so California's technology and entertainment industries have become ones of idea generation, from prototyping to developing the capital model. The work to produce a product, be it an iMac or iPad computer, a cellphone or a TV episode of *Lost*, takes place outside the state.

Briefly examining the top five imports and exports (Table 12.1) from data provided by the Port of Los Angeles tells a picture of the cyclical nature of California's economic history. As Chapter 7 described with respect to the Mexican rancho era, one can make a current connection between our imbalanced trade with Asian countries (i.e., dominated by China) and the hide and tallow trade. During the hide and tallow trade era, raw resources were exported and produced-finished items from those resources were imported for consump-

TABLE 12.1 The Top Five Imports and Exports Coming Through the Port of Los Angeles at San Pedro for 2008.

Imports	Exports
Furniture	Paper, Paperboard and Wastepaper
Apparel	Scrap Metal
Automobile Parts	Grains, Wheat & Soybean Products
Electronic Products	Fabrics & Raw Cotton
Toys	Pet and Animal Feed

tion by a society controlled by a wealthy class (i.e., the rancho dons) supported by a service class. Table 12.1 describes current similar condition as 165 years ago. The future has a past, and California's current fiscal problems and ties to the global economy with respect to the major economic downturn (2008–present)—the likes of which this state has not witnessed since the Great Depression of the 1930s—has seen the same industries and geographic regions taking the strain yet again.

12.4 CALIFORNIA'S ERA OF EXPANDING THE MIND AND RESTLESS YOUTH ENERGY

"Made up my mind to make a new start. Going to California with an aching in my heart. Someone told me there's a girl out there. With love in her eyes and flowers in her hair." Led Zeppelin, *Going to California*, 1971.

Many regard the 1960s as a time of peace, love, and flower power. There were also dark flowers in the peace garden of the Age of Aquarius that the youth developed during this era. California became a center for much of the '60s' liberal, exciting, and distrustful period, however this decade had its roots in the disaffected college age youth of the late 1940s and 1950s.

A group of youths known as the **Beat Generation** represented the postwar "lost" generation later named "beatniks." The founding members were young bohemian writers originating from the East Coast who rejected their parents' obsession with security (i.e., largely a legacy of World War II weariness and nuclear war fears generated from the Cold War). The central arguments of the beatnik culture were based on a rejection of mainstream American values: an "organization man" existence in the eight to five workday, the accumulation of material goods and life in suburban tract-home sameness with its potential boredom, divorce, stifling of creativity and self-dependence on a system that sucks away the spirit. The main proponents—Jack Kerouac (i.e., writer of *On the Road*—chronicled a trip to California), Allen Ginsberg and Neal Cassady—became obsessed with drug experimentation, alternate forms of sexuality and an interest in Eastern spirituality that all focused on the liberation of the mind. They also followed the same mode of thought that drove countless others to the West Coast: freedom and breaking ties with the Eastern establishment that they felt shackled their creativity. An obsession with California developed and centered in San Francisco and later in Los Angeles and the Big Sur region. By the 1960s the Beat culture underwent a transformation and gave way to the **"hippies"** with their counterculture movement born and established in California's urban centers.

The 1960s counterculture movement in California coincided with the development of the rock'n roll and pop music scenes shared by the U.S. and Britain. The U.S. music scene was largely derived from African-American ghetto blues, which also originally inspired the British scene. The continual push by the youth during this period to do the opposite of their conservative parents and perceived societal repression played into the Cold War and the escalating violent engagement in Vietnam. Following on the beatnik generation, this era's youth wanted even more freedom and personal expression—freedom from traditional modes of authority and consumerism. This led to increased psychedelic drug experimentation, new forms of musical expression, a new code of equality within race relations, flexible sexual mores and the advancement of women's rights. California and especially the San Francisco scene became the vanguard for the countries counterculture movement, later renamed the hippie movement. Important people instigating the movement include Ken Kesey, leader of the LSD-fuelled group the Merry Pranksters. He became the originator of the famous Trips Festivals that got the San Francisco acid-rock scene started with the following musical groups: the *Grateful Dead, Quicksilver Messenger Service, The Doors, Jefferson Airplane, Santana*, etc. The famous figures of this era and hippies in general all had desires to journey to the Far East, but many of these journeys started out as journeys to the West Coast. For many in the mystic sixties the search for nirvana began with a voyage west of the Mississippi, to the land of orange grove memories, the Gold Rush, movies and sunshine—California

Dreamin'. Interestingly, the hippie movement had connections to the utopian settlements around California from the 1880s (see Chapter 9). During that earlier era one of the main utopian/spiritualist proponents, the Theosophical Society, believed that California was the future development region for the "Pacific root race" which was to replace the dominant Aryans with a more spiritually/planetary enlightened culture. This idea had become popular and dominant again with the hippies in especially their main place of origin, the Haight-Ashbury district of San Francisco.

This period is also punctuated by tense events generated by Vietnam War protesters consisting of university student standoffs and riots that occurred at U.C. Berkeley (Peoples Park riots, 1969) and U.C. Santa Barbara (Isla Vista riots, 1970). The lead-up to these vocal and sometimes violent protests derived from students in California universities gaining student free speech rights on university campuses, a freedom that students and faculty now all take for granted in the lecture halls across the nation. This was an extension of the civil rights movement won earlier in the decade.

The federal **Civil Rights Act** passed in 1964, but California had its own problems when voters passed Proposition 14 that same year. Since the end of World War II, real estate agents and developers in the suburbs had discriminated against minorities. In 1963, the **Rumford Fair Housing Act** passed to stop property owners from denying people housing because of ethnicity, religion, sex, marital status, physical handicap or familial status. An overwhelming percentage, however, voted for Proposition 14 the following year, which was cleverly worded to reverse the Rumsford act. This vote subsequently created anger and a feeling of injustice within the African-American minority, who dominated the major inner cities. Finally, the inner city situation exploded in south-central Los Angeles with the **Watts Riots** in 1965 (Figure 12.13). The anger and frustration towards a California system rigged against the African-American community, which led to a high jobless rate, poor housing conditions and underperforming schools was too much to bear. The response in the African-American community was militant with the rise of the **Black Panthers** in Oakland 1966, which later spread to

FIGURE 12.13 *The burning streets of the Watts riots. New York World Telegram photo. (Courtesy of The Library of Congress)*

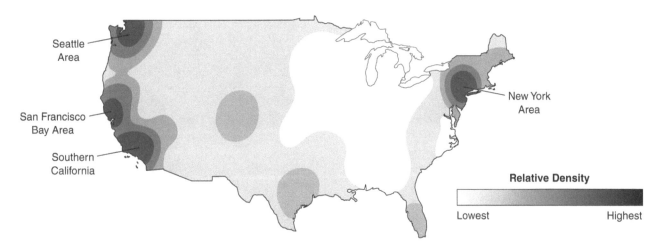

FIGURE 12.14 *Map based on analysis by Richard Florida from "Who's Your City," the U.S. coastal populations are "Open to Experience," especially in California. (Data source: Jason Rentfrow and Kevin Stolarick, Adapted map originally by Paulo Raposo)*

all the major inner city African-American communities in the country. The Black Panthers used Marxist-Maoist rhetoric and armed militancy within inner city African-American strongholds to promote "Black Power" and self defense from police brutality. The group was flawed, however, and the movement disbanded from pressure by the FBI after ten years. Therefore, it was rather unfortunate to witness the same rioting in the same geographic area (south-central Los Angeles, Watts/Compton) during the aftermath of the 1992 Rodney King vs. Los Angeles police beating incident court case (the officers were acquitted). The second time around only illustrated that the same issues and problems still had not been corrected. Since 1992 the California inner cities have received attention for their economic needs.

The culminating events of 1960s California were the Human Be-In and **Summer of Love** festival 1967 in San Francisco. So many youth came from outside California and the U.S. to these festivals that the "hippie" values they took away from San Francisco and Berkeley made a lasting impression upon the world and especially upon California that has existed to this day. Outsiders still see California and Californians guided by their desire to "experience" and be an open culture (Figure 12.14). This cultural trait has led the region to produce impacts on philosophy, morality, music, art, lifestyle, religion, and fashion.

The important events and figures of this period marking California places and culture with each passing year, moves deeper into misinterpreted myth. This era, however, is still deeply etched in California's psyche as well as to the world outside its borders.

> "The '60s are gone; dope will never be as cheap, sex never as free, and the rock and roll never as great." Abbie Hoffman, (American radical activist and writer, Berkeley, 1936–1989)

Bibliography

Almaguer, T. (1994). *Racial Fault Lines: The Historical Origins of White Supremacy in California*. University of California Press, Berkeley, CA.

Avila, E. (2006). *Popular Culture in the Age of White Flight: Fear and Fantasy in Suburban Los Angeles*. University of California Press, Berkeley, CA.

Binford, G. (2005). Sunshine technopolis: southern California's dimming future. *Futures* (37): 345–348.

Bottles, S. (1987). *Los Angeles and the Automobile: The Making of a Modern City*. University of California Press, Berkeley, CA.

Brasil, E. (1992). Up Country: It's divide and rule. *San Francisco Examiner*, February 9.

Bricker, D. (2000). Ranch Houses Are Not All the Same. In *Preserving the Recent Past 2*, edited by D.Slaton and W.G. Foulks, Historic Preservation Education Foundation, National Park Service, and Association for Preservation Technology International, Washington, D.C. [online] http://www.nps.gov/nr/publications/bulletins/suburbs/Bricker.pdf

California Legislative Analyst's Office (2007). *California Tribal Casinos: Questions and Answers*. California's Nonpartisan

Fiscal and Policy Advisor. [online] http://www.lao.ca.gov/2007/tribal_casinos/tribal_casinos_020207.aspx.

Calisphere (2008a). *California Cultures.* University of California [online] http://www.calisphere.universityofcalifornia.edu/calcultures/

Calisphere (2008b). *A world of California Primary Sources.* University of California [online] http://www.calisphere.universityofcalifornia.edu/

Camarillo, Albert. 1984. *Chicanos in California: A History of Mexican Americans.* Sparks, NV: Materials for Today's Learning, Inc.

Clark, W.A.V. (1998). *The California Cauldron: Immigration and the Fortunes of Local Communities.* Guilford Press, New York, NY.

Collins, K.E. (1979). *Black Los Angeles: The Maturing of the Ghetto, 1940–1950.* R & E Research Associates, San Francisco.

Davis, M. (2000). *Magical Urbanism: Latinos Reinvent the U.S. City.* Verso, New York, NY.

Davis, C. and D. Igler (2002). *The Human Tradition in California.* Scholarly Resources, Wilmington, DEL.

DeGraaf, L.B., Mulroy, K., and Q. Taylor (2001). *Seeking El Dorado: African Americans in California.* Autry Museum of Western Heritage and University of Washington Press, Los Angeles, CA.

Deverell, W. and D. Igler (2008). *A Companion to California History.* Wiley-Blackwell, New York, NY.

DeWitt, H.A. (1999). *The Fragmented Dream: Multicultural California.* Kendall/Hunt Publishers.

DeWitt, H.A. (1999). *The California Dream.* 2nd edition. Kendall/Hunt Publishers.

Diamond, S.J. (1993). Split personality. *Los Angeles Times,* July 25.

Dilsaver, L.M., Wyckoff, W., and W. Preston (2000). Fifteen events that have shaped California's Human Landscape. *The California Geographer,* 40: 3–78.

Donley, M.W., Allan, S., Caro, P., and C.P. Patton (1979). *Atlas of California.* Pacific Book Center, Culver City, CA.

Durrenberger, R.W. and R.B. Johnson (1976). *California Patterns on the Land.* 5th edition, Mayfield Publishing Company, Mountain View, CA.

Foster, M.S. (1975). The model-t, the hard sell, and Los Angeles's urban growth: the decentralization of Los Angeles during the 1920s. *The Pacific Historical Review* 44(4): 459–484.

Gregor, H.F. (1963). Spatial Disharmonies in California's Population Growth. *Geographical Review* 53: 100–22.

Hanson, V.D. (2007). *Mexifornia: A State of Becoming.* 2nd Edition, Encounter Books.

Hornbeck, D. (1983). *California Patterns: A Geographical and Historical Atlas.* Mayfield Publishing Company, Mountain View, CA.

Kling, R., Olin, S., and M. Poster (1995). *Postsuburban California: The Transformation of Orange County since World War II.* University of California Press, Berkeley, CA.

Lotchin, R.W. (1992). *Fortress California 1910–1961: From Warfare to Welfare.* Oxford University Press, New York, NY.

McIntire, E. (1998). California's Changing Landscapes. *Journal of the West,* 37(3):44–54.

McWilliams, C. (1973). *California, the Great Exception.* Peregrine Smith, Santa Barbara, CA.

Matthews, G. (1999). The Los Angeles of the North: San Jose's Transition form Fruit Capital to High-Tech Metropolis. *Journal of Urban History,* 25: 459–476.

Meinig, D.W. (1979). Symbolic Landscapes: Some Idealizations of American Communities. In: *The Interpretation of Ordinary Landscapes: Geographical Essays,* D.W. Meinig, ed., pp. 164–192. Oxford University Press, New York, NY.

Michaelson, J. (2008). *Geography of California.* Course at UC Santa Barbara, Dept. of Geography. [online] http://www.geog.ucsb.edu/~joel/g148_f08/.

Mitchell, D. (1996). *The Lie of the Land: Migrant Workers and the California Landscape.* University of Minnesota Press, Minneapolis, MN.

Morello-Frosch, R., Pastor, M. Jr., Porras, C., and J. Sadd (2002). Environmental Justice and Regional Inequality in Southern California: Implications for Future Research. *Environmental Health Perspectives* 110(suppl 2): 149–154.

Morrison, P.A. (1971). *The Role of Migration in California's Growth.* Rand Corporation, Santa Monica, CA.

Nash, G. (1990). *World War II and the West: Reshaping the Economy.* University of Nebraska Press, Lincoln, NE.

Pitti, S.J. (2004). *The Devil in Silicon Valley: Northern California, Race, and Mexican Americans.* Princeton University Press.

Pryde, P.R., ed. (1979). *San Diego: An Introduction to the Region.* 2nd edition, Kendall/Hunt, Dubuque, IA.

Rawls, J.J. and W. Bean. 2008. *California: An Interpretative History.* 9th Edition, McGraw-Hill, Boston.

Rice, R., Bullough, W., and R. Orsi (2001). *The Elusive Eden: A New History of California.* 3rd edition, McGraw-Hill.

Sanchez, George J. 1993. *Becoming Mexican American: Ethnicity, Culture, and Identity in Chicano Los Angeles, 1900–1945.* New York: Oxford University Press.

Sanchez-Korrol, Virginia V. 1983. *From Colonia to Community.* Westport, Conn: Greenwood Press.

Saxenian, A. (1985). The Genesis of Silicon Valley. In: *Silicon Landscapes,* P. Hall and A. Markusen, eds., pp. 20–34. Allen and Unwin, Boston, MA.

Self, R.O. (2003). *American Babylon: Race and the Struggle for Postwar Oakland.* Princeton University Press, NJ.

Social Explorer (2009). *Los Angeles County 1940–2000: Race Map*. [online] http://www.socialexplorer.com/pub/maps/home.aspx.

Social Explorer (2009). *United States Census Demographic Maps: California 1850–2007*. [online] http://www.socialexplorer.com/pub/maps/home.aspx.

Social Explorer (2009). *United States Religion Maps 1980–2000*. [online] http://www.socialexplorer.com/pub/maps/home.aspx.

Starr, K. (2005). *California: A History*. The Modern Library, New York, NY.

State of California, Department of Finance (2009). *California Current Population Survey Report: March 2007*. Sacramento, CA. [online] http://www.dof.ca.gov.

State of California, Department of Finance (2009). New state projections show 25 million more Californians by 2050; Hispanics to be state's majority ethnic group by 2042. Sacramento, CA. [online] http://www.dof.ca.gov

Steiner, R. (1981). *Los Angeles: The Centrifugal City*. Kendall/Hunt, Dubuque, Iowa.

Takaki, R. (2008). *A Different Mirror: A History of Multicultural America*. 2nd edition, Back Bay Books.

Thompson, W.S. (1955). *Growth and Changes in California's Population*. The Haynes Foundation, Los Angeles.

Vance, J.E. (1972). California and the search for the ideal. *Annals of the Association of American Geographers* 62(2): 185–210.

Weber, David. J. (1979). *Foreigners in Their Native Land: Historical Roots of the Mexican-American*. Albuquerque: University of New Mexico Press.

CHAPTER 13

California: Paradise or Gritty Reality?

Key terms

1965 Immigration and Nationality Act	Coyotes	Operation Gatekeeper
AB32	Cultural regions	Proposition 187
California Environmental Quality Act	Gaming industry	Refugee Act of 1980
	Gujuartis	Secure Fence Act of 2006
California Land Conservation Act of 1965	H-1B visa	South Asian Indian
	Hmong	Split state initiative
Chain migration	Immigration and Naturalization Service	State of Jefferson
Chicanos		
Community Supported Agriculture	Maquiladora	
	Marijuana	
	NAFTA	

Introduction

CALIFORNIA'S GEOGRAPHIC FUTURES... THE GEOGRAPHY OF THE FUTURE HAS A PAST

"Current events form future trends." Gerald Celente, *Trends Research Institute*, 2008.

California's population continues to increase through native birth and immigrations. The state's estimated population at the end of 2010 was 37.2 million with a projection of just fewer than 60 million by 2060. Before 1980 internal U.S. migration drove the population increases, but after 1980 foreign immigration has largely driven population growth. California's dramatic population increases have roots in the U.S. Immigration Acts of 1965 and 1986. In 1965 the governmentally removed the barriers to immigrants coming into the country from non-European countries.

The **1965 Immigration and Nationality Act** abolished the national origins and quota systems-based policy put in place by the 1924 Immigration Act. This new act based the immigration numbers on global hemisphere of origin, it instituted education and occupational criteria, and it allowed U.S. citizens to sponsor the immigration of family members, which set the stage for **chain migration** (see Chapter 1 for review). In 1986 an amendment to the 1965 act increased the number of immigrants allowed into the country and provided amnesty to illegal immigrants that had been

249

in the country prior to 1982. Finally, the last immigration act passed was in 1990, which further increased the numbers that could be let in. However, there was a greater emphasis on skilled and degreed workers (see page 26 in the *California Atlas*). All of these immigration acts had a profound influence on California's population increases. In summary, before 1980, 60 percent of the immigrants came from Europe, while after 1980, 80 percent of the immigrants started to come from Asia and Latin America (see page 26 in *California Atlas*).

Table 13.1a reveals the top ten counties that will potentially dominate in population size by 2050 and Table 13.1b presents those counties that will have the largest change in population between 2000–2050. They reveal an all too familiar pattern in California's population history. First, Southern California will continue to swell in numbers, along with the Sacramento metropolis driven by the Interstate 80 and Highway 50 transportation corridors to the San Francisco Bay area, then the San Joaquin Valley's main hubs of Fresno and Bakersfield and finally the San Francisco East Bay communities. Second, the data reveals tremendous growth in the interior Central Valley, the Sierra Nevada foothills and the Southern California Inland Empire of Riverside. The trend will be towards developing the interior of the state by sacrificing farmland and eventually restructuring the state's perceived northern versus southern divide into a coastal versus an interior one, based on cost of living (Figures 13.1 and 13.2).

TABLE 13.1a

County	2000 Census Rank	2010 Census Rank	2050 Projected Rank	Region
Los Ángeles	1	1	1	SoCal
Riverside	6	4	2	SoCal
San Diego	3	2	3	SoCal
Orange	2	3	4	SoCal
San Bernardino	4	5	5	SoCal
Santa Clara	5	6	6	Bay Area
Sacramento	8	8	7	Sacramento Metro
Kern	14	11	8	San Joaquin Valley
Alameda	7	7	9	Bay Area
Fresno	10	10	10	San Joaquin Valley

TABLE 13.1b

County	Projected Percent Change, 2000-2050	Region
Yuba	232.2	Sacramento Valley
Madera	231.7	Sacramento Valley
Sutter	225.3	San Joaquin Valley
Kern	216.4	San Joaquin Valley
San Joaquin	213.5	San Joaquin Valley
Merced	208.5	San Joaquin Valley
Riverside	203.5	SoCal
Placer	197.8	Sacramento Metro
Tulare	177.6	San Joaquin Valley
Mono	177.3	Central Sierra

CHAPTER 13 *California: Paradise or Gritty Reality?* **251**

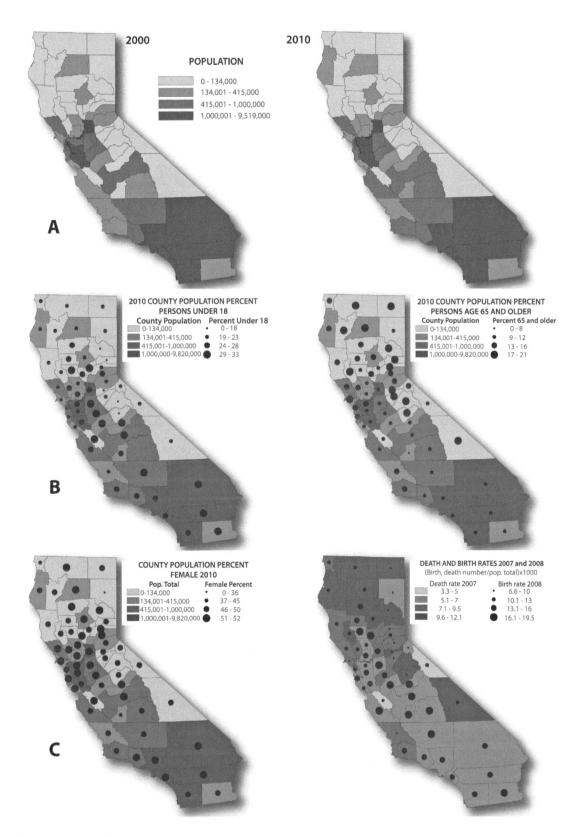

FIGURE 13.1 *U.S. Census county level demographics: (a) population changes 2000 vs. 2010; (b) population percent under 18 and age 65 and older; (c) population percent female and death/birth rates for 2007 and 2008 combined. (Data source: U.S. Census 2000 and 2010; Center for Disease Control; Courtesy Curtis Page).*

252 **CHAPTER 13** *California: Paradise or Gritty Reality?*

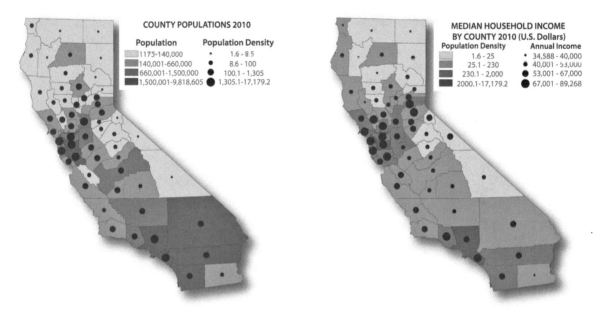

FIGURE 13.2 *U.S. Census county level socio-economics: population density per square mile and the median household income. (Data source: U.S. Census 2010; Courtesy Curtis Page).*

13.1 Modern Geographic Cultural Patterns

According to the California Department of Finance (2012) the state will pass the 40 million mark in 2018, and exceed 50 million by 2048 and will remain an ethnic "minority"-majority state until 2042 when projections reveal that Hispanics will constitute the majority of Californians at 52% of the state's population. Whites will comprise 26%; Asians are expected to be 13%; African-Americans will be 5%; and Multirace (mixed) persons, 2% (see page 27 in *California Atlas*). Both Native Americans and Hawaiian and Pacific Islander groups will each make up less than 1% of the total.

The population projections also reveal county patterns with Trinity County predicted to have the highest percentage of Whites, and Hispanics will be the highest in Imperial County—strikingly opposite ends of the state. Asians will have their highest population represented in Alameda County; Pacific Islanders will have their highest numbers in Santa Clara County, and African-Americans will show a move from the Los Angeles inner city to the Inland Empire (i.e., San Bernardino and Riverside counties) of Southern California, with their highest numbers in San Bernardino County. Finally, Native Americans will continue to have their largest percentage in Alpine County, while the largest share of people declaring multirace (mixed) ancestry will be in Inyo County. In summary, of the two major ethnic groups, Whites (Figure 13.3) will be the majority in twenty-three counties, while Hispanics will be the majority in twenty-two counties.

13.1.1 NATIVE AMERICANS

After nearly 250 years of interaction with outside cultures, which initially led to the near destruction of their population and culture, this ethnic group has managed to survive on the California landscape. Remarkably, in the 2000 and 2010 census California contained the largest Native American population in the nation with over 300,000 individuals recorded (i.e., 1 percent of the total California population). This population rivals the original encountered by the Spanish in 1769, but there is a catch. The general minimum percentage a person must have of Native American ancestry is 6.25 percent (this can vary by tribal requirements) to declare oneself as Native American in the census count. This low percentage allows for many people with mixed heritage to declare Native American and, in addition, many Native Americas from tribes outside of California have moved

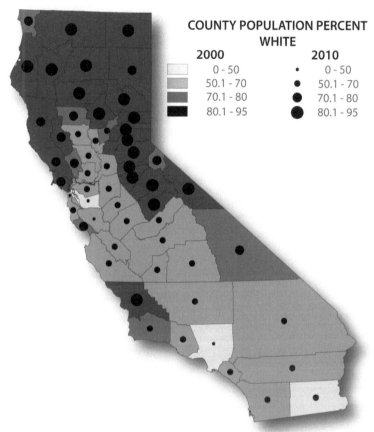

FIGURE 13.3 *U.S. Census county population data for Whites. (Data source: U.S. Census 2000 and 2010; Courtesy Curtis Page)*

to the state over the last century as they got pushed westward and/or followed work. By consulting Figure 13.4 one can see though that the majority of Native Americans largely live in far Northern California and in the Basin and Range province, while the original mission belt, Central Valley, and desert areas remain sparse. Since the development of the **gaming industry** (casinos) on tribal lands (Rancherias) in the 1980s, financial resources have helped with health care, disease control (alcoholism), and education. As of 2012, 57 out of 109 tribes operate 60 casinos in 27 counties in the state with their concentration in Southern California (Figure 13.5). The casinos and their profitability have created a tremendous power structure amongst the gaming tribes in the state, which has allowed them to become a respected political force.

13.1.2 AFRICAN-AMERICANS

After the 1992 Rodney King riots, a significant shift occurred in the African-American population in Southern California. South-central Los Angeles started to see an influx of predominantly Asian and Hispanic immigrants, and African-American families moved to resettle in the Inland Empire counties of Riverside and San Bernardino, and the San Joaquin Valley. The urban African-American communities in the Northern California urban centers of Oakland, Vallejo, Richmond and San Francisco revealed a further shift of African-Americans to the East Bay and into the Sacramento metropolitan region. We can view these shifts in the population as a joining of the "White flight" scenario to the suburban areas as a middle class structure was able to grow within the population. Analyses of the 2000 versus 2010 census results have shown these trends (Figure 13.6). The overall population trend, however, between 2000 and 2010 has witnessed a decline from 6.7 to 6.2 percent of the total state population.

13.1.3 ASIANS

By the 1990s, 40 percent of Asians in the U.S. came primarily from the East Asian realms of China, Hong Kong, Taiwan, Korea, Philippines, and south and southeast Asia via India, Thailand and Laos. The

254 CHAPTER 13 *California: Paradise or Gritty Reality?*

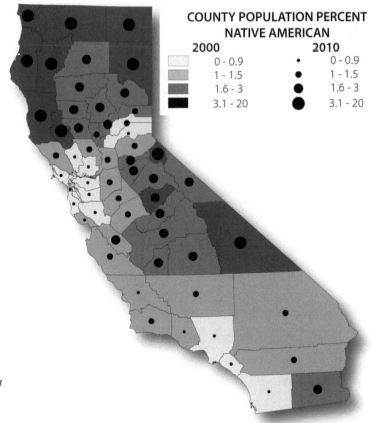

FIGURE 13.4 *U.S. Census county population data for Native Americans. (Data source: U.S. Census 2000 and 2010; Courtesy Curtis Page)*

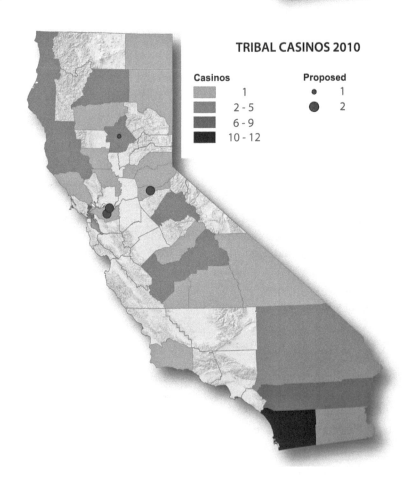

FIGURE 13.5 *Tribal casinos as of 2012 (Data source: California Legislative Analyst Office; Courtesy Curtis Page).*

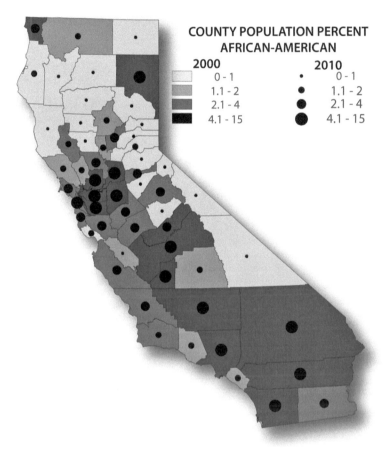

FIGURE 13.6 *U.S. Census county population data for African-Americans. Note that the high percentage of African-American men in prisons located in Del Norte and Lassen Counties account for their large outlier population percentages. (Data source: U.S. Census 2000 and 2010; Courtesy Curtis Page)*

majority of the people from these groups became residents of California. The overall concentration of Asian Americans resides in the San Francisco Bay Area, Los Angeles and Orange Counties and the lower Sacramento Valley (Figure 13.7). There has been a significant increase in overall Asian population between 2000 and 2010 from 10.9 to 13.0 percent as total of the state population.

A special category of Southeast Asian immigrants to California includes the **Hmong** whom the federal government helped to settle in the U.S. as refugees from Laos (Southeast Asia) after they helped the CIA during the Vietnam War. **The Refugee Act of 1980** helped Hmong families escape the harsh realities of Thailand's refugee camps and come to the U.S., with California being one of the top destinations. They largely live throughout the Central Valley and on the California north coast (i.e., Humboldt and Del Norte Counties). Fresno is the metropolitan city that has their highest numbers.

In general, **South Asian Indian** immigration increased (Figure 13.7) with two different groups: the Indian **Gujuartis** cultural group and Indian high technology professionals. First the Gujuartis have become mainly involved as hotel owners in the U.S. and especially along interstate highways in California. They began to acquire and found unbranded motels in the late 1970s (branded later in the 1990s) after being expelled or fleeing regime changes in Africa (especially Uganda after Idi Amin's brutality). Ex-British colony African countries had been many Guajarati's home since the late 1800s after having been sent their as indentured agricultural labor when the British controlled India. Through subsequent Indian immigration and acts of contagion, the hotel/ motel business has been very attractive and noticeable on the cultural landscape for this group. In another immigration wave after 1990 in response to the high-technology industry with its large population clusters in the Silicon Valley (Santa Clara County), well educated Indians were actively recruited. This has been helped by the **H-1B** nonimmigrant visas for specialty occupations, where in 2008, fifty-five percent have gone to Indians for employment in the technology sector.

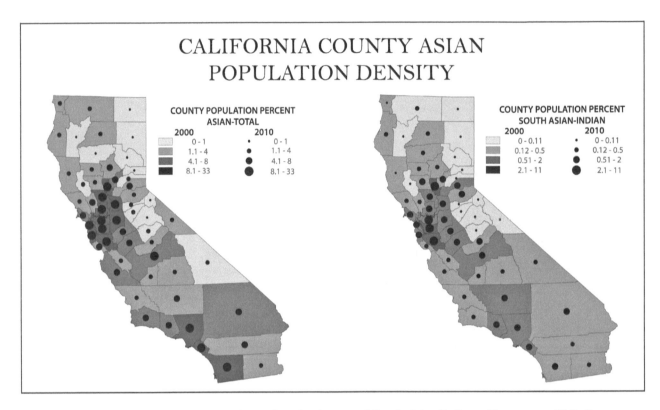

FIGURE 13.7 *U.S. Census county population data for Asians and South Asian Indians. (Data source: U.S. Census 2000 and 2010; Courtesy Curtis Page)*

13.1.4 HISPANICS

This cultural group is vast and it makes sense not to focus on one ethnic origin (i.e., Mexican) over another (i.e., Guatemalan) because its general population is so culturally diverse. In the 2000 census, 32.4 percent of California's total population was in the Hispanic category, then in the 2010 census this rose to 37.6 percent. By 2025 the Hispanic population could potentially equal the state's non-Hispanic White population; then by 2042 they will represent the majority racial group. Despite its large population size, the group as a whole, finds it difficult to achieve a sense of community, as many of the individual cultural groups stick together and tend not to mingle with other cultural groups under the broader Hispanic census group. The only truly unifying element is the Spanish language and Catholic religion.

The proportions of Hispanics in counties throughout the state still reflect the pattern from the 1930s (Figure 13.8), with the largest concentrations in Los Angeles and Imperial Counties, the San Joaquin Valley and the Central Coast. This group is still highly correlated with the agricultural and services industries. California and the Southwest are also clearly part of the Hispanic speaking world, even though California and the U.S. do not officially recognize the language (Figure 13.9).

Since the 1960s the Mexican-American segment of the Hispanic population has self-described itself as **Chicanos**. This has largely been a way for the second and third generation Mexican-Americans to distance themselves from the older generations whom they often derided. If we look deeply into the Hispanic tradition in California we can see that there are several groups. The largest Hispanic group is "American of Mexican ancestry"; the next is Mexican-American; then there are the varied Hispanic ethnic groups linked to illegal immigration; and finally there are the Chicanos.

The most recent history for this group revolves around the issue of illegal alien status. This came to a head in 1994 in the form of the **Proposition 18**7 "Save Our State" ballot initiative, which politicians designed to deny all but emergency services to illegal aliens. All other social services or access to state education was to be denied to all people of an illegal alien status. In addi-

CHAPTER 13   *California: Paradise or Gritty Reality?*   257

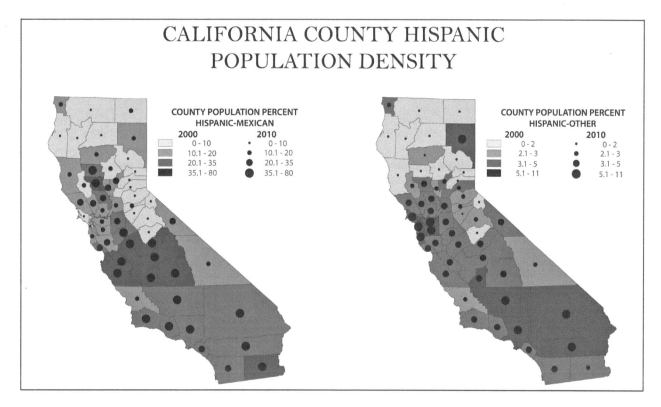

FIGURE 13.8   *U.S. Census county population data for Hispanics: Mexican and classified Hispanic other. Note that the high percentage of Hispanic men in prisons located in Del Norte and Lassen Counties account for their large population percentages. (Data source: U.S. Census 2000 and 2010; Courtesy Curtis Page)*

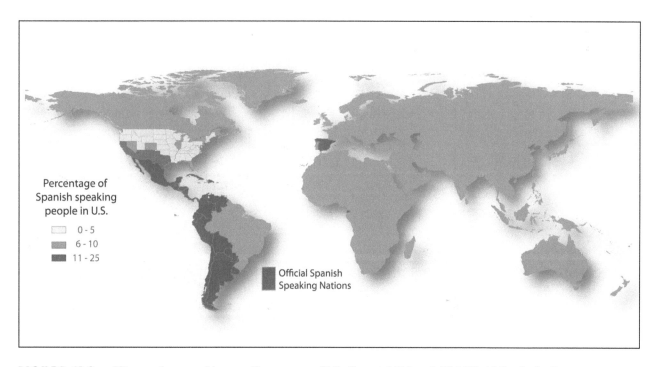

FIGURE 13.9   *Hispanophone world map. (Data source: U.S. Census 2010 and CIA World Factbook; Courtesy Curtis Page)*

tion, state and local agencies were required to report the identification of an illegal immigrant to the U.S. **Immigration and Naturalization Service** (INS). This proposition passed overwhelmingly across the state with the vast majority of counties voting yes; however, the San Francisco Bay Area counties of San Francisco, Marin, San Mateo, Sonoma, Santa Cruz, Santa Clara and Alameda, as well as Yolo County, voted no. In 1998 the U.S. District Court declared the proposition unconstitutional, but it remains the defining moment in the immigration debate that has only heated up in recent years.

The illegal immigrant issue is most closely associated with citizens from Mexico coming into the U.S., with the largest group residing in California. Southern California has had the longest connection with Hispanic heritage in the state, and since 1910 there has been a contagion effect further developing the *barrios* in East Los Angeles. The more illegal immigrants that come to the state builds a "network effect" furthering immigration to join families together in a central area and therefore developing cultural communities. Large California cities claim that illegal immigrants have been helpful to their economies and in 2003 they instructed their law enforcement to not check on a resident's status or report illegal identities to the INS. San Diego, Los Angeles, and San Francisco have become "sanctuary cities" which has effectively blurred the lines between federal immigration law and a city making an amnesty immigration law. In the same year as Proposition 187, several other events transpired which has created the current tense illegal immigration issue:

1. Passing of the North American Free Trade Agreement (**NAFTA**) which did not work as planned to keep Mexican citizens employed in Mexico;

2. The **Maquiladora** program which was linked to NAFTA to allow factories to manufacture products in the Mexican border towns (Figure 13.10) thereby keeping Mexican citizens employed in Mexico; and

FIGURE 13.10 *A small fence separates densely populated Tijuana, Mexico, right, from the United States in the Border Patrol's San Diego Sector. Construction is underway to extend a secondary fence over the top of this hill and eventually to the Pacific Ocean. (Courtesy of the Department of Defense)*

3. **Operation Gatekeeper** which was designed to tighten the border but it resulted in increased illegal immigration through the use of smugglers known as ***coyotes*** or *polleros*.

All of the above has led to a series of INS border programs that have had various levels of success. They include:

- Operation Endgame (2003–2012)
- Operation Front Line (2004–2005)
- Operation Return to Sender (2006–2007)
- Operation Scheduled Departure (2008)

Each of these operations was designed in various ways to send illegal immigrants back to their countries of origin.

In 2006 major illegal immigration control and policy came to a head with protest marches by immigrants in major cities around the country in response to a proposed 700 mile border fence and threatened mass deportations under The Border Protection, Anti-terrorism, and Illegal Immigration Control Act of 2005. This bill died in the U.S. Senate but it resulted in the immigration reform protests in April and May of 2006 of which the largest gathering was in Los Angeles. Despite the protests, the **Secure Fence Act** of 2006 was signed into law which has allowed the building of a double-reinforced fence for 700 miles between the U.S. and Mexico. Immigration whether illegal or legal will continue to be an intense topic in California into the future.

13.2 CALIFORNIA'S MODERN CULTURAL REGIONS REVEALED

The picture of California is not the same for everyone in the state—no, there are definitely differences if not highly contrasting **cultural regions** (see Chapter 1). Both extremely wealthy and extremely poor regions are within its boundaries, represented for example by the San Francisco Bay area versus the southern Cascades and the northeast corner of the state (see Figure 13.2). To regionalize a piece of geography, to lay down boundaries, is never an easy task as the regionalization has to make sense to the reader of the map in combination with an apt description of the region in text. The following sub-sections outline a generally accepted view of how California can be regionally defined as they pertain to a set of regionally unique socio-economic processes and environmental characteristics that have been locked into place for these regions since before WWII. These distinctive regions can generally be defined by vast metropolitan areas, agricultural landscapes, mountains, and lands of water and forest abundance to lands of aridity. It is appropriate in this last chapter to describe their character with the hindsight developed from the supporting twelve chapters this textbook has presented thus far.

13.2.1 URBAN EMPIRES ON THE PACIFIC

While the majority of the state's cultural regions can be defined based on landforms, the major metropolitan landscapes of the state are defined by the sheer density of people, buildings/homes and thousands of miles of roads that stitch them together to make them function as dynamic units. Both the Southern California megalopolis, extending from Santa Barbara to San Diego and inland to San Bernardino and Riverside, and the San Francisco Bay urban complex where Santa Rosa, Gilroy and Walnut Creek triangulate a center on the bay, define very different urban empires. These are California's contribution not only to the U.S., but also to the Pacific Rim economic region.

13.2.1.1 Southern California

This region is defined by the gentle bending of the California coast from south-facing Santa Barbara in the north to west-facing San Diego in the south. The landscape is strongly defined by the tectonic development of the Transverse and Peninsular ranges forming the hinterland boundary for the region as one leaves the coast for the interior. In many outsiders' eyes, this region defines California—sun, warmth, beaches, exotic plant landscapes, Spanish mission accents, Hollywood's cinema mystique, the sprawl of tract homes and skyscraper buildings. Unfortunately, the rapid growth and change that this region has undergone have led to air pollution, commuter traffic nightmares (especially characteristic of the 405 freeway), overcrowding and an unstoppable demand for outside resources in water, power and food to sustain this demographically challenged region.

The Los Angeles basin is probably one of the most demographically diverse regions in the U.S., with a 2010 census count of 9.7 million people. The city of Los Angeles (Figure 13.11) is the core of this region with 88 politically independent satellite cities alone surrounding the city in Los Angeles County proper. Within the county, Long Beach would be considered the next largest but much smaller sibling city (500,000 persons) to Los Angeles's 3.8 million people. However, while Los Angeles might be the region's identifier, the region really represents a near continuous stretch of infrastructure (housing, industry, freeways) ranging from Santa Barbara (88,400) south through Ventura (106,400) and the Oxnard Plain (198,000) to the San Fernando Valley, the Los Angeles Basin, the inland empire cities of Ontario (164,000), San Bernardino (210,000) and Riverside (304,000) (with the counties of San Bernardino and Riverside boasting 2.1 million people each). The region continues down the 405 freeway to the mini-cities of Orange County (3.0 million people) where the landscape once covered in oranges is now covered by Anaheim (336,000, home of the Magic Kingdom, Disneyland), Santa Ana (324,500), Irvine (212,000), Costa Mesa (110,000) and Newport Beach (85,000). The bottom-end of this region is defined by San Diego (1.3 million), which is the second largest city in the region and in the state (Figure 13.12). San Diego has developed and survived based on its association with the Navy and its recreation focus. All of the satellite cities around San Diego also benefit from the military and the area's sub-tropical climate for luring retirees and biotechnology companies that are affiliated with the University of California San Diego. Even accounting for all the racial diversity in Southern California, it is still easily recognized as a vast megalopolis of large cities and satellite communities woven together by a

FIGURE 13.11 *Downtown Los Angeles skyline with smaller buildings and homes in the foreground and the San Gabriel Mountains in the background. (© Jose Gil, 2012. Used under license from Shutterstock, Inc.)*

FIGURE 13.12 *Downtown San Diego skyline with San Diego Bay in the foreground. (© Albert Michael Cutri, 2012. Used under license from Shutterstock, Inc.)*

vast transportation network supporting a populace that prefers to live in a distinctly drier version of the already mild, human friendly Mediterranean climate.

13.2.1.2 The San Francisco Bay Area

Bridges, water, skyscrapers, Ivy League universities, vineyards, earthquakes, landslides, hippies, hipsters, Silicon Valley, Google, Facebook, eBay, Apple—the list can go on in many ways to describe this region's character and contribution not only to California but to the world. The region is diverse physically and culturally, and it is distinctly defined by the urbanization around the bay, which then expands into the surrounding hillsides and into the connected lowland valleys of Napa, Sonoma, Santa Rosa and Santa Clara, as well as the basins containing Livermore and the 680 freeway cities of the East Bay—Pleasanton, Concord, Danville, Walnut Creek, and Martinez.

San Francisco (805,000) is the hub (Figure 13.13), and has been since the Gold Rush. Oakland (391,000) across the bay is a competing second (only by proximity), but can hardly compare with the growing domination of San Jose (945,000), the Silicon powerhouse to the south in the Santa Clara Valley. Despite San Francisco's smaller size compared to San Jose, it still continues to represent the cultural and financial core of this region. As described in earlier chapters, San Francisco has the maturity, density and architecture in many places to rival older East Coast cities from which it was originally fashioned. San Francisco is not unlike New York City where the majority of the inhabitants find themselves densely packed in a vertical city composed of apartments, and divided into culturally and historically unique districts (i.e., North Beach, Bernal Heights, Chinatown, the Mission, etc.). As far as character goes, San Francisco as the cultural seat leans liberal in its thinking and politics compared to the Southern California metropolis. Even the conservative leaning Sonoma, Napa and East Bay region are considered more liberal than their Orange County sibling to the south. It is hard to pin down what drives the cultural characteristic of the bay to be different from Southern California, and it may not matter much as the differences were much stronger in the historical past than they are today.

The rest of the Bay Area can be comfortably divided into the much larger East Bay, consisting of the 680, 580 and 880 freeway cities divided by the Berkeley-Oakland Hills, and the Southern Bay, represented by the spine of the peninsula and the Santa Clara Valley with the surrounding mountainous hinterlands of the Santa Cruz and Diablo mountains. Finally, there is the North Bay defined as the mountain and valley region north of the Golden Gate and the Carquinez Strait.

The East Bay 680 corridor cities consist of Pleasanton (70,000), Dublin (46,000), San Ramon (72,000), Danville (42,000), Walnut Creek (64,000), Lafayette (24,000), Concord (122,000) and Martinez (36,000). The far eastern edge of the region in the north

FIGURE 13.13 *Aerial view of downtown San Francisco with Berkeley-Oakland Hills in the background. (© Jenny Solomon, 2012. Used under license from Shutterstock, Inc.)*

is represented by the Delta cities of Antioch (102,000) and Pittsburg (64,000), while Livermore (81,000) in the south is in an outlier basin connected via the 580 freeway. The cities along the 880 corridor facing the San Francisco Bay are represented by Fremont (214,000), Hayward (144,000), San Leandro (85,000), Oakland, Berkeley (112,500), Richmond (104,000) and a string of oil industry-based mini-cities. These include Pinole (18,000), Hercules (24,0000 and Rodeo (8,600). The split between these two corridors representing the East Bay are quite different in character. The interior 680 corridor cities represent recent post-WWII residential bedroom-overflow and edge city communities of sprawling tract homes, which have removed pressure from those communities that are more densely packed and costlier to live in that surround the bay. The 880 cities are much older and represent the evolving agriculture, industrial/oil refinery and military development that at one time was a hallmark of the East Bay shoreline, but is now high density housing and restored industrial areas transformed into housing, such as around Alameda (74,000).

The South Bay is a corridor of bedroom communities linked between the 280 and 101 freeways down through San Mateo into the Silicon Valley region of Palo Alto (64,500) and the San Jose metro-complex (see Figure 1.3). This is the fastest growing sub-region of the San Francisco Bay Area, but, like the Los Angeles basin in the south, the Santa Cruz and Diablo Mountains control its sprawl. Thus, development becomes narrowed towards the bottom end of the valley into Morgan Hill (38,000) and Gilroy (49,000).

The North Bay is characterized by its pastoral and wooded agricultural rurality interspersed with European style wineries in the Napa and Sonoma Valleys (Figure 13.14), by the wealthy and picturesque communities of Marin County in contrast to the industrial Vallejo-Benicia corridor, and by Santa Rosa, the last large residential community on the 101 freeway. Despite its great complexity, the North Bay is primarily known for its viticulture landscapes in the finger-like valleys coming off the bay that trend in a north-south direction represented by Napa and Sonoma. Marin County has also become synonymous with affluent secluded communities, such as Mill Valley (14,000), Sausalito (7,000) and San Anselmo (12,000). Often Marin County is given the title as one of the richest places in the U.S. to live, mostly due to its lack of affordable housing (Figure 13.15).

FIGURE 13.14 *Aerial view of the Sonoma Valley wine country. (© FloridaStock, 2012. Used under license from Shutterstock, Inc.)*

FIGURE 13.15 *Aerial view of the picturesque and very rich housing options in Marin County. (© Galina Barskaya, 2012. Used under license from Shutterstock, Inc.)*

13.2.2 THE GREAT CENTRAL VALLEY

As outlined in Chapter 10, few places in the U.S. or the world can compete with the agricultural output of this region. Flat, fertile soils and a federally engineered and supported water supply have made this region incredibly productive. However, in the last 20 years the agricultural productivity has been coming under pressure from housing tract crops that have been eating away at the farms on the edge of the cities of this region. Now with the hopes and hype of the High-Speed Rail Authority working on a route that will link San Diego to Sacramento and the San Francisco Bay Area, in addition to the ever dominating Interstate 5 transportation spine, the Great Central Valley is under pressure in its southern San Joaquin and Sacramento metropolitan corridor sections. This region is ground zero for the rural restructuring that is taking over the Western U.S.

13.2.2.1 Sacramento Metro-Delta Corridor

The corridor can loosely be linked to Interstate 80, starting geographically in Vacaville and ending in Rocklin. The dominating city is Sacramento (466,000), the capital of the state and the cultural, educational and recreational center of the region (Figure 13.16). Sacramento is still considered an inland harbor like its southern counterpart, Stockton, as they both are uniquely tied to the Sacramento-San Joaquin Delta complex (Figure 13.17). In the last decade this region has become a bedroom community for the San Francisco Bay Area at the corridor's western end around Vacaville (92,000), Dixon (18,000) and Davis (65,600), and especially west of the Sacramento River for the overspill in housing needs for Sacramento in places like Elk Grove (153,000) and Roseville (119,000). The movement east into the Sierra Nevada foothills follows the Interstate 80 and Highway 50 corridors quite closely. In fact, places like Auburn (13,000), Placerville (10,000), Grass Valley (13,000), Nevada City (3,000) and Jackson (4,700)—the old Gold Rush Belt of mining towns—have seen a spurt in growth from the economic expansion of Sacramento. The region is becoming more ethnically diverse and the sprawl has no end, as in the last decade the Interstate 5 and Highway 99 transportation corridors are being used to extend housing north and south of Sacramento. They are eating away at valuable farmland and Highway 70 has become the new northern corridor out of Roseville that is carving out housing in the oak woodlands of the Sierra Nevada Foothills.

FIGURE 13.16 *Sacramento skyline from the Sacramento River. (© Andy Z), 2012. Used under license from Shutterstock, Inc.)*

FIGURE 13.17 *The industrial Port of Stockton utilizing the Sacramento-San Joaquin Delta. (© Terrance Emerson, 2012. Used under license from Shutterstock, Inc.)*

13.2.2.2 San Joaquin Valley

Long described by Interstate 5 travelers as "boring," the San Joaquin Valley, its agricultural productivity and its many cities linked to the Highway 99 corridor, is anything but simply "boring." While the Interstate 5 is the main artery of the valley, it is situated in such a way that roadside travelers never get to see the true San Joaquin Valley, since the I-5 is nestled up against the western side of the valley and was built long after the railroad and Highway 99 created a series of large cities and mini-cities along the eastern side of the valley. This is the region of big agribusiness, Gold Rush and Depression-era memories, perennial agricultural water woes and a sharply expanding multi-cultural society. It is represented by the Delta inland port of Stockton (291,000) in the north (Figure 13.17), continues through the mini-metropolis of Fresno (494,000) in the center of California and ends with Bakersfield (347,000) anchoring the southern end up against the Tehachapi Mountains. This region is the most agriculturally productive in the state, but it also represents a strong industrial segment centered on Stockton, Modesto (201,000) and Fresno, and it revivals the Central Coast and Southern California in oil production. In the last decade, however, the region has become a boon to housing developers and those workers who could not afford to live in the San Francisco Bay area or Southern California. Places like Stockton, Modesto and Tracy (83,000) represent the overflow of people from the Bay Area that commute along the 580 corridor to work, and this has placed pressure on the valuable farmland in the region. In the south, Bakersfield has been one of the fastest growing cities in the state (Figure 13.18), as it absorbs residents leaving the San Fernando Valley and the top end of the Los Angeles basin who are looking for less expensive housing and less crowding, but who are willing to commute the I-5 corridor over the Grapevine to get to work in Los Angeles County.

13.2.2.3 The Sacramento Valley

The well-watered lowland pastoral landscape north of Sacramento to Redding (90,000) is defined by the sinuous layout of the Sacramento River as it makes it way through Redding-Anderson at the narrow top end of the valley until it opens up at Red Bluff to continue through the broad southern section of the Sacramento Valley. From Red Bluff (14,000) south, the valley's western and eastern sides are quite different. The eastern side of the valley is composed of small towns and one regional center, Chico (86,000), which is linked via the Highway 99 corridor. This pattern is similar to the development that occurred in the San Joaquin Valley. In fact, the same minimal population development has occurred along the Interstate 5 corridor on the west

FIGURE 13.18 *New residential housing developments outside of Bakersfield are pushing against the foothills of the Sierra Nevada Range. (© Richard Thornton, 2012. Used under license from Shutterstock, Inc.)*

side of the valley. The westside towns of Corning (7,600), Orland (7,000), Willows (6,000), Williams (5,000), Maxwell (1,100) and Colusa (6,000) have remained as small farming communities as the railroad and Highway 99 were in place long before the development of I-5 in the 1960s. Redding and Chico represent the two major regional center urban nodes, in a valley whose primary landscape function largely consists of orchards, rice, cattle range, wetlands, and the production area for 85% of California's water needs.

13.2.3 MOUNTAINOUS, ARID-SCAPES, AND BETWEEN THE URBAN EMPIRES

The remaining cultural-environmental regions are special because they are so dominated by their physical environment: strong topographic diversity and climatic extremes. The people and developments in these regions are deeply tied to extractive natural resources and unique agriculture, but there are minimal transportation connections due to the constraints of the physical landscape, which keeps the populations low.

13.2.3.1 Central Coast

This is the region between the two urban empires, the old mission belt that lies along the El Camino Real or the old Spanish cart trail that is now the U.S. 101 freeway, and the Pacific Coast Highway 1. This is the region that is traveled between San Francisco and Los Angeles. It is a land made up of small towns and some regional centers, but there is no dominant settlement. This is because there have never been enough flat landscape valleys in the region for places to grow, except for Salinas (150,000) and Santa Maria (99,000), which have become overflow cities for the Silicon Valley and the Santa Barbara south coast respectively. The landscape is strongly controlled by major tectonic uplift courtesy of the neighboring San Andreas Fault, which has left small narrow valleys for agriculture and settlement. The region ends in the north at Santa Cruz (60,000) and in the south at Lompoc (42,000)-Santa Ynez Valley, with Pt. Conception and Gaviota Pass (U.S. 101 tunnels) being the border with Southern California. The region is commonly referred to as the "fog belt" represented by the small coast port (or formerly) towns of Santa Cruz, Monterey (60,000), Carmel (8,000), Cambria (6,000), Morro Bay (10,000), Avila-Pismo Beach (17,000), and Lompoc. There is also a scattering of inland cities represented by Salinas, King City (13,000), Paso Robles (28,000), Atascadero (28,000), San Luis Obispo (45,000), Santa Maria, and Solvang (5,200). The theme for most of these cities and towns is the reliance on wineries and truck crop agriculture, as well as recreational opportunities for stressed out urban empire citizens needing a change of pace found in this pastoral landscape.

13.2.3.2 The Wet Northwest

North of Santa Rosa, the San Francisco Bay Area's last major urbanization node, and to the west of the Sacramento Valley and the southern Cascade region, lies the Northwest region. To many in this region it represents an isolated stretch of California that is not well connected to the rest of the state except by a two-lane version of U.S. 101 and two very windy east-west highways originating in the Sacramento Valley: Highways 20 and 299. There are no true metropolitan hubs in the region, at least in comparison to the other regions discussed so far. The Eureka-Arcata complex on Humboldt Bay just reaches 44,000 persons, with most towns, such as Crescent City (7,600), Ft. Bragg (7,200), Ukiah (16,000) and Clearlake (15,000), coming up distant seconds and thirds scattered across this vast, very wet (see Chapter 3 for details) and mountainous territory. While at one time it relied on redwood trees and fisheries for its survival, the region has fallen on hard times since the 1990s and now represents diversification into wineries, vacation homes for the rich along the beautifully rugged coastline and cannabis "pot" growing. It is identified as the emerald triangle **marijuana** production region. The very isolation of this region and the difficulty to get around have allowed it to regionally benefit from federally identified illegal activities, as medical marijuana has been given the socially acceptable "okay" to be produced for sale in the state since 1996.

13.2.3.3 The Lonely Southern Cascade and Northeast

While some would argue that the Southern Cascade and Northeast should not be lumped together, they both function in a similar manner and have a similar population group made up of small towns that are widely spread apart. This is California's forgotten corner. Transportation system is very sparse, although Interstate 5 runs on its western border from the Lake Shasta Reservoir to the Oregon border. However, I-5s presence has never allowed the area to develop considerably since it was put in during the 1960s. This region is little known by most Californians, but it is very dramatic with its volcanic landscapes crowned by Mt. Shasta and Mt. Lassen, as well as the flat lava landscape of the Modoc plateau. The "major" settlements of this region consist of Yreka (7,700), Mount Shasta (3,400), Alturas (2,800) and Susanville (18,000). The isolation of this region in conjunction with a minimal tourism base has created a populace that lives by politically conservative principles and strong local independence. In many respects, the region is similar to Wyoming or northern Colorado, in look and in major industry—beef cattle ranching. While there is some limited agriculture around Tule Lake, Yreka and Susanville, for the most part the region is made up of retirees, those employees associated with the two prisons in Susanville and with forestry operations and watershed management.

13.2.3.4 The Sierra Nevada

John Muir's "range of light" is primarily entrusted to the federal government in the form of national forests, national parks and Bureau of Land Management parcels. The higher elevation portions of the region have always been sparsely populated (with the exception of around Lake Tahoe). Focus has always been on the Sierra Nevada foothills during the Gold Rush with the development of small mining camps that became towns like Oroville in the north ranging to the Tehachapi Mountains in the south. There are only three appreciably major transportation links that cross the range east to west: Highway 70 (which represents the range's northern boundary), Highway 50 and Interstate 80. The terrain only allows for a vast network of hiking trails to cross the range in a north-south fashion: the John Muir trail and the Pacific Crest Trail.

Since the mid-1990s, the Sierra Nevada has been in environmental trouble after 150 years of damming rivers, logging old-growth forests, cattle ranching and removing oak woodlands for the creeping housing developments that have expanded both the old mining towns and have sprawled out of the Sacramento metroplex and the Lake Tahoe area. In many ways, the love that so many metropolitan based Californians have for the Sierra Nevada has caused much of the problems because of the increased extraction of natural resources and an exodus to retire in "wilderness" areas or to have recreation homes away from their city lives. Environmental historian William Cronon has noted (1994), "People who no longer earned their livings on the land, whose homes and workplaces were located in immense metropolitan districts, saw in the western wilderness a much-loved alternative to the complicated lives of quite desperation that they both cherished and maligned."

Settlements of any real size begin in the central Sierra Nevada at Mariposa (2,200) and continue sporadically through the old Gold Rush territories and gentrified towns of Sonora (4,900), San Andreas (2,700), Jackson (4,600), Placerville (10,300), Nevada City (3,000) and Oroville (15,500). The higher Sierra hosts settlement only in the north section, which includes Quincy (4,000), Truckee (16,000) and the settlement fringe that hugs the shores of Lake Tahoe (Figure 13.19) represented by Tahoe City (1,500) and South Lake Tahoe (21,400). For the most part, this region is maintained economically by the remnants of cattle ranching, by forestry operations, by numerous recreational opportunities and by overflow and the re-inflation of old Gold Rush towns by retirees.

13.2.3.5 The Mojave Desert and Trans-Sierra

The open spacious desert region to the east of the mountains bordering Southern California and the southern Sierra Nevada, but north of the Salton Sea-Imperial Valley, is the realm of the Mojave Desert and the areas of Death Valley, Owens Valley and Mono Lake in the Trans-Sierra. The region is vast, and while similar in its climate characteristics (see Chapter 3), the settlement and functional use of the region is varied. The majority of the landscape is under ownership by the federal government in the form of national forest lands, national parks, national preserves, Bureau of Land Management parcels and military training grounds. There are several main transportation arteries that crisscross this region: Interstates 10, 40 and 15, and U.S. 66 and 395. The settlement of the region is concentrated on the backsides of the San Gabriel and San

FIGURE 13.19 *Scenic view of the south shore of Lake Tahoe. (© Maisna, 2012. Used under license from Shutterstock, Inc.)*

Bernardino Mountains, reinforcing the fact that inhabitants of this region are strongly connected to overflow from Southern California, with the exception of the Owens Valley area. For many Californians, and for those coming from east of the state's border, this is a region one must just cross to get "anywhere," and several of the small settlements in the core of the Mojave region thrive on pass through travelers. These include Needles (4,800), Baker (740) and Barstow (22,600). The Antelope and Mojave River Valleys are the true hosts to any sizeable inhabitants: Victorville (116,000), Palmdale (152,800) and Lancaster (156,600). Those aforementioned cities are strongly connected to Los Angeles and San Bernardino as bedroom communities for commuters (Figure 13.20), as the economic prospects in this region are sparse, with the exception of military bases, prisons and limited industrial mining. In the Trans-Sierra there are two major settlements, Bishop (3,800) and Lone Pine (2,000), which have always served as ranching, limited mining, and tourism headquarters for the area. The U.S. 395 highway is the lifeline to the region (Figure 13.21), allowing the surrounding eastern Sierra Nevada, White Mountains and ancient saline Mono Lake to be explored for their beauty, mystery, and vastness.

13.2.3.6 The Colorado Desert and Salton Sea

This region is truly a desert: arid, hot and a host to cactus. But the region is deceiving, as the import of exotic water via the Colorado River, fertile soils, a very mild winter climate and proximity to Southern California has created both a playground and retirement area for the rich and one of the nation's most productive agricultural regions. Far from being just a desert, this region has strangely become an oasis, both by accident (i.e., the Salton Sea) and by sheer desire (i.e., agriculture and golf courses). These characteristics, however, are at opposite ends of the region. There are recreation and housing in the northern end and agriculture in the south, creating two different cultural landscapes. The region is connected via Interstate 8 out of San Diego in the south and Interstate 10 in the north out of San Bernardino. In the northern end of the region, Palm Springs (44,500) and Indio (76,000) play host to the visiting rich at play and the retired (and rich) at rest or play on the numerous golf courses servicing the region (Figure 13.22). It's not all play, however, as Indio is the business and agricultural center for the Coachella Valley, which is most famous for its date palm and orange orchards. In the south, nestled between the

FIGURE 13.20 *Aerial view of Palmdale's sprawl in the Mojave Desert. (© trekandshoot, 2012. Used under license from Shutterstock, Inc.)*

FIGURE 13.21 *Empty road of the U.S. 395 highway near Lone Pine with the Eastern Sierra Nevada and Mt. Whitney in the background. (© Sarah Fields Photography, 2012. Used under license from Shutterstock, Inc.)*

FIGURE 13.22 *Aerial view of Palm Springs and Palm Desert with numerous golf courses. (© Tim Roberts Photography, 2012. Used under license from Shutterstock, Inc.)*

FIGURE 13.23 *Aerial view of the Salton Sea with the Coachella Valley in the north and the Imperial Valley to the south. (© Thomas Barrat, 2012. Used under license from Shutterstock, Inc.)*

Salton Sea (Figure 13.23) and the border with Mexico, the Imperial Valley plays host to its only major economic activity, some very large twelve months of production agriculture. The settlements of El Centro (42,500), Calexico (38,500) and Brawley (25,000) are the service centers for the agricultural enterprise and also offer travelers services on Interstate 8 between San Diego and Yuma, Arizona.

13.3 CALIFORNIA: EDEN OR WASTELAND?

What are we left with in California? Environmentally the golden state had become tarnished and on the road to ecosystem decay by the 1960s. All the new residents did not spread evenly across the landscape but instead intensified the crowded areas along the coast—the urban nodes of Southern and Northern California. They came to achieve their California dream—affluence in a gentle climate. However, what happened during this process amounted to crowded coastal cities with large planned tracthome suburbs of "McMansions" and transportation corridors clogged with ride-alone commuters in private cars as the state moved into the 21st century. The state that invented car culture (especially Southern California), however, found the era of real estate booms and expanding commuter culture coming to an end in the current economic climate (c. 2007–present).

The environment has become damaged from air pollution, water quality problems (some left over from the Gold Rush eras use of mercury), deforestation and habitat fragmentation leading to the largest numbers of endangered species in the continental U.S., and limited wilderness—all accelerated by greed and ignorance. The future of California agriculture is also in the balance. The **California Land Conservation Act** (Williamson Act) of 1965 tried to stem the loss of farmlands to tract home sameness by reducing a farm parcels property tax, but the lure of riches from land developers has been hard to resist. It did not help that developers pushed through "prodevelopment" reforms in the 1980s, which made the act largely ineffective. In the absence of effective regional or statewide growth regulation, runaway development has shifted to outlying areas in a process of "leap-frogging" statutory boundaries in a process of exurbanization.

Nevertheless, with all the above said, California has become the leader with respect to landmark environmental legislation in the form of the Natural Communities Conservation Protection Act (2002) and the **California Environmental Quality Act** and climate change legislation (**AB32**, 2007). There has also been promotion of organic agriculture and **community supported agriculture** (CSAs) through the embracing of the "slow food" movement, localvore foodsheds, and therefore a growing locally evolving rejection of agribusiness methods and/or corporate processed foods and farming with GMOs (genetically modified organisms). Future challenges for the state's agriculture and urbanization sectors will come from climate change effects on water supply but also from loss of species-specific croplands due to the higher temperatures.

13.3.1 NORCAL VS. SOCAL OR COASTAL VS. INTERIOR?

"Driven by demographic, cultural and political shifts, a rift is growing between California's huge coastal cities and its increasingly conservative interior, replacing the north-south split that has defined state politics for more than 150 years." John Wildermuth, *San Francisco Chronicle*, October 6, 2002.

The state of mind in California up until the twenty-first century had been about being from "NorCal" or "SoCal"—any other division did not matter. The state received its first notions of this in a drive for a literal spilt in 1941 when a secession movement began linking the far northern counties of California with the southern counties of Oregon to form the **State of Jefferson** (Figure 13.24). The internal rebellion qui-

FIGURE 13.24 *The location of the State of Jefferson. (Courtesy Curtis Page)*

eted with the bombing of Pearl Harbor and subsequent entry of the U.S. into the war. The State of Jefferson and the unique mountainous landscapes largely populated by White rural peoples and Native American rancherias have never forgotten their feelings of being neglected by both state capitals of Salem, OR and Sacramento. This feeling has only increased as the fortunes of the natural resource industries (i.e., lumber, fisheries) on which this region has historically relied on have deteriorated based on mismanagement and environmental regulation. In 1992, another **split-state initiative** developed within the California State Assembly, but before it could go on the ballot it died in the Senate rules committee. It would have posed a non-binding question to the voters as to whether California should be split into three states (Figure 13.25). Historically there have been twenty-six attempts to date to look at splitting this culturally complex state. Voters have voted on other local ballot initiatives to split counties with the latest having occurred in 2006 in Santa Barbara County's north versus south county split initiative. It failed at the ballot box.

The current economic, political and cultural scenario in California has been pointing at the true split being a coastal versus interior grouping of counties. The interior has become more conservative and is not as wealthy, while the coastal counties hold the largest populations, and have a more affluent and liberal leaning agenda as large urbanized centers. It will be fascinating to watch this new power geography play itself out in the coming years.

FIGURE 13.25 *The split state initiative of 1992. (Courtesy Curtis Page)*

13.4 Living the "La Vida Loca" in Mediterranean California

California, a dream, a nightmare—a sense of place for tens of millions made famous at one time as a mythical island on the far side of the European known world. There has always been a sense of the Gold Rush mentality in the state, a sense of gambling for greater affluence, recreation and relaxation, and a view that it is the place where the American dream of freedom of thought and pursuit of happiness in excess could become possible. If only one had the desire to come West, young man or woman, to the land where both dreams and broken promises intermingle, and economic bubbles spectacularly rise and devastatingly burst. California asks a person to participate in the game of life to its fullest; however, it would not be life if there were not temporality involved, which is sorrow—loss. California's myriad of cultures act themselves out on a wonderful landscape set, a diverse and dynamic physical geographic backdrop, one that is a wonder to live in.

Bibliography

Almaguer, T. (1994). *Racial Fault Lines: The Historical Origins of White Supremacy in California*. University of California Press, Berkeley, CA.

Avila, E. (2006). *Popular Culture in the Age of White Flight: Fear and Fantasy in Suburban Los Angeles*. University of California Press, Berkeley, CA.

Binford, G. (2005). Sunshine technopolis: southern California's dimming future. *Futures* (37): 345–348.

Bottles, S. (1987). *Los Angeles and the Automobile: The Making of a Modern City*. University of California Press, Berkeley, CA.

Brasil, E. (1992). Up Country: It's divide and rule. *San Francisco Examiner*, February 9.

Bricker, D. (2000). Ranch Houses Are Not All the Same. In *Preserving the Recent Past 2*, edited by D.Slaton and W.G. Foulks, Historic Preservation Education Foundation, National Park Service, and Association for Preservation Technology International, Washington, D.C. [online] http://www.nps.gov/nr/publications/bulletins/suburbs/Bricker.pdf

California Legislative Analyst's Office (2007). *California Tribal Casinos: Questions and Answers*. California's Nonpartisan Fiscal and Policy Advisor. [online] http://www.lao.ca.gov/2007/tribal_casinos/tribal_casinos_020207.aspx.

Calisphere (2008a). *California Cultures*. University of California [online] http://www.calisphere.universityofcalifornia.edu/calcultures/

Calisphere (2008b). *A world of California Primary Sources*. University of California [online] http://www.calisphere.universityofcalifornia.edu/

Camarillo, Albert. 1984. *Chicanos in California: A History of Mexican Americans*. Sparks, NV: Materials for Today's Learning, Inc.

Clark, W.A.V. (1998). *The California Cauldron: Immigration and the Fortunes of Local Communities*. Guilford Press, New York, NY.

Collins, K.E. (1979). *Black Los Angeles: The Maturing of the Ghetto, 1940–1950*. R & E Research Associates, San Francisco.

Davis, M. (2000). *Magical Urbanism: Latinos Reinvent the U.S. City*. Verso, New York, NY.

Davis, C. and D. Igler (2002). *The Human Tradition in California*. Scholarly Resources, Wilmington, DEL.

DeGraaf, L.B., Mulroy, K., and Q. Taylor (2001). *Seeking El Dorado: African Americans in California*. Autry Museum of Western Heritage and University of Washington Press, Los Angeles, CA.

Deverell, W. and D. Igler (2008). *A Companion to California History*. Wiley-Blackwell, New York, NY.

DeWitt, H.A. (1999). *The Fragmented Dream: Multicultural California*. Kendall/Hunt Publishers.

DeWitt, H.A. (1999). *The California Dream*. 2nd edition. Kendall/Hunt Publishers.

Diamond, S.J. (1993). Split personality. *Los Angeles Times*, July 25.

Dilsaver, L.M., Wyckoff, W., and W. Preston (2000). Fifteen events that have shaped California's Human Landscape. *The California Geographer*, 40: 3–78.

Donley, M.W., Allan, S., Caro, P., and C.P. Patton (1979). *Atlas of California*. Pacific Book Center, Culver City, CA.

Durrenberger, R.W. and R.B. Johnson (1976). *California Patterns on the Land*. 5th edition, Mayfield Publishing Company, Mountain View, CA.

Foster, M.S. (1975). The model-t, the hard sell, and Los Angeles's urban growth: the decentralization of Los Angeles during the 1920s. *The Pacific Historical Review* 44(4): 459–484.

Gregor, H.F. (1963). Spatial Disharmonies in California's Population Growth. *Geographical Review* 53: 100–22.

Hanson, V.D. (2007). *Mexifornia: A State of Becoming*. 2nd Edition, Encounter Books.

Hornbeck, D. (1983). *California Patterns: A Geographical and Historical Atlas*. Mayfield Publishing Company, Mountain View, CA.

Kling, R., Olin, S., and M. Poster (1995). *Postsuburban California: The Transformation of Orange County since World War II*. University of California Press, Berkeley, CA.

Lotchin, R.W. (1992). *Fortress California 1910–1961: From Warfare to Welfare.* Oxford University Press, New York, NY.

McIntire, E. (1998). California's Changing Landscapes. *Journal of the West,* 37(3):44–54.

McWilliams, C. (1973). *California, the Great Exception.* Peregrine Smith, Santa Barbara, CA.

Matthews, G. (1999). The Los Angeles of the North: San Jose's Transition form Fruit Capital to High-Tech Metropolis. *Journal of Urban History,* 25: 459–476.

Meinig, D.W. (1979). Symbolic Landscapes: Some Idealizations of American Communities. In: *The Interpretation of Ordinary Landscapes: Geographical Essays,* D.W. Meinig, ed., pp. 164–192. Oxford University Press, New York, NY.

Michaelson, J. (2008). *Geography of California.* Course at UC Santa Barbara, Dept. of Geography. [online] http://www.geog.ucsb.edu/~joel/g148_f08/.

Mitchell, D. (1996). *The Lie of the Land: Migrant Workers and the California Landscape.* University of Minnesota Press, Minneapolis, MN.

Morello-Frosch, R., Pastor, M. Jr., Porras, C., and J. Sadd (2002). Environmental Justice and Regional Inequality in Southern California: Implications for Future Research. *Environmental Health Perspectives* 110(suppl 2): 149–154.

Morrison, P.A. (1971). *The Role of Migration in California's Growth.* Rand Corporation, Santa Monica, CA.

Nash, G. (1990). *World War II and the West: Reshaping the Economy.* University of Nebraska Press, Lincoln, NE.

Peters, G.L., Lantis, D.W., Steiner, R., and A.E. Karinen (2001). *California.* 4th Edition. Kendall/Hunt Publishing Co., Dubuque, IA.

Pitti, S.J. (2004). *The Devil in Silicon Valley: Northern California, Race, and Mexican Americans.* Princeton University Press.

Pryde, P.R., ed. (1979). *San Diego: An Introduction to the Region.* 2nd edition, Kendall/Hunt, Dubuque, IA.

Rawls, J.J. and W. Bean. 2008. *California: An Interpretative History.* 9th Edition, McGraw-Hill, Boston.

Rice, R., Bullough, W., and R. Orsi (2001). *The Elusive Eden: A New History of California.* 3rd edition, McGraw-Hill.

Sanchez, George J. 1993. *Becoming Mexican American: Ethnicity, Culture, and Identity in Chicano Los Angeles, 1900–1945.* New York: Oxford University Press.

Sanchez-Korrol, Virginia V. 1983. *From Colonia to Community.* Westport, Conn: Greenwood Press.

Saxenian, A. (1985). The Genesis of Silicon Valley. In: *Silicon Landscapes,* P. Hall and A. Markusen, eds., pp. 20–34. Allen and Unwin, Boston, MA.

Self, R.O. (2003). *American Babylon: Race and the Struggle for Postwar Oakland.* Princeton University Press, NJ.

Social Explorer (2009). *Los Angeles County 1940–2000: Race Map.* [online] http://www.socialexplorer.com/pub/maps/home.aspx.

Social Explorer (2009). *United States Census Demographic Maps: California 1850–2007.* [online] http://www.socialexplorer.com/pub/maps/home.aspx.

Social Explorer (2009). *United States Religion Maps 1980–2000.* [online] http://www.socialexplorer.com/pub/maps/home.aspx.

Starr, K. (2005). *California: A History.* The Modern Library, New York, NY.

State of California, Department of Finance (2009). *California Current Population Survey Report: March 2007.* Sacramento, CA. [online] http://www.dof.ca.gov.

State of California, Department of Finance (2009). New state projections show 25 million more Californians by 2050; Hispanics to be state's majority ethnic group by 2042. Sacramento, CA. [online] http://www.dof.ca.gov

Steiner, R. (1981). *Los Angeles: The Centrifugal City.* Kendall/Hunt, Dubuque, Iowa.

Takaki, R. (2008). *A Different Mirror: A History of Multicultural America.* 2nd edition, Back Bay Books.

Thompson, W.S. (1955). *Growth and Changes in California's Population.* The Haynes Foundation, Los Angeles.

Vance, J.E. (1972). California and the search for the ideal. *Annals of the Association of American Geographers* 62(2): 185–210.

Weber, David. J. (1979). *Foreigners in Their Native Land: Historical Roots of the Mexican-American.* Albuquerque: University of New Mexico Press.